WATER INDEX

DESIGN STRATEGIES FOR DROUGHT, FLOODING AND CONTAMINATION

SETH McDOWELL

UNIVERSITY *of* VIRGINIA
SCHOOL OF ARCHITECTURE

Editor
Seth McDowell, Assistant Professor

VERDANT PHREATOPHYTES PLANTS FED BY A MYRIAD OF SPARKLING SPRINGS WERE REPLACED BY A XERIC LANDSCAPE OF SAGEBRUSH AND SAND DUNES [229] GROWING POPULARITY OF WASHING BEGAN TO THREATEN THE SANITARY ARRANGEMENTS OF THE PREINDUSTRIAL CITY BY FLOODING CESSPITS AND DILUTING THE NITROGEN CONTENT OF HUMAN MANURE [120] HALF OF THE WORLD'S CITIES ARE LOCATED IN DEPLETED WATERSHEDS [261] HEALTH DEPENDED UPON SANITATION [125] HOT WEATHER AND THE USE OF THOUSANDS OF WATERCLOSETS CREATED AN UNGODLY STENCH LASTING TWO YEARS [131] HOUSES ON STILTS SEEM TO PREFER TO AVOID THE WATER THAN TO ENGAGE WITH IT. [50] IF A CITY CAN ALLOCATE 10% OF ITS TOTAL AREA AS A GREEN SPONGE AREA FOR STORMWATER MANAGEMENT, IT CAN VIRTUALLY SOLVE THE STORMWATER PROBLEM THAT IS COMMONLY SEEN IN CONTEMPORARY CITIES [161] IF NEW ORLEANS FAILS TO ADAPT, THE CITY WILL BE GONE IN A HUNDRED YEARS [49] IF THIS DIKE SHOULD BREACH, THE DAMAGE HERE WOULD BE IMMEASURABLE [41] INCAPABLE OF HANDLING THE AMOUNTS OF EFFLUENT LEAVING THE HOMES AND BUSINESSES [129] INCREASED POLLUTION LOAD THREATENED THE PURITY OF THE WATER SUPPLY [132] INCREASE IN IMPERMEABLE PAVED SURFACES [159] INCREASING POLLUTION OF THE RIVERS AND STREAMS OF THE COUNTRY IS AN EVIL OF NATIONAL IMPORTANCE [132] INSTANTANEOUS SEA LEVEL RISE (AND SOMETIMES SEA LEVEL DROP) CAUSED BY EARTHQUAKES IS A SURPRISINGLY WIDESPREAD PHENOMENON [31] IN THE MEDIA, NEWS ABOUT STORMS AND FLOODING ELSEWHERE IN THE WORLD CROP UP IN QUICK SUCCESSION [39] ISSUES OF WATER QUALITY AND SUPPLY ARE MAKING FRONT-PAGE NEWS [215] KEEPING THE CITY DRY, OR SEPARATING THE HUMAN-MADE ENVIRONMENT FROM ITS NATURAL ENDOWMENT, HAS BEEN THE PERPETUAL BATTLE FOR NEW ORLEANS [48] LAGGING LEVELS OF CONNECTION TO MODERN WATER SUPPLY AND SANITATION SYSTEMS IN THE CITIES OF THE GLOBAL SOUTH [121] LAND SPECULATORS NEEDED MORE IRRIGATION WATER TO TRANSFORM THE DESERT INTO THE VISION OF LUSH GARDENS ADVERTISED IN MAGAZINES [217] LET'S NOT, UNDER THE PRETEXT OF SECURITY, BUILD DIKES THAT ARE MERELY BEAUTIFUL [45] LOCAL GOVERNMENT WAS TOTALLY UNPREPARED FOR THE STORM [34] MAMMOTH SHIFTS IN POPULATION MAY BE REQUIRED FOR THESE CITIES [36] MOST PARKS AND ROAD GREENERY ARE IRRIGATED WITH DRINKING WATER [255] NETHERLANDS IS DROWNING IN LAWS AND REGULATIONS [47] NEW ORLEANS WILL FOREVER BE ASSOCIATED WITH HURRICANE KATRINA [47] NO CLEAR PLACE TO BEGIN A DESIGN [129] NOW THE TIME HAS COME FOR TODAY'S POPULATIONS TO DEAL WITH THE HUGE IMPACTS OF AN EXPANDING OCEAN [31] ONE-FOURTH OF THE DIKES DO NOT EVEN MEET THE OUTDATED STANDARDS [41] OWENS VALLEY FACES STAGNATION, DESICCATION, AND TOXIC DUST STORMS [225] PEOPLE ACCEPTED CHRONIC DYSENTERY AND OTHER ENDEMIC DISEASES AS NORMAL [127] PLUMBING INNOVATIONS WITHIN THE HOME REMAINED HIGHLY UNEVEN IN DIFFERENT NATIONAL AND CULTURAL CONTEXTS [121] POLLUTION IN THE LAKES HAS BEEN A PRIMARY BY-PRODUCT OF OVER A CENTURY OF URBANIZA-

TION [185] PROBLEMS OF THE CITIES WILL TRUMP THOSE OF RURAL OR TOURISTIC AREAS [29, 36, 272] PRODUCTION OF ARTIFICIAL FERTILIZERS WAS BECOMING MORE WIDESPREAD [120] PUBLIC ACTIVITIES SUCH AS WASHING WERE INCREASINGLY RESTRICTED TO THE PRIVATE SPHERE [120] PURE WATER SUPPLY WAS NOT JUST A CONVENIENCE, BUT A NECESSITY FOR GOOD HEALTH [133] PUTREFYING SEWAGE CAUGHT IN THE TIDAL REACH OF THE RIVER [131] RAPID GROWTH OF 19TH-CENTURY CITIES QUICKLY OVERWHELMED THE HISTORIC RELIANCE ON WELLS, WATER VENDORS, AND OTHER SOURCES AND LED TO THE INTRODUC- TION OF CENTRALIZED WATER SUPPLY SYSTEMS [119] RELUCTANT TO AC- KNOWLEDGE A NEED FOR SEPARATE SANITARY AND STORM SEW- ERS [130] RESIDENTS THOUGHTLESSLY ASSUME THAT THEIR GARDEN PARADISE MERELY COMES FROM "TURNING ON THE TAP" [216] RISING SEAS WILL AFFECT NOT JUST THE BUILT ENVIRONMENT, BUT THE NATU- RAL WORLD AS WELL. [32] ROOTED IN THE HISTORIC AMERICAN CONFLICT BETWEEN RURAL VIRTUE AND URBAN INTELLECTUALISM [219] SAFETY IS INVISIBLE TO THE CITIZEN [40] SAFETY STANDARDS FOR THE DELTA WORKS AND THE DIKES DATE FROM THE YEARS IMMEDIATELY FOLLOW- ING THE SECOND WORLD WAR [41] SEVERAL CHOLERA EPIDEMICS TO RAVAGE GREAT BRITAIN TOOK 60,000 LIVES, MANY AMONG THE POOR [127] SEWAGE OF NEARLY THREE MILLIONS OF PEOPLE HAD BEEN BROUGHT TO SEETHE AND FERMENT UNDER A BURNING SUN, IN ONE VAST OPEN CLOACA LYING IN THEIR MIDST [131] SILENT VICTIM OF THE CITY'S DESTRUCTIVE THIRST [209, 225] SOME OFFICIALS REMAIN UNCON- VINCED THAT THE THREAT TO THE NILE DELTA IS REAL [35] SOUTHERN CALIFORNIA HAS ALWAYS BEEN EXTREMELY RELUCTANT TO DISCUSS ITS BASIC WEAKNESS [215] SOUTH LOUISIANA, LIKE THE NETHERLANDS, MUST ADAPT TO THE THREATS INHERENT TO LIVING IN A SUBSIDING DELTA [49] SPREAD OF PIPED WATER AMONG THE COMFORTABLE DIRT- IED THE ENVIRONMENT OF THE POOR [133] STORMS ARE, IN A SENSE, OF- TEN THE ADVANCE GUARD OF SEA LEVEL RISE [32] STORMS ARE THE AC- TIVE AGENT OF SEA-LEVEL-RISE DESTRUCTION [32] TAINTED WATER WAS THE MEDIUM OF TRANSMISSION [127] THE ANIMALS UNIQUE TO THESE ECOSYSTEMS WILL HAVE NO PLACE TO MIGRATE AND THUS MAY PER- ISH. [32] THE CAPITAL WILL HAVE TO BE MOVED 112 MILES (180 KM) TO HIGHER ELEVATIONS [36] THE CHANGING RELATIONSHIP BETWEEN WA- TER AND THE HUMAN BODY IN THE MODERN CITY [117] THE DAYS OF BUILDING MASSIVE AQUEDUCTS IN THE UNITED STATES ARE OVER. [231] THE DUTCH LIVE SIGNIFICANTLY BELOW SEA LEVEL [40] THE ISLANDS MAY "COLLAPSE" OR DISAPPEAR, POSSIBLY WITHIN THE NEXT FEW DECADES [33] THE MISSISSIPPI DELTA IS A SMALL TAIL ATTACHED TO A GREAT BIG DOG [50] THE MOST DENSELY POPULATED COUNTRY IN EU- ROPE, WHICH GENERATES MORE THAN 60% OF ITS ECONOMY BELOW SEA LEVEL, HAS NO NATIONAL EVACUATION PLAN [41] THE PHYSICAL EN- VIRONMENT EXERCISED A PROFOUND INFLUENCE OVER THE WELL-BE- ING OF THE INDIVIDUAL [125] THE PLACE OF WATER WITHIN THE 19TH-CEN- TURY CITY REFLECTS AN AMBIGUITY BETWEEN THE STRATEGIC NEEDS OF THE MODERN STATE AND THE DEVELOPMENT OF REFORMIST DI-

MENSIONS TO URBAN POLITICAL DISCOURSE. [118] THE PLACE WILL BE-COME A BATTLEGROUND BETWEEN ENGINEERS AND SEA LEVEL RISE [33] THE PROBLEM OF WATER IN SOUTHERN CALIFORNIA IS A CULTURAL PROBLEM [215] THERE ARE NO DISTANT WATERSHEDS REMAINING FOR THE CITY TO TAP [231] THERE IS NOWHERE FOR THE EVERGLADES TO GO [33] THE RITUAL OF WATER IS NO LONGER A PUBLIC ACTIVITY [216] THIS LABYRINTH REMAINS HIDDEN FROM VIEW [216] THIS LAST PHASE IN THE MODERNIZATION OF WATER INFRASTRUCTURE REMAINS ONLY PARTIALLY COMPLETED [121] THIS LOW-LYING DELTA IS VULNERABLE TO CLIMATE CHANGE, WHICH INVOLVES NOT ONLY RISING SEA LEVELS BUT ALSO MORE FREQUENT AND MORE INTENSE RAINFALL. [40] URBAN FLOODING CAUSED BY STORMWATER HAS BECOME A GLOBAL ISSUE [159] VIETNAM WILL SUFFER ENORMOUSLY IN TERMS OF LAND AREA LOST AND PEOPLE DISPLACED FROM A 3.3-FOOT (1 M) SEA LEVEL RISE [34] WATER-INTENSIVE LAWNS AND ARTIFICIAL PONDS IN PARKS PUT EVEN MORE STRESS ON THE LIMITED AND INEFFICIENTLY MANAGED WATER RESOURCES [255] WATER, LIKE OTHER FACETS OF URBAN NATURE, WAS INCORPORATED INTO AN INCREASINGLY RATIONALIZED AND SCIENTIFICALLY MANAGED URBAN FORM [117] WATER REMAINS ASSOCIATED WITH FEAR [49] WHILE BATHING IN A HOUSEHOLD SPRING, THERE IS LITTLE TO REMIND ONE OF THE WATER'S SOURCES [221] WHOLE TOWNS, EVEN NATIONS, WILL DISAPPEAR [31] YOU CAN ALSO LIMIT THE DAMAGE BY USING THE LAND DIFFERENTLY, BY BUILDING THE HOUSES DIFFERENTLY [45] ADAPT INDEX [84, 158, 250] ABOVE-THE-WETLAND EXPERIENCE [161, 164] A FLEXIBLE LANDSCAPE THAT IS GROWN, SHAPED, AND DEFINED OVER TIME AS A RESULT OF THE OPERATION OF AN EXPANDABLE AND MOBILE CELLULAR NETWORK [179] A FLEXIBLE SYSTEM CAN STIMULATE AND INTERVENE IN THE EXISTING NATURAL PROCESSES OF THE WET LANDSCAPE [179] ALLOWS FOR THE CO-HABITATION OF THESE NETWORKS, BUT ALSO LEVERAGES EXTRA ECONOMIC, ECOLOGICAL, AND CULTURAL VALUES [179] A PARK PERFORMING MANY FUNCTIONS [161] A PERFORMATIVE WATERSCAPE [111, 183] ARTIFICIAL ISLANDS THAT GROW OVER TIME [181] BALANCE BETWEEN THE NEEDS OF HIGH-QUALITY SPACE FOR CITIZENS, RIVER DYNAMICS, CROP EXPLOITATION OF LOCAL VARIETIES RECOVERED BY ORGANIC FARMING [85] BECOME PART OF A PUBLIC SPACE [87] BIODYNAMIC PRODUCTION [51, 85, 272] BIOLOGICAL MODEL OF AMD TREATMENT [173] BURGEONING AQUACULTURAL INDUSTRIES [197] COLLECTION PODS COLLECT CO_2 FROM THE TUNNELS OF NEW YORK CITY [181] COLLECTS, CLEANSES, AND STORES STORMWATER AND INFILTRATES IT INTO THE AQUIFER [161] COMBINES WATER STORAGE WITH THE IMPROVEMENT OF THE QUALITY OF URBAN PUBLIC SPACE [99] CONSTRUCTED WETLAND [137, 167, 168, 275] COUPLING OR BUNDLING OF TRANSPORTATION INFRASTRUCTURE WITH WATER-BASED URBAN PROGRAMMING FOR PRODUCTION, RECREATION, LIVING, PROTECTION, RESEARCH, AND REHABITATION [181] CYCLICAL STRATEGIES FOR WATER STORAGE AND CONVEYANCE [185] DOUBLE AS A PARK [173] DYNAMIC PLACE [99] ECOLOGICAL INFRASTRUCTURE [239, 255, 278] ECOLOGICAL WASHING MACHINE TO CLEANSE THE ACRID GOLDEN POISON INTO A

VITAL, GREEN ORGANISM [173] ECOSYSTEM SERVICES APPROACH TO UR-
BAN PARK DESIGN [161] FLOODABLE FOREST [51, 85, 87, 272] FORGING A NEW
COURSE [93] FRAMEWORK FOR AN EMERGENT AQUACULTURE INDUS-
TRY [199] GIVING FORM TO THE ALCHEMY OF ACID-TO-ALKALINE [173] HY-
BRID INFRASTRUCTURE THAT DYNAMICALLY TUNES THE INPUT AND OUT-
PUT OF EXISTING INFRASTRUCTURE, MODULATING AND ALLOCATING
WATER THROUGHOUT THE CITY [93] HYBRIDIZATION OF INDUSTRIAL AND
ECOLOGICAL PROCESSES [199] INCREMENTAL IMPLEMENTATION BY VAR-
IOUS PUBLIC AND PRIVATE ENTITIES [181] INFRASTRUCTURES ARE RE-
SPONSIVE TO ECOLOGICAL, ENVIRONMENTAL, ECONOMIC AND SOCIAL
VARIATION OVER TIME [201] INTEGRATES LARGE-SCALE URBAN STORM-
WATER MANAGEMENT WITH THE PROTECTION OF NATIVE HABITATS,
AQUIFER RECHARGE, RECREATIONAL USE, AND AESTHETIC EXPERI-
ENCE [159] INTEGRATES THE HYDROLOGICAL CYCLE INTO A MULTIFUNC-
TIONAL OPEN SPACE SYSTEM [255] INTENSE PARTICIPATORY TRAJEC-
TORY WITH THE LOCAL COMMUNITY [99] LANDSCAPE AS A SPONGE [159,
160] LATENT PROCESSES INTEGRALLY EMERGE OVER TIME [199] LAYERED
SOCIAL AND ECOLOGICAL STRUCTURES [210, 239, 261, 263, 279] LIVING LABO-
RATORY [173, 239, 251, 253, 279] MORE PROGRESSIVE AND ADAPTIVE [25, 93] MORE
THAN JUST A GIANT SCIENCE PROJECT [173] NEW PARADIGM OF DECEN-
TRALIZED URBAN WATER MANAGEMENT SYSTEMS [185] NEW TEMPO-
RARY RIVER COURSE FOR HIGH WATERS [87] NEW TYPOLOGY OF WATER
MANAGEMENT [51, 93, 273] OVERALL STRATEGIES OF DETOXIFICATION,
CONTAINMENT, COEXISTENCE, AND CONVERSION [199] PASSIVE TREAT-
MENT SYSTEM TO TACKLE AMD [173] PATHS ARE KEPT HIGH [87] POSITIVE
ENVIRONMENTAL AMENITY IN THE CITY [161] POSSIBILITY FOR A NEW
PUBLIC [201] PRODUCTIVE, ADAPTIVE, AND MULTIFUNCTIONAL [93] PRO-
DUCTIVE AND LOW-MAINTENANCE LANDSCAPE [167] PROTOTYPICAL IN-
TERVENTIONS THAT RECONFIGURE STORMWATER AND GREYWATER
INFRASTRUCTURE [210, 261, 263] PROVIDING NATIVE HABITATS FOR BIODI-
VERSITY AND OFFERING RECREATIONAL USES, [167] REBALANCE THE
POSITION OF STRENGTH WITH WHICH HUMAN HAS BEEN LINKED WITH
THE ENVIRONMENT [87] RECONNECTED WATER NETWORKS AND TRANS-
FORMED THE AREA INTO AN URBAN STORMWATER PARK THAT WILL PRO-
VIDE MULTIPLE ECOSYSTEM SERVICES. [161] REDESIGN WATER INFRA-
STRUCTURE SO THAT MONOFUNCTIONAL SYSTEMS ARE TRANSFORMED
INTO RESILIENT SOCIO-ECOLOGICAL CYCLES [261] RE-ESTABLISHMENT
OF THE WETLANDS [251] REGENERATIVE LIVING LANDSCAPE [167] RE-
INTRODUCE SWAMPS INTO THE PUBLIC REALM AS A VITAL PART OF A
SMART WATER GRID [93] REINTRODUCTION OF SWAMPS AS A VITAL PUB-
LIC SPACE [93] REPROGRAMMED THROUGH AN EMBEDDED LOGIC TO
SUPPORT A MORE SUSTAINABLE AQUACULTURAL INDUSTRY [197] RE-
VEAL LATENT ECOLOGICAL PROCESSES [261] SECULAR "SOFT" INTELLI-
GENCE OF THE FARMERS [87] SMART NETWORK OF SWAMPS [93] SPACE
OF CULTURAL PRODUCTION—A SPACE FOR CITIZEN ENGAGEMENT AND
INTERACTION ON PRODUCTIVE TERMS [201] SPACE OF PRODUCTIVE EN-
GAGEMENT [199] SPATIO-ECONOMIC TRANSFORMATION [138, 179, 276] STORM-
WATER COULD INFLUENCE THE SQUARE [99] STORMWATER-FILTRATING

AND-CLEANSING BUFFER ZONE [161] SUSTAINABLE DESIGN TO TREAT POLLUTED RIVER WATER AND RECOVER THE DEGRADED WATERFRONT IN AN AESTHETICALLY PLEASING WAY [167] SYSTEM THAT VALUES WATER AS AN INDISPENSABLE RESOURCE IN A NEW SITUATION [93] THE CONFLUENCE OF INDUSTRY AND HYDROLOGICAL ECOSYSTEMS [197] THE MEMBRANE BEHAVES LIKE THE HUMAN SKIN [269] THE RIPARIAN FOREST IS WIDENED [87] THE SINGLE-FAMILY HOME AS PART OF A DECENTRALIZED MICRO-WATERSHED MANAGEMENT SYSTEM [185] THE THIRD USER, THE RIVER, HAS OCCUPIED THE PARK [87] THIS FLOATING PARK CAN ALSO REMEDIATE THE WATER OF METALS AND OTHER TOXINS [183] TO RESTORE THE DYNAMISM OF A NATURAL MEANDER [85] TRANSFORM MONOFUNCTIONAL INFRASTRUCTURE INTO MULTIFUNCTIONAL COMMUNITY WATERSHEDS [261] TRANSFORM THE BROWNFIELD INTO A PUBLIC PARK [167] TRANSITION BETWEEN NATURE AND CITY [161] UNIQUE PUBLIC SPACE [161] VISIBLE AND ENJOYABLE [99] WATER CAN BE OBTAINED FROM AIR, WHICH IS STILL AN UNPRIVATIZED RESOURCE [269] WATER IS LANDSCAPE IN THE PARK [85, 87] WATER IS NEITHER CREATED NOR DESTROYED, BUT TRANSFORMED [269] WATER IS THE FOCAL POINT OF THE SITE [252, 253] WATER WILL BE IMAGINED AS A NEW PUBLIC SPACE [181]

DEFEND INDEX [52,139,240] A BATTLE BETWEEN CITY AND COUNTRY [65] A HEROIC MACHINE TO A NONCHALANT ROUTINIZING OF THE MACHINIC LANDSCAPE [65] BETWEEN MACHINE-EQUIPPED MODERN SOCIETY AND PRIMORDIAL NATURE [63] BROKERING DEALS WITH NEIGHBORING STATES TO "BANK" THEIR WATER [241] COLLECTIVE EFFORT TO "EDUCATE" BY TAMING A WILD NATURE WITH HEROIC TECHNICAL, JOURNALISTIC, AND GRAPHIC LANGUAGES [61] COMPLEMENTING THE NATURALIZATION OF AN ARTIFICIAL ECOLOGY, EVEN A "ZOO-IFICATION" OF THE MACHINE APPEARS. [67] EFFECT OF THIS PROJECT ON THE REGION HAS BEEN ENORMOUS [53] EVERY BIT THE HEROIC MODERNIST IDEAL, THE PRECISION-CALIBRATED DEVICE AND ITS MACHINIC LANDSCAPE WOULD TAME A THREATENING RIVER BEAST [59] EXCEPTIONALISM [63] GAINING GROUNDWATER RIGHTS FROM OTHER PARTS OF THE STATE, [241] GLORIFICATION OF THE MACHINE [63] HALLMARK OF MODERNIST INFRASTRUCTURAL PRODUCTION [59] HEROIC IMAGERY [51, 65, 272] HUNDREDS OF MILES OF BOOMS WERE LAID OUT, ENCIRCLING THE OIL [141] IDEOLOGIES OF RATIONALIZATION AND CONTROL [61] IT'S NOT JUST FOR RECREATION [65] LAYING HUNDREDS OF MILES OF PIPE [241] MACHINIC PHANTASMAGORIA ON DISPLAY [63] MEN AND CUTOFFS TRANSFORM SNARLING WILDCAT INTO PURRING KITTEN [61] PATCHES OF OIL WERE DISPOSED OF WITH A CONTROLLED BURN [141] PERCHED ALWAYS ON HIGH GROUND, LEVEE, OR CAUSEWAY [63] PLANS TO CREATE A SERIES OF MASSIVE INFRASTRUCTURAL PROJECTS TO INSULATE THE COAST [53] PREFERRING THE SCALE AND CONTROL [53] PUBLIC FASCINATION WITH A "GROTESQUE" NATURE, TECHNOPHILIC FETISHISM OF A NEW NATURE-QUELLING MACHINE [61] PURSUIT OF SECOND NATURE [63] SENSATIONALIZED AS A SUDDEN AND URBAN CATASTROPHE [65] SEVENTEEN DESALINATION PLANTS HAVE BEEN PROPOSED ALONG THE CALIFORNIA COAST [261] SLOWLY EVOLVED OVER TIME, FROM A SIMPLE SERIES OF

DAMS TO A COMPLEX SYSTEM OF DAMS, SLUICES, STORM WALLS, AND STORM GATES [53] SOME 7,000 SHIPS WERE DEPLOYED TO CONTAIN THE ESCAPED OIL [141] SPECTACULARIZED IMAGERY OF THE SPILLWAY'S OPERATION [63] SPILLWAY BECOMES A ZOO FOR THE SPECTACLE OF AN ARTIFICIAL ECOLOGY [67] STARK MODERNIST GEOMETRIES OF THE STRUCTURE STAND OUT AGAINST A "RAGING" RIVER [61] TAKES ADVANTAGES OF THE CONTRIBUTIONS FROM THE FLOOD WHILE CONTROLLING THE RIVER. [69] THE RIVER-MACHINE SPECTACLE [65] THESE CITIES ARE FORMED OF AN ENVELOPING SEAWALL [69] THE SPECTACLE OF HUMAN LABOR IN THE PROCESS OF CONTROL OVER THE BEAST [63] THE WILD RIVER BEAST TAMED BY THE PROWESS OF MODERN MACHINE ENGINEERING [63] TO MAINTAIN THE GREATEST DEGREE OF CONTROL [53] URBANITES TO GAZE UPON NATURE-AS-SPECTACLE [63] WATER CHANNELED BY STRICT ORTHOGONAL MACHINE GEOMETRIES [63] WATER IS IMPORTED OVER VAST DISTANCES [263] WATER PURIFICATION PLANT AND PARK [145]

RETREAT INDEX [72,148,246] ACTIVATING DEMOLITION SITES AFTER THEIR POST-DESTRUCTION ABANDONMENT, [149] ADJUST BY MOVING OR COMPLETELY DEMOLISHING AND RECONSTRUCTING THE NUMEROUS CULTURAL AND INSTITUTIONAL BUILT RESOURCES ALONG THE SHORE [81] A FLOATING PRODUCT THAT CAN BE ADDED TO ANY CITY, SIMILAR TO ADDING AN APP ON YOUR SMARTPHONE [249] A NETWORK OF DEMOLISHED PALESTINIAN RESIDENTIAL SITES SIT ABANDONED [149] A SERIES OF FORMAL AND SPATIAL CHANGES [81] BLURS THE DISTINCTIONS BETWEEN "WASTE" AND "RESOURCE" AND BETWEEN "LAND UTILIZATION" AND "LAND ABANDONMENT" [149] CONFRONT THIS EVER-CHANGING LANDSCAPE [81] HILLS ARE THE CULMINATING FEATURE OF THE GOVERNORS ISLAND MASTER PLAN AND WILL RISE 30 TO 80 FEET ABOVE SEA LEVEL [75] HYDROLOGICAL AND URBAN GRADIENT. [157] IMPORTING HUNDREDS OF THOUSANDS OF CUBIC YARDS OF FILL MATERIAL [25, 75] LIFTED THE MAJORITY OF THE ISLAND OUT OF THE FLOOD ZONE [75] MATCHING THE EROSION OF THE CLIFF ON WHICH THE BUILDING SITS [81] NEW ISLAND ELEVATION THAT LIFTS THE ROOT ZONES OF THE NEW PLANTING AWAY FROM BRACKISH WATER AND ABOVE PROJECTED FLOOD LEVELS [25, 75] PLAN FOR THE CONTINUING LONG-TERM INCREASE IN MEAN SEA LEVEL AND FOR THE MORE FREQUENT AND VIOLENT STORMS THAT ARE EXPECTED TO ACCOMPANY CLIMATE CHANGE ALONG THE EASTERN SEABOARD [73] REDISTRIBUTION OF LAND-USE PATTERNS AND TRANSPORTATION NETWORKS WITH RESPECT TO HYDROLOGICAL SYSTEMS [157] REORGANIZED PILES OF CONCRETE DEBRIS ARE SORTED ON-SITE TO CREATE A NEW WASTEWATER FILTRATION SYSTEM [149] RESILIENT IN THE FACE OF RISING WATERS [73] REZONES LAND USES ACCORDING TO HYDROLOGICAL ADAPTABILITY [157] SHIFTED LAND-USE PATTERNS FROM FORESTLANDS AND AGRICULTURE TO SUBURBS [155] TOWARDS A NEW ERA OF WATERFRONT RECREATION [73] TRANSPORTATION NETWORKS BECOME ECOLOGICAL FRAMEWORKS [157]

WATER INDEX

DESIGN STRATEGIES FOR DROUGHT, FLOODING AND CONTAMINATION

TABLE OF CONTENTS

Left:
Hieronymus Bosch, *The Hell and the Flood*, panel from *Deliverance from the Deluge* (circa 1450–1516). Oil, 27.2 x 15.4". Museum Boijmans Van Beuningen, Rotterdam.

"For this purpose [we] chose that substance which is normally liquid, colorless, incompressible and horizontal in small quantities; having a curved surface, blue in depth and with edges that tend to ebb and flow when it is stretched; which Aristotle terms heavy, like earth; the enemy of fire and renascent from it when decomposed explosively; which vaporizes at a hundred degrees, a temperature determined by this fact, and in a solid state floats upon itself—water, of course!"
Alfred Jarry, from *Exploits and Opinions of Dr. Faustroll, Pataphysician*

A HYDROLOGICAL TRAGEDY IN THREE ACTS: INTRODUCTION

Seth McDowell

Water, our indispensable resource, is the subject of this book highlighting critical design projects from around the world that radically engage issues of drought, flooding, and water quality. The projects in this collection reveal design strategies that respond to the mounting global water crisis facing the earth in the 21st century, largely resulting from climate change and population growth. Humanity's relationship with water is strained, and those who design the built environment demand that relief reside in the choreography between the city, architecture, landscape, and that natural resource—water.

In the wake of an escalating global crisis with water, this book is a critical inventory and analysis of innovative architecture and design solutions to address the rising, disappearing, and contamination of water. As fear of ecological disaster ferments in contemporary architectural discourse, design competition briefs, conference topics, and journal themes optimistically call for designers to reconcile or reimagine the relationship between water, architecture, and the city. Anxiety is elevated by the onslaught of extreme weather in the form of superstorms, hurricanes, tsunamis, landslides, floods, and droughts whose frequency and intensity continue to increase. Couple the ever-present exposure to disaster with scientific data that suggests a future characterized by climate change and population growth, and we have the ingredients for a full-fledged paranoia: the perfect motivation for absurd, expansive, and radical building projects. *Water Index* examines three hydrological tragedies (flood, contamination, and drought) through strategies that offer methods for controlling, escaping, or adapting to this vital natural resource.

The book is a collective vision of the future, providing solutions for every continent and spanning the disciplines of urban design, landscape architecture, and architecture. The overarching goals of *Water Index* are (1) to create an enduring manual and critical manifesto for water development and design in the 21st century and (2) to acknowledge crisis-initiated design as an important trajectory for architectural discourse. The more immediate objectives are (a) to recommend concrete, actionable strategies for water management in diverse international contexts, (b) to reach both local and global audiences with these wide-ranging design solutions, and (c) to suggest a precedent for multidisciplinary approaches to thinking about the relationship between water and the built environment. *Water Index* highlights a moment when designers have linked formal concerns with social, ecological, and political agendas offering solutions for expanding global problems.

This book is an index of a vast transformation occurring in architecture, landscape architecture, and urban design—a transformation that is particularly evident in the fundamental relationship between human habitation and water. It is the transformation from the conquering paradigm, in which construction becomes a mechanism for controlling nature, to a more adaptive and responsive agenda in which construction is situated as a latent system, fluctuating in response to the forces of ecology. Computational and simulation tools now enable designers to precisely evaluate the symbiotic relationship between constructed and natural environments, prompting more responsive iterations in the technological conquest over nature. This evolving ecological precision encourages solutions of a calculated language. We have transitioned from a mode of controlled nature to the condition of calibrated ecologies.

The terms, projects, and essays in this book echo this moment of calibrated ecology. While the imagery is dominated by blue and green, do not be seduced; these are often highly artificial environments, if not full-fledged mutants, whose ecosystems are generated through structured design operations. In an age of genetic engineering, design has adopted the ability to manipulate the DNA of natural systems in order to achieve an interdependent relationship between the constructed and the natural. For example, wave attenuation and water quality can be rectified by a biotic process of implanted oyster farming, as seen in the project "Parallel Networks" by Ali Fard and Ghazal Jafari (page 179), rather than by a hard control system of seawalls or filtration plants. In this project, a material intervention is introduced as a living system and positioned within an ecosystem as a latent yet productive infrastructure.

This approach of calibrated, adaptive design relies on process. Water is a variable system that has generated a preference for design solutions articulated as procedures. These solutions incorporate time, phasing logics, seasonal fluctuations, and changing datum. The ebbs and flows of water prompt an elongated time frame for design. Water-sensitive projects are not calculated for a single moment or ideal; rather, they have become responses to a series of possible futures or "what if" scenarios that span across time and varied conditions. This is the basis of an adaptive aquatic architecture: architecture embedded with a responsive intelligence that reacts to hydrological disruptions, fluctuations, and mutations.

For, water has instituted a new materiality. Embedded in the conditions of process is this ephemeral, living material. The presence of water initiates growth systems that are formulated by environmental stimuli and controlled by design parameters. Many of the projects in this book examine material conditions that grow or mature, initiating remediation on water during their development, in a scripted choreography. Emerging aquatic systems act as material catalysts—placed in, on, or around water, they facilitate, rather than dictate, an action.

This ephemeral, adaptive, reactionary position for design in relation to water has surfaced largely due to the fact that the 21st century is facing an elevating crisis with water. The seas are rising, the deserts are expanding, and the cities are growing at alarming rates. The underlying agenda for *Water Index* is to employ the dystopian threats surrounding water as a provocation. We must construct new mechanisms for subsistence. Yet, in a state of expanding fragility, what is the format of deliverance for these hydrological economies and ecologies? Is it sufficient to build life support systems that offer momentary alleviation by means of controlling and channeling water for specific, temporal needs and desires? Or is it possible to reframe the age-old problems in order to formulate new solutions? The aim of this index is to position a collection of design strategies that range from the historic practice of control to the contemporary practice of adaptation.

Hieronymus Bosch's "The Hell and the Flood", a panel from the triptych *Deliverance from the Deluge* painted on the inside of an altarpiece, reveals an image of a post-apocalyptic landscape. The water has subsided, and rotting corpses of drowned sinners litter the land. This Old Testament narrative depicts water as a device for ethical cleansing. Yet, if the moral connotations are set aside, the Deluge in Genesis 8 is essentially a story of man's technological adaptation to imposing natural phenomena. The ark, as a response for survival, has become the paradigm for humanity's response to ecological disaster: construct a mechanism for deliverance. This book is essentially a catalog of mechanisms that enable the control of, escape from, and adaptation to water.

Three hydrological crises are examined in *Water Index*, through the lenses of three strategies for deliverance: "The Rising", "The Contaminated", and "The Disappearing " frame three acts that present fundamental complications for humanity's dependence on the natural resource. Within each of these "acts," design responses are examined in relation to a pallet of tools or mechanisms that enable relief. Each water crisis is examined in accordance with strategies for defense, retreat, and adaptation.

The first strategic operation examined in each

act relates to the method of defense. DEFCON,[1] the term used by the United States Armed Forces to describe the alert state known as a defense readiness condition, appropriately characterizes this militaristic approach to controlling water and water's relationship with populations. The archetype for this strategy is the dam, the levee, or the aqueduct—engineered structures of a monumental scale that seek to defend against the given crisis. These are structures of the state, artifacts of modernity with the agenda of conquering, capturing, and distributing water for the purpose of progress. These tools for defense largely operate outside of a public realm; they are seen as mere infrastructure and are positioned at a distance from civic activities. In Act One: The Rising, Travis Bost examines the 1935 Bonnet Carré Spillway in New Orleans in his essay "The Spectacle of Water and Machine," highlighting with great depth the covert objectives entrenched in the "iconography of machine technology" found in this strategy for defense.

The second strategy examined in each act is one of retreat. The archetype for this approach is the ark. It is a tool for escaping the crisis, a tool that enables mobility. This approach allows people, structures, and ecologies to migrate or relocate, prompting a repositioning relative to water. Throughout the book this strategy is interpreted in various ways, from a nomadic exercise as seen in Ben Gregory and Ed Ford's "Connections on Uncertain Ground" (Act One: The Rising) to the practice of establishing a new datum for inhabitation similar to West 8's project for Governors Island Park (Act One: The Rising), where imported fill and new topography are deployed as tools for elevating dry, safe ground.

The third and final strategy examined in each act is adaptation. This strategy references the archetype of the step-well or bioswale as a mechanism that accepts, if not celebrates, the given hydrological instability. As opposed to the strategies of defense or retreat, adaptation allows water to invade the spaces of inhabitation, and the architectural and landscape interfaces then react to resolve or remediate the stressed water condition. Thus, urban space, landscape, and architecture are calibrated to respond to the fluctuations of an unstable aquatic ecology. As previously mentioned, this strategy is heavily situated in contemporary practice, where design intent is often framed as a response to ecological fluctuations. This approach can be seen in projects like Turenscape's Shanghai Houtan Park (Act Two: The Contaminated), where landscape is deployed

as a living system that introduces a regenerative process in a former industrial brownfield.

This book is the result of a yearlong focus on water at the University of Virginia's School of Architecture (UVA SARC), which was initiated by Dean Kim Tanzer as an exercise in bridging a common interest of four academic departments: Urban Planning, Architectural History, Architecture, and Landscape Architecture. During the 2012-2013 academic year, UVA SARC investigated the relationships between water, urban design, landscape, architecture, history, art, music, literature, and the environmental sciences. This examination included public lectures by Kate Orff, founding partner of SCAPE Landscape Architecture, whose water-related credentials include Oyster-Tecture, a project for the Museum of Modern Art (MoMA) 2009 exhibition Rising Currents and the project Living Breakwaters, the winning submission for the 2014 HUD Rebuild by Design Competition and the 2014 Fuller Challenge. Adam Yarinsky, founding partner of Architecture Research Office (ARO), also lectured on MoMA's Rising Currents exhibition and ARO's participation and submission. Katherine Wentworth Rinne, author of The Waters of Rome, presented a public lecture entitled "Plumbing Rome" that discussed the historic relationship between Rome's public space and the city's water infrastructure.

The year of water-centric activities at UVA SARC also included a series of symposiums and workshops that activated a discourse about the past, present, and future relationship between the built environment and water. The India Initiative Symposium, led and coordinated by SARC Professors Peter Waldman and Phoebe Crisman, brought together a panel of distinguished architects and educators to discuss "Water as a Spatial Generator of the Emergent Megacity and the Enduring Village in India." The panel included Tod Williams, founding partner of Williams-Tsien Architects; Pankaj Vir Gupta, architect, Delhi; David Turnbull, professor at The Cooper Union and founding partner of Harrison-Turnbull Architects; and Dr. Vikram Prakash, professor of architecture and urban planning at University of Washington, St. Louis.

UVA SARC collaborated with landscape architect Adriaan Geuze, founding partner of West 8 and the 2012 Robertson Chair Visiting Professor, for a ten-day, all-school design workshop that examined the Rivanna River's civic and ecological role for the city of Charlottesville, Virginia. This design vortex brought together over 400 students from all departments

Above:
The University of Virginia's School of
Architecture all- school vortex: *River, City,
and County*. Final Reviews with WEST 8's
Adriaan Geuze, 2012 Robertson Chair
Visiting Professor.

Below:
2010 *Rising Currents* exhibition at the
Museum of Modern Art, New York City.

within the school, forming 30 multidisciplinary teams focused on generating radical ideas for celebrating the interface between riparian ecology and the city. This effort included guidance from the local community and the City of Charlottesville planning officials.

Lastly, "After the Deluge: Reimagining Leonardo's Legacy" was a four-session symposium supported by a UVA Arts Council Grant and coordinated by Kim Tanzer and George Sampson that presented talks focused on water's relationship to the "practical imagination." In the 15th century, Leonardo da Vinci fused an artist's persuasive understanding of water with an engineer's reasoned response in his *Deluge* series of drawings and his proposal to contain the flooded Arno River. The symposium's four dialogues between diverse perspectives explored the relationship between art and science with keynote presentations by artist/biologist Brandon Ballengée, computational composer Matthew Burtner, Princeton University PhD candidate Leslie Geddes, and artist/environmentalist Margaret Ross Tolbert.

UVA SARC's interest in water and the urgent issues surrounding the resource echoes a phenomenon that emanates on a global scale. We are in a moment when designers and planners are exceedingly eager to devote all available professional and creative resources to the most relevant and urgent matters of the time, including water relations. In this scenario, crisis prompts design motives and design acts as a catalyst for recovery. This paradigm has been highlighted by Barry Bergdoll, former Philip Johnson Chief Curator of Architecture and Design at MoMA, and two MoMA exhibitions that situate designers in the midst of two global disasters: sea-level rise and economic recession. Both *Rising Currents: Projects for New York's Waterfront* [2] in 2010 and *Foreclosed: Rehousing the American Dream* [3] in 2011 asked designers to rigorously analyze social, political, environmental, and economic problems and devise solutions that are both systemic and spatial. These two exhibitions became extremely charged while becoming vastly popular due to their ambition and relevancy. The projects signify that scientific and economic anxieties have infiltrated the world of architecture and architecture is grasping to remain relevant in a world fixated on uncertainty.

This book does not aim to be an ambulance chaser; however, it must be acknowledged that its relevance is only increasing as our complications with water continue to become more severe and frequent. At the time of the book's first formulation [4] the news was dominated

by tragic water-related catastrophes: Hurricane Sandy drastically disrupted the northeastern United States, taking over 150 lives and causing over $50 billion in damage; Hurricane Isaac hit the Gulf Coast exactly seven years after Hurricane Katrina; the summer of 2012 brought the worst drought to the United States in a quarter of a century, while the Mississippi River also shrank to its lowest levels in a quarter century; flooding brought chaos to the Philippines; and scientists at the US National Snow and Ice Data Center reported in August 2012 that the amount of ice covering the sea surface had shrunk to its lowest recorded level.

In order to emphasize urgency, each act of the book begins with a Call to Action. These calls are critical, previously published texts by select historians, environmental scientists, urban geographers, or theorists whose expertise reside in water issues. Tracy Metz outlines how Dutch culture has been dominated by the fear of floods and how the Disaster of 1953 still haunts the national psyche in "Catastrophe," an excerpt from her 2012 book *Sweet and Salt* (Act One: The Rising). Matthew Gandy explores a combination of developments during the 19th century that culminated in the emergence of the "bacteriological city" as an identifiable urban form in his essay, "The Bacteriological City and Its Discontents" (Act Two: The Contaminated). Then, in Act Three: The Disappearing, William Moorish dissects Los Angeles' stressed relationship with water supply in the essay "The Urban Spring: Formalizing the Water System of Los Angeles." These essays frame the problems or shed light on the specifics of flooding, drought, or contamination. They provide a historic narrative and situate our current fears and fixations within a lineage of hydrological dystopias.

There are many books and films focusing on water which document the struggles we are facing in the 21st century. When starting this book, we faced the daunting task of outlining a trajectory that would offer new information on a topic that is already over-published. Our hope is that this collection of work will provide a mirror for the design community and that in this mirror we can see a reflection that is both admirable and frightening. While architecture, landscape architecture, and urban design are making great, worthy strides in tackling issues of water and climate change, I cannot help but feel that there are fashionable theatrics involved in this pursuit. As we shuffle through these pages and gaze into this collective reflection, I think we must ask ourselves, Are we on a hydrological snark hunt?

Notes ---

1. The defense readiness condition (DEFCON) is an alert state used by the United States Armed Forces. The DEFCON system was developed by the Joint Chiefs of Staff and unified and specified combatant commands. It prescribes five graduated levels of readiness (or states of alert) for the U.S. military, which increase in severity from DEFCON 5 (least severe) to DEFCON 1 (most severe) to match varying military situations.

2. MoMA and P.S.1 Contemporary Art Center collaborated to address one of the most urgent challenges facing the nation's largest city: sea-level rise resulting from global climate change. "Though the national debate on infrastructure (was at the time) focused on "shovel-ready" projects that would stimulate the economy, the joint effort enabled new research and fresh thinking about the use of New York City's harbor and coastline." An architects-in-residence program at P.S.1 (November 16, 2009–January 8, 2010) brought together five interdisciplinary teams to re-envision the coastlines of New York and New Jersey around New York Harbor and to imagine new ways to occupy the harbor itself with adaptive "soft" infrastructures that would be sympathetic to the needs of a sound ecology. These creative solutions were intended to dramatically change residents' relationship to one of the city's great open spaces.

3. *Foreclosed: Rehousing the American Dream* was an exploration of new architectural possibilities for cities and suburbs in the aftermath of the foreclosure crisis. During summer 2011, five interdisciplinary teams of architects, urban planners, ecologists, engineers, and landscape designers worked in public workshops at MoMA P.S. 1 to envision new housing and transportation infrastructures that could catalyze urban transformation, particularly in the country's suburbs. Responding to *The Buell Hypothesis*, a research report prepared by the Buell Center at Columbia University, teams—led by MOS, Visible Weather, Studio Gang, WORKac, and Zago Architecture—focused on a specific location within one of five "megaregions" across the country to come up with inventive solutions for the future of American suburbs. This installation presented the proposals developed during the architects-in-residence program, including a wide array of models, renderings, animations, and analytical materials.

4. *Water Index* was formulated and produced from fall 2012 to fall 2014.

ACT ONE: THE RISING

For the last 2.5 million years, the earth has participated in an ecological dance with the massive continental ice sheets covering the northern hemisphere. This dance entails a constant oscillation between ocean levels and frozen water deposited in the glaciers—a give-and-take ballet that has instigated fluctuations in water level by more than 500 feet (150 m). "Shorelines have moved landward or seaward tens of miles as a result."[1] Rising water level is not a recent phenomenon; it has been an unceasing operation over the history of the earth. What is urgent for us today is that water is approaching major centers of population and capital. We humans are the third, and somewhat awkward, participant in this environmental dance. For the first moment in earth's history, sea level rise is encountering a densely developed edge condition, putting millions of lives, thousands of buildings, and hundreds of cities at risk.

This is the situation of "The Rising": water is breaching established thresholds and invading the city, architecture, and constructed landscape. The breach is no longer an anomaly, but rather a common occurrence, and is forcing us to rethink the built environment's relationship with water. Is it possible to build mechanisms to keep the rising water out? If so, how long will these devices protect us? Should we flee, surrendering home and culture to the forces of nature? If so, where do we go? Is it possible to imagine an amphibious architecture? Can we build environments that have the ability to withstand or adapt to fluctuating water levels?

The first act of this index of water propositions will investigate a liquid invasion on the built environment. Two Call to Actions set the tone, highlighting the urgency by framing sea level rise and flooding within a historical context. Orrin Pilkey and Rob Young's "People and the Rising Sea," taken from a chapter in their 2009 book *The Rising Sea*, situates sea level rise within a global, historical context. This essay recalls the "catastrophic submergence" of several ancient cities along the shores of the Mediterranean, including Alexandria, a city built by Alexander the Great, strategically located as a central harbor. Pilkey and Young underscore that rising water is associated with disaster when it occurs instantaneously; such instantaneous sea level rise is a "surprisingly widespread phenomenon" linked to earthquakes, tsunamis, hurricanes, or storms. Thus, exceptional climatic or geological events great-ly heighten the built environments' vulnerability to water.

"The countries with the biggest problems are the atoll na-tions; deltaic countries such as Egypt, the Netherlands, and Bangladesh; and countries with large, low-lying, heavily de-veloped coastal plains such as the United States, Brazil, and China,"Pilkey and Young write.[2] Their essay, "People and the Rising Sea," takes a global survey of vulnerability, touching on the conditions of Vietnam, Myanmar, Bangladesh, Egypt, the Netherlands, Singapore, Indonesia, and the United States.

Following this global account of the effects of the rising sea, another Call to Action is supplied by Tracy Metz in her essay "Catastrophe," taken from her 2012 book *Sweet and Salt: Water and the Dutch*, co-written with Maartje van den Heu-vel.[3] This account specifically examines a nation and culture dominated by the fear of flooding. Metz traces the effect that the 1953 North Sea Flood has had on the Dutch psyche and questions the "sense of security" that is provided by the large infrastructural projects of the Delta Works. By revealing the deep contradictions within the gigantic scale and ambitions of the Delta Works, the essay emphasizes that while on paper the country appears to be prepared for the next major flood, the realities of aging infrastructure, antiquated standards, and inaccessible evacuation plans suggest that the Dutch are in a very precarious situation. Metz associates the issues in the Netherlands with the disaster in New Orleans due to Hur-ricane Katrina, revealing that this false sense of security is a liability for many cities and populations across the globe.

The Calls to Action thus promote an unrelenting sense of urgen-cy: populations, capital, and cultures are at a high risk of being engulfed by the bodies of water surrounding them. Every city, building, and landscape residing in a delta or low-lying coastal plain must develop a strategy for dealing with the management of rising water, especially during extreme weather. Following these Calls to Action are three strategic responses: a response of defense, a response of retreat, and a response of adaptation.

Three projects provide examples of defense against the rising: the Delta Works in the Netherlands, Travis Bost's account of

New Orleans' Bonnet Carré Spillway in his essay "The Spectacle of Water and Machine," and the speculative project Hydropolis by Marion Ottmann, Margaux Leycuras, and Anne-Hina Mallette. These three projects share a common strategy of constructing large mechanisms to keep rising water levels away from people and the built environment. The archetypes for this strategy are the dam, the levee, the dike, the pump, the storm wall, the storm gate, the weir, and the spillway—engineered structures of a monumental scale that seek to protect against water's encroachment. These are structures of the state, artifacts of modernity, with the agenda of conquering, capturing, and redistributing the natural fluctuations of water for the purpose of sheltering capital, culture, and people.

Travis Bost's in-depth examination of the 1935 Bonnet Carré Spillway in New Orleans details the covert objectives entrenched in the "iconography of machine technology" found in this strategy for defense. Although these tools for defense largely operate outside of a public realm, the monumental scale of these solutions produces an engineered spectacle. The "stark modernist geometries" of the Bonnet Spillway, a common attribute of 20th-century flood-proofing infrastructures, render the ideologies of rationalization and control against the "dangerous beast" of unpredictable waterways.

The strategy of defending against rising waters concludes with the Hydropolis project, a modular system of cities embedded within the Nile in Egypt. This project complicates a straightforward explanation of defending against rising water because it proposes to locate the city in the middle of the rising water source. However, in an act of defense, these modular cities are fortified with a system of seawalls, dikes, and floodgates. The cities form rings and the rising water flows around their idealized geometry. Meanwhile, within the city's seawall a productive, calibrated ecology emerges from a system of controlled canals and lakes. The proposal is a strange, utopian hybrid of the modernist strategy for control and the contemporary fascination with adaptive ecology.

"The Rising" then moves to a second response to rising water levels within the territory of inhabitation: retreat. The projects using this strategy move people, buildings, and communities away from the threat. It is an operation of evacuation, of mobility, of migration. The archetype for this approach is the ark, a tool for escaping the crisis, a tool enabling mobility and prompting a repositioning relative to water.

This strategy is interpreted by two formats in the condition of rising water. One approach is a nomadic exercise as seen in Ben Gregory and Ed Ford's "Connections on Uncertain Ground." Here, architectural tectonics enable a building constructed on the receding Cape Cod seashore to periodically disassemble and retreat. It is an architecture that slowly migrates with a landscape in flux.

Another interpretation of retreat is found in West 8's project for Governors Island Park in New York City. This project withdraws from the water by establishing a new datum for inhabitation. The occupied ground is literally lifted out of the flood zone using imported fill to create a new topography: "By importing hundreds of thousands of cubic yards of fill material, West 8 has created a new island elevation that lifts the root zones of the new planting away from brackish water and above projected flood levels." These two projects reveal that retreat is both a vertical and a horizontal operation.

The final strategy for addressing threats of rising water comes in the form of adaptation. This strategy references the archetype of the step-well, a mechanism that accepts, if not celebrates, hydrological instability. As opposed to defending or retreating, adaptation allows water to invade the spaces of inhabitation, at which point the architectural and landscape interfaces act to negotiate and sustain the presence of water. Public space, landscape, and architecture are calibrated to respond to the fluctuations of an unstable aquatic ecology. This strategy is heavily situated within contemporary practice, where design intent often parallels tidal fluctuations.

Two realized projects exploiting adaptation as strategy are aldayjover arquitectura y paisaje's Aranzadi Park in Pamplona, Spain, and De Urbanisten's Watersquare Benthemplein located in Rotterdam. Aranzadi Park presents a landscape that is calibrated to interact with the flood, and water is treated as a living material within a shared space. De Urbanisten's Watersquare takes a similar approach, allowing the flood to invade the public space. However, this project is situated within the density of the city as a hardscape that transforms from public square to public basin with the weather. Both projects reveal that water can occupy civic space if the physical and natural interfaces are designed to sustain the dynamics of hydrology.

The third project in this operation of adapting to rising waters is Isaac Cohen, Kate Hayes, and Jorge Sieweke's "Swamp Thing: Smart Grid, Smarter Water Management in New Orleans, LA," which examines the complex delta region surrounding New Orleans and the Mississippi River. The project's narrative is driven by the desire to unleash the Mississippi River, allowing the wild beast to "forge a new course, away from New Orleans." This project exemplifies the adaptive approach as it radically abandons static control systems and deploys "more progressive and adaptive" methods for managing water. Here, infrastructure is linked to ecology, and the swamp becomes a multifunctioning urban mechanism.

In the following pages you will embark on the situations and solutions of rising water levels. These projects address the rising as an operation related to sea level rise, river dynamics, and urban hydrology. They range from fantastic speculations to constructed realities. While some provide very specific, technical solutions, others are included to generate larger, conceptual discussions. The hope is that this index of possibilities for the future can catalyze a discourse about the rising waters and inspire design methodologies to deal with water's encroachment.

Notes ---

1. Orrin H. Pilkey and Rob Young, *The Rising Sea* (Washington, DC: Island Press/Shearwater Books, 2009).

2. Orrin H. Pilkey and Rob Young, "People and the Rising Sea," in *The Rising Sea* (Washington, DC: Island Press/Shearwater Books, 2009) 132.

3. Metz, Tracy, and Maartje van den Heuvel. Sweet & Salt : Water and the Dutch. Rotterdam: NAi Publishers , 2012.

ELING TO THE DELTAS [34] LOWEST-LYING AREAS OF THE CITY [48] LOW-LYING [23, 24, 34, 35, 40, 41, 48] MAINTENANCE BACKLOG OF THE EXISTING DIKES AND BARRIERS [45] MEGASTORMS [34] MELTING OF BEACH PERMAFROST [32] MONSOON RAINS [33] MORE FREQUENT AND MORE INTENSE RAINFALL [40] NATURAL DISASTER [43, 48] NO NATIONAL EVACUATION PLAN [41] NOWHERE FOR THE EVERGLADES TO GO [33] OCEAN LEVELS [23] OFFICIALS REMAIN UNCONVINCED THAT THE THREAT TO THE NILE DELTA IS REAL [35] ONCE-IN-TEN-THOUSAND-YEARS STORM [40] OUTDATED STANDARDS [41] PESSIMISTIC SCENARIOS [36] POPULATION GROWTH [42] PRECIPITANT OF DISASTER [32] PREDICTIONS HAVE BEEN EXAGGERATED [35] PRESSURE ON THE MECHANICAL SYSTEMS [49] PRETEXT OF SECURITY [45] PROBLEMS OF THE CITIES WILL TRUMP THOSE OF RURAL OR TOURISTIC AREAS [36] RAPIDLY RETREATING SHORELINES [31] REFUGEE [31] RISK FOR NATURAL DISASTERS [40] RIVER DYNAMICS [25, 85, 87] SAFETY IS INVISIBLE TO THE CITIZEN [40] SALINE INTRUSION [69] SALINIZATION [36] SALINIZATION OF WELLS [34] SEA LEVEL RISE [23, 25, 31, 32, 33, 34, 35, 36] SEASONAL DYNAMICS OF FLOOD WATER [85] SENSE OF SECURITY [23, 24, 40] SHIFTS IN POPULATION [36] SHORELINE EROSION [32, 33, 34] SHORELINE RETREAT [34] SHORELINES [23] STATIC CONTROL SYSTEMS [25] STORMS [32, 77] STORM SURGE [32, 34, 36, 73, 75, 77] STORM SURGE AND STORM WAVE POTENTIAL [36] STORM SURGES [33, 34, 36, 49] STORMWATER [75, 77, 94, 99, 103] STORM WAVE POTENTIAL [36] SUBMERGED BY MIDCENTURY [33] SUBMERGED CITY [31] SUBMERGED SANDBAR [33] SUBMERGED SANDBARS [32] SUBSIDENCE [31, 32, 34, 53] SUBSIDING DELTA [49] SUSCEPTIBILITY TO CLIMATE CHANGE [40] TECTONIC FORCES AND STORMS [31] TECTONIC SEA LEVEL CHANGE [32] TERMS OF THREAT [43] THE ACTIVE AGENT OF SEA-LEVEL-RISE DESTRUCTION [32] THREATENING RIVER BEAST [59] TIDAL AMPLITUDES [36] TSUNAMIS [23] UNCERTAIN GROUND [24] UNDERMINED FOUNDATIONS [81] UNPREDICTABLE WATERWAYS [24] UNPREPARED FOR THE STORM [34] URBAN HYDROLOGY [25] VAST COASTAL CITIES AT LOW ELEVATIONS. [31] VIOLENT NATURE [61] VULNERABILITY [23, 33, 35, 36, 45] VULNERABLE CITIES [35, 36] VULNERABLE DELTAS [34] WATER AS THOUGH IT WERE THE ENEMY [49] WATER DAMAGE [46] WEATHER TSUNAMI [39] WIDESPREAD PHENOMENON [23, 31] WORST CASE PREDICTION [65]

ACT ONE: THE RISING

CALL TO ACTION

The following text is taken from the book chapter "People and the Rising Sea" from *The Rising Sea* by Orrin H. Pilkey and Rob Young.

PEOPLE AND THE RISING SEA

From *The Rising Sea*, 2009

Orrin Pilkey and Rob Young

People have had to deal with changing sea levels for tens of thousands of years. Some of the earliest humans must have lived next to the rapidly retreating shorelines that followed the end of the last ice age more than ten thousand years ago—a period when sea level was sometimes rising more than 7 feet (2m) in a century. Society then was not encumbered with beachfront condominiums and vast coastal cities at low elevations. Now the time has come for today's populations to deal with the huge impacts of an expanding ocean, but it will be much more difficult for us to move back. Sea level rise will have an impact on land use of every sort, on parks, natural areas, subways, communications, sewers, roads, railroads, and buildings by the thousands. Whole towns, even nations, will disappear. A refugee problem of enormous magnitude is likely.

Sea Level Rise and the Ancients

A 2008 National Public Radio program entitled "Rising Sea Levels of Alexandria" brought into focus the long history of sea level rise that has left ancient parts of the Egyptian city submerged well offshore from the seawalls that protect today's Alexandria. The city sits on the edge of the Nile Delta, a site Alexander the Great chose two thousand years ago for its excellent potential as a centrally located harbor. The harbor entrance was guarded by the Pharos Lighthouse, one of the seven wonders of the ancient world, the ruins of which were discovered on the Mediterranean seafloor in 1994.

Alexandria is one of several submerged ancient cities along the shores of the Mediterranean, all victims of some combination of global ocean expansion, land subsidence, and relative sea level rise caused by local tectonic forces and storms. The evidence suggests that the Mediterranean Sea simultaneously engulfed the two cities Menouthis and Herakleion, along with Alexandria, sometime around 740 AD (the age of the youngest coins found in the rubble). The fact that the pillars and statues in all three cities largely lie in the same orientation on the seafloor is an indication that an earthquake destroyed them. The fact that the pillars and statues exist at all is evidence of catastrophic submergence of the city during an earthquake. A surf zone cannot slowly move across the remains of a city during a slow sea level rise without pulverizing the ruins.

A famed submerged city of the New World is Port Royal, Jamaica, founded in 1654 and submerged in a 1692 earthquake. By the 1670s, wild, freewheeling, and diverse Port Royal was a larger and more important merchant port than the staid, puritanical city of Boston, with a character about it that was decidedly missing from Boston. In the 1692 tremblor, most of the storehouse and port facility district fell into the harbor, ultimately killing perhaps five thousand people, well more than half the town's population.

Instantaneous sea level rise (and sometimes sea level drop) caused by earthquakes is a surprisingly widespread phenomenon. In addition to the examples of sunken ancient Mediterranean cities, there are many more recent instances of tectonic sea level change. Hundreds of miles of Alaskan shoreline south of Anchorage, for example, suddenly dropped from 1 to 4 feet (0.3 to 1.2 m) in the 1964 Good Friday earthquake, and along Colombia's Pacific coast, 50-mile-long (80 km) reaches of shoreline drop 2 or 3 feet (0.6 or 0.9 m) with every small earthquake on the local continental shelf. During the Great Tumaco Earthquake in 1979, the resulting tsunami killed two hundred and fifty people (almost all the residents) in the remote Colombian fishing village of San Juan de la Costa. It was a double whammy. Simultaneously with the earthquake, the barrier island sank, causing sea level to rise instantaneously by perhaps 1 meter (3.3 feet) and the erosion rate to accelerate, destroying most of the village's remaining buildings within a few years.

Sea Level Rise in America's Past

Some American towns and villages, too, have gone to sea in centuries past. These seaside villages existed in a time when people who lived along hazardous shoreline reaches did not expect government help. Many communities are threatened with the same fate today, but we have more resources to pour into the sea, for better or worse, to try to hold shorelines in place using seawalls and other measures. The most important lesson from these past episodes, as we'll see in the cases that follow, is that storms are the active agent of sea-level-rise destruction, and just as in the past, they are almost always the immediate precipitant of disaster. In our modern era of rising sea level, storms will become more important as their impacts push farther into the interior of the beach communities, across barrier islands, and into coastal cities.

Last Island (Isle Derniere), Louisiana (1856)

Ocean-facing barrier islands make up most of America's shoreline along the eastern and Gulf coasts, extending in an almost continuous chain from Long Island's southern shore to the Mexican border. In the 19th century, Last Island (Isle Derniere) was a retreat for the wealthy and privileged of Louisiana because, among other things, it provided a place to escape from the yellow fever epidemic that hit New Orleans in 1853. According to the U.S. Geological Survey, the 20-mile-long (32 km) Last Island was, at the time, a single barrier island with a mature maritime forest. The average elevation of the island was then probably about 5 feet (1.5 m) or less, with both significant

shoreline erosion and relative sea-level-rise problems due mainly to subsidence, or sinking of the delta.

In 1856, there were around a hundred homes on Last Island. In the Last Island Hurricane of August 10 of that year, a storm surge destroyed a multistory hotel and submerged the town's other buildings, killing at least 190 of the 400 vacationers there. Many of the bodies were carried more than 6 miles (10 km) inland of the island. Those who survived had clung to pieces of wreckage and to a grounded schooner that had arrived too late to carry them off the island. Hurricane parties seem to be an essential part of the legend of all hurricane disasters, and the Last Island Storm is no exception: a raucous party was said to have been underway at the hotel when the storm swarmed ashore. A single cow was the only animal to survive.

Storms are, in a sense, often the advance guard of sea level rise; in the absence of such a large storm, Last Island might have remained habitable as a resort, but only for a time. Since the Last Island Storm, Isle Derniere has experienced a 3.3-foot (1 m) sea level rise due to a combination of the expanding ocean and delta subsidence and has been cut into many smaller islands. The hurricanes of the last two decades have put the finishing touches on the island chain, turning them into mostly submerged sandbars, although there have been efforts to maintain them through beach nourishment.

Sea Level Rise and Nature Preserves

Rising seas will affect not just the built environment, but the natural world as well. Some of the natural environments that will be impacted, such as the Sundarbans, the Great Rann of Kutch, and the Everglades, are unique environments, and the animals unique to these ecosystems will have no place to migrate and thus may perish.

The Bering Land Bridge National Preserve consists of 2.7 million acres and sits astride the Arctic Circle. The Shishmaref Island Chain, part of this preserve, sits along the Chukchi Sea at the tip of Alaska's Seward Peninsula. Sea level rise in this and other Arctic preserves operates in tandem with greatly increased storm impacts because of longer periods of ice-free conditions on the ocean and melting of beach permafrost.

The Sundarbans of both India and Bangladesh is part of the Ganges Delta. It contains one of the world's largest mangrove forests and is the home of the endangered Bengal tiger. It is also home to a number of other endangered species, including marine turtles, crocodiles, and freshwater dolphins. Two million people live on the food (fish, crabs, mollusks, and honey) and firewood from the Sundarbans. The very fact that this is a mangrove forest is an indication of its close proximity to sea level and the great peril it faces from sea level rise. According to World Bank estimates, a 24-inch (60 cm) sea level rise would inundate the whole area and a 3.3-foot (1 m) rise would destroy the Sundarbans.

The Great Rann of Kutch in Gujarat, India, is the largest wildlife reserve in India and a World Natural Heritage Site. It is home to the last two thousand endangered Indian wild asses and to one of the world's largest colonies of greater and lesser flamingoes. The entire habitat, much of it a seasonal salt marsh flat during the monsoon rains, with scattered small islands where the wild asses survive, will likely be submerged by midcentury.

The Cape Hatteras National Seashore in North Carolina is a chain of thin, low barrier islands with a low sand supply. Sea level rise is already narrowing the width of the islands (by shoreline erosion on both sides). Thanks to a lifetime of studies by Stan Riggs, professor of geology at East Carolina University, these may be the best-understood barrier islands in the world. Riggs believes the islands may "collapse" or disappear, possibly within the next few decades. With each increment of sea level rise, the possibility of collapse increases. He anticipates that in a storm of sufficient duration and intensity, a large number of new inlets could open up, simultaneously isolating the eight small tourist villages within the National Seashore. If Riggs is right, a large portion of the Outer Banks will become a long, submerged sandbar quite similar to Louisiana's Isles Dernieres after Hurricane Katrina.

Sadly, the North Carolina Department of Transportation has not heeded this warning. A multimillion-dollar bridge spanning Oregon Inlet, a part of the National Seashore, will be rebuilt in the exact location of the current Bonner Bridge even though it is likely to go to sea within the life span of the bridge.

The Everglades cover more than 1.3 million acres in south Florida. As sea level rises, salt water will intrude on the Everglades and its fauna and flora will change in profound ways. The changes in birds, fish, and various plants are likely to be adverse ones, Everglades National Park Superintendent Dan Kimball has noted. Absent human influence on the planet, the changes would simply be the natural fluctuations that have occurred over millions of years as the level of the seas changed. But with humans involved and the demands of agriculture, tourism, and recreation, the changes are indeed likely to be adverse because there is nowhere for the Everglades to go. The multibillion-dollar Everglades restoration that is underway is designed for a 1-foot (0.3 m) sea level rise in the next century. Since a much larger sea level rise, certainly in excess of 3 feet (0.9 m) in that period, is a strong possibility, it is questionable what "restoration" will mean in the Everglades in the long term.

The Ria Formosa Nature Reserve is the Portuguese equivalent of the Cape Hatteras National Seashore. The reserve is a chain of seven barrier islands in the Algarve (in southern Portugal) near the Spanish border. The area has some unique problems that will make a response to sea level rise difficult. More than two thousand houses on the islands are built on public property and occupied by relatively wealthy squatters, who have even insisted on the right to build seawalls to protect their property. If the government is unable to remove the houses in the future, the place will become a battleground between engineers and sea level rise, to the great detriment of the reserve. Although the built-up environments of the coastal areas of the world may capture the lion's share of attention, the natural areas along the ocean shorelines that have been wisely preserved for future generations are clearly also in trouble.

Sea Level Rise and the Peril of Nations

Each coastal nation has unique problems to solve in a time of rising sea level, depending not just on the physical and biological nature of its coasts but also on its resources, politics, and culture. The impact on biodiversity, fisheries, and tourism along heavily developed coasts will be vastly different than in remote, lightly developed regions.

In broad perspective, the principal nation-scale impacts of sea level rise are likely to be the following:

- Loss of agricultural and nonagricultural land
- Flooding
- Increased vulnerability to storm surges
- Accelerated erosion of shorelines and artificial beaches
- Increased salinization of surface and groundwaters
- Increased flood heights of tidal rivers
- Loss of biodiversity (loss of marshes/mangroves/coral reefs)
- Loss of aquaculture, fishery, marina infrastructure
- Tourism decline as beaches erode and resorts are threatened

Land loss itself is generally viewed as the least of the problems brought on by rising sea levels. Of course, the major exception to this view is Tuvalu and other atoll nations that have nowhere to go. More important are the more indirect events that will adversely affect a nation's economic and social well-being, such as patterns of shoreline retreat relative to threatened buildings; loss of tourist infrastructure; loss of coastal infrastructure including seawalls, marinas, utilities, and roads; the salinization of wells; the destruction of sewage and drainage systems with consequent health problems; and loss of the nearshore ecosystem, including edible resources.

The countries with the biggest problems are the atoll nations; deltaic countries such as Egypt, the Netherlands, and Bangladesh; and countries with large, low-lying, heavily developed coastal plains such as the United States, Brazil, and China. The elevation of deltas is always low to begin with, and natural compaction of river muds leading to subsidence adds to the rate of sea level rise. In addition, delta subsidence is increased by extraction of water and oil and by the absence of replenishing sediment trapped by upstream dams. Furthermore, the continental shelves off most deltas are flat and gently sloping, a geometry that leads to storm surges that are particularly high and laterally extensive.

Vietnam

Among developing nations, according to a 2007 World Bank report, Vietnam will suffer enormously in terms of land area lost and people displaced from a 3.3-foot (1 m) sea level rise. The country has two major deltas, the Red River and the Mekong. These densely populated areas are, like the Nile Delta, the breadbaskets of the country. A 3.3-foot (1 m) rise will flood 1,930.5 square miles (5,000 km) of the Red River Delta and 7,722 square miles (20,000 km) of the Mekong Delta. Such a rise is expected to displace more than a tenth of the nation's people, gobble up 12% of its land area, and reduce food output by 12%.

Adding to the problem will be loss of sediment traveling to the deltas from likely future dam construction. China in particular is eyeing the Mekong River as ripe for damming. Loss of sediment will increase the subsidence rate because land will no longer build up; thus, the sea-level-rise rate will increase (just as is happening in the Mississippi Delta today) and shoreline erosion rates will grow in proportion.

Myanmar

The Irawaddy River in Myanmar splits into innumerable distributaries that end up discharging into the Andaman Sea on a 200-mile-wide (320 km) delta front. The Irawaddy Delta's rice production has long provided the breadbasket of Myanmar. In May 2008, the catastrophic Cyclone Nargis carved a corridor of destruction across the entire breadth of the lower Irawaddy Delta. The cyclone could not have followed a more disastrous route, flooding the deltaic lowlands, which are often less than 3 feet (0.9 m) above sea level, with a 12-foot (3.7 m) storm surge. Perhaps 100 thousand people died in the storm.

Myanmar's ruling military junta drew worldwide condemnation for its inept response to the human tragedy that was Cyclone Nargis. In Myanmar, the local government was totally unprepared for the storm, even issuing an initial estimate of 390 dead (off by three orders of magnitude). Few warnings were given before the storm arrived, no evacuations were attempted, and no plans were in place for shelters. The widely unpopular and insecure Myanmar military regime couldn't bring itself to quickly loosen its clamp on foreigners, even to allow aid workers immediate access to the huge numbers of homeless and starving people. The saga of this storm response is a vision into the future of vulnerable deltas in incompetently governed countries at a time of accelerating sea level rise. Cyclones like Nargis and other megastorms will have increasingly disastrous impacts on delta residents as sea levels continue to rise.

Bangladesh

The "Gift of the Rivers", as Bangladesh is appropriately known, is about 90% floodplain or delta and is the poster child of sea-level-rise impact. Around 15% to 17% of the country's produce is located at elevations within 3.3 feet (1 m) of sea level. Estimates of numbers of people that will be heavily affected by a 3.3-foot (1 m) sea level rise range from 13 million to 30 million, and at least 15 million likely will be forced to become environmental refugees.

Asia's two biggest rivers, the Ganges and the Brahmaputra, meet in Bangladesh and form the Meghna River, which finally brings the water across the world's largest delta to the sea through a complex pattern of distributaries. The delta's future problems will come from a combination of sea level rise and expected increase in cyclone intensity, which in turn will cause higher storm surges. The 1970 and 1991 storm surges caused by cyclones killed perhaps 500,000 and 140,000 people, respectively, while floods in 1992 and 1998 inundated more than half the country's land area. As sea level rises, the elevation change at the coast will reduce the gradient of the lower river and will thus cause the delta to drain more slowly and floods to penetrate farther inland.

Changes in agriculture and water quality and quantity will be the most obvious everyday manifestations of sea level rise in Bangladesh. Rice production will decrease dramatically. Already, on the outermost habitable portions of the delta, increased saltiness of the soil has changed crop yields, and a significant part of the rice crop has been lost.

The storm preparation of the democratic Bangladeshi government stands in sharp contrast to that of the Myanmar military regime. Bangladesh has responded to past catastrophes with coastal building codes, construction of 2,500 concrete shelters high atop pilings, installation of warning systems, organization of a 32,000-person rescue group, and well-publicized evacuation plans. These efforts have greatly reduced the vulnerability of this low-lying country.

Another ray of hope in Bangladesh, however faint, is the diversion of river silt into "soup bowl" depressions on the outer Ganges Delta. Like the transfer of sediment on the Mississippi Delta, the addition of new material adds much-needed elevation.

Egypt

The Nile Delta, perhaps the oldest intensively cultivated region on earth, makes up only 2.5% of Egypt's land area, but the World Bank projects that 9% of the country's population will be displaced by a 3.3-foot (1 m) sea level rise. The 3.1-mile-wide (5 km) stretch paralleling the delta coast is mostly below 7 feet (2 m) in elevation and is protected from storm floods and storm wave action by a chain of barrier islands and dune fields along the shoreline. But because of a combination of sea level rise and loss of sand trapped behind the Aswan Dam across the Nile, this protective band of sand is rapidly being lost.

Some officials remain unconvinced that the threat to the Nile Delta is real. Mostafa Saleh, head of Environmental Quality in Egypt, for example, believes that flooding predictions have been exaggerated to draw international attention to Egypt's problem. Saleh was quoted in 2008 in the *Middle East Times* as saying, "If sea levels rise by 1m that would bring the water inland by about 40 miles, so it is not necessarily a large portion of the delta"! Saleh's comments reflect a head-in-the sand attitude about sea level rise, not uncommon among politicians in many countries.

The Netherlands

The Netherlands has spent more money and intellectual capital preparing for sea level rise than any other country. Its research institutions are constanty studying the economic, social, oceanographic, engineering, and political aspects of sea level rise. There is also widespread recognition among the citizenry that they are experienced in protecting themselves from the sea and are ready and willing to spend much national treasure to continue to do so. Miles and miles of levees, dikes, nourished beaches and seawalls, groins and jetties, and giant tide gates to control the in-and-out flow of water from the land—all attest to the world's most advanced skill in coastal engineering. Engineering is the solution, say the Dutch, and unlike Bangladesh, they have the money to implement solutions.

The Netherlands is a relatively small and relatively wealthy country. The engineering solutions the Dutch have found for protecting their society from sea level rise would be too costly and too environmentally damaging in most other countries. Many Louisiana politicians have looked to the Netherlands as a model for how the Mississippi Delta coast could be protected. But using this model in Louisiana would be an environmental disaster. It would also be unnecessary. The United States is a big country with plenty of room for its sea-level-rise refugees.

Singapore

Most of the business district, airport, and port facilities of this small island-city-state at the southern tip of the Malay Peninsula lie less than 7 feet (2m) above sea level. Considerable land has been reclaimed from the sea, and it is particularly vulnerable to rising waters. Like the Netherlands, this is a wealthy, efficient country. But unlike the Netherlands, Singapore has little experience with large-scale coastal engineering. Local politicians assume that extensive diking will save the city, for a while.

Indonesia

Among Indonesia's seventeen thousand islands, the most endangered is Java, where more than 110 million people live. Its capital, Jakarta, with a population of 8.5 million, on the north coast of Java, is Indonesia's largest city and among the most vulnerable cities in the world to sea level rise. Indonesian scientists predict the city's airport will be inundated by 2035. In November 2007, the road to the airport was breached by a storm, an event expected to occur with increasing frequency in the future. The same November storm produced storm surges that stepped over the seawalls of Jakarta along the northern shores. By 2050, about 24% of the city will be gone, according to local estimates (assuming no coastal engineering). Some believe the capital will have to be moved 112 miles (180 km) to higher elevations in the city of Bandung.

Sea Level Rise and the Cities

Just as the expanding ocean will affect coastal nations differently, each city that is vulnerable to sea level rise has a different physical situation, including tidal amplitudes, storm surge and storm wave potential, and area of the city that is low enough to be affected by rising waters. The nature of the culture and the resources available to the city will determine the nature of the response. And of course each city will have unique vulnerabilities, such as the subway systems of New York, Boston, London, and Washington, DC. For the world's vulnerable coastal cities, the effects of sea level rise will include the following:

- Blockage of city storm drainage, sewage treatment facilities and subways
- Salinization or pollution of domestic water supplies
- Flooding
- Increase in the extent and penetration of storm surge
- Loss of protective barrier islands that rim many coastal plains
- Infrastructure loss—water, electrical power, roads, railroads, port facilities
- Requirement for dikes, levees, seawalls, and relocation of buildings.

According to a recent study led by Professor Robert Nicholls of Middlesex University and his colleagues and sponsored by the Organization for Economic Co-operation and Development (OECD), because of expected population increases, as many as 150 million people in the world's major cities—more than three times the 40 million currently endangered—may need to rely on engineering structures such as dikes for survival by 2070 if the sea level rise reaches 20 inches (50cm) by then, a very conservative estimate.

The OECD study ranked what it considered to be the ten most vulnerable cities in the world as measured by susceptibility of property to flooding (but not storm surge impacts), a useful proxy for evaluating the potential for damage (though not loss or disruption of life) from sea level rise: Miami (the most endangered), New York/Newark, New Orleans, Osaka/Kobe, Tokyo, Amsterdam, Rotterdam, Nagoya, Tampa/St. Petersburg, and Virginia Beach. Half of the top ten cities are American, and Miami tops this list (and every other list of vulnerable cities).

Using the same criteria, the OECD ranked the vulnerability of major American cities as: Miami, New York/Newark, New Orleans, Tampa/St. Petersburg, Virgina Beach, Boston, Philadelphia, San Francisco/Oakland, Los Angeles, and Houston. The top seven vulnerable cities are all East Coast and Gulf Coast communities on the rims of coastal plains. Such rankings should be viewed as order-of-magnitude approximations but are nonetheless useful.

When it comes to the economics of response to sea level rise, it is likely that the problems of the cities will trump those of rural or touristic areas. Preservation of Manhattan will certainly be seen as a higher national priority than preservation of Wrightsville Beach, North Carolina; Ocean Shores, Washington; or hundreds of other American beachfront tourist towns. In the United Kingdom, preservation of London will be higher priority than preservation of Margate and North Blackpool beach resorts. Preservation of the city of Dhaka, Bangladesh, will undoubtedly be higher priority than saving agricultural land on the outer Ganges Delta.

The impact of sea level rise on low-elevation coastal cities over the next century will range from catastrophic to minor depending on the resources a city or nation is willing and able to expend. In addition, much depends on the willingness of more developed countries to aid the developing nations. The 2007 UN Bali conference on global warming assumed that the wealthier nations would build seawalls for the developing countries, but this is questionable. At some point—probably a 3.3-foot (1m) rise in sea level—the lower elevation rims of vulnerable coastal cities such as Dhaka, Bangladesh; Ho Chi Min City, Vietnam; Lagos, Nigeria; Barranquilla, Colombia; Rangoon, Myanmar (now known as Yangon); and Abidjan, Ivory Coast, will be abandoned. For some of these areas, perhaps a century from now, mass exodus may prove the only recourse. The same sea level rise will be less damaging in Inchon, South Korea; Vancouver, British Columbia; London, England; Tokyo, Japan; and Miami, Florida, because of available resources in these wealthy countries (if the nations choose to hold the line). But sooner or later, if some of the more pessimistic scenarios become reality, mammoth shifts in population may be required for these cities as well.

References ---

Cline, W.R. 2007. *Global Warming and Agriculture: Impact estimated by country*. Peterson Institute, 201.

Dasguopta, S., Laplante, B., Meissner, C., Wheeler, D.,and Yan, J. 2007. The impact of sea level on developing countries: A comparative analysis. World Bank Policy Research Working Paper 4136, 51.

Emanuel, K.A. 1987. The dependence of hurricane intensity on climate. *Nature* 326: 483-85.

Francis, J.A., and Hunter, E. 2006. New insight into the disappearing Arctic Sea ice. *EOS: Transactions of the American Geophysical Union* 87: 509-11.

Gibbons, S. J.A., and Nicholls, R.J. 2006. Island abandonment and sea level rise: An historical analog from the Chesapeake Bay, USA. *Global Environmental Change* 16: 40-47.

Jacob, K., Gornitz, V., and Rosenzweig, C. 2007. Vulnerability of the New York City metropolitan area to coastal hazards, including sea-level rise: Inferences for urban coastal risk management and adaptation policies. In L. McFadden, R. Nicholls, and E. Penning-Rowswell (eds.), *Managing Coastal Vulnerability*. Elsevier, 139-56.

Kerr, R.A. 2006. Global warming may be homing in on Atlantic hurricanes. *Science* 314: 910-11.

Kister, C. 2004. *Arctic Melting:*

How global warming is destroying one of the world's largest areas. Common Courage Press, 224.

Komar, P. 1997, *The Pacific Northwest Coast*. Durham, NC: Duke University Press, 195.

Milliman, J.D. and Haq, B.U. (eds), 1996, *Sea Level Rise and Coastal Subsidence: Causes, consequences and strategies.* Kluwer Academic Publishers, 369.

Nicholls, R.J., et al. 2007. Ranking of the world's cities most exposed to coastal flooding today and in the future. OECD, Environmental Working Paper No. 1, Executive Summary, 12.

Nishioka, S., and Harasawa, H. (eds). 2000. Global warming:

The potential impact on Japan. *Climatic Change* 47: 213-15.

Pilkey, O.H., and Dixon, K.L. 1996. *The Corps and the Shore*. Island Press, 272.

Riggs, S.R., Cleary, W.J., and Snyder, S.W. 1995. Influence of inherited geologic framework upon barrier beach morphology and shoreface dynamics. *Marine Geology* 126: 213-34.

Sarwar, G.M. 2005. Impacts of sea level rise on the coastal zone of Bangladesh. Master's thesis, Lund University, Sweden, 38.

Wind, H.G. (ed). 1987. *Impact of Sea Level Rise on Society*. A.A. Balkema Publishers, 191.

ACT ONE: THE RISING

CALL TO ACTION

CATASTROPHE

From *Sweet and Salt: Water and the Dutch* , 2012

Tracy Metz

"The War" and "The Disaster".
Occupation and floods

Next to the Second World War, the Disaster in 1953 was an event of a severity unequaled for the Dutch during the 20th century. During the night of 31 January to 1 February 1953, the spring tide and a northwesterly storm combined to flood a large part of Zeeland, western Brabant and the islands of South Holland. More than 1,800 people were killed, along with many thousands of animals; more than 100,000 people lost their homes.

The 1953 North Sea Flood is still the worst, but certainly not the last incident in terms of flooding (or risks of flooding) in the Netherlands. As I write this, a fast-moving, intense northwesterly storm is raging over the country. A cold front hundreds of kilometers long is moving across the land, rapidly and violently pushing up barometric pressure and sweeping up the shallow water along the coast. At IJmuiden, the sea has risen by 1.5 m in minutes and fallen again just as quickly; the weather website www.weer.nl is talking about a "weather tsunami." In the north, a dike in the Frisian village of Grou has breached and the inhabitants and their livestock are being evacuated; the Groninger Museum is rushing to move the exhibition on fashion designer Azzedine Alaïa out of the lower levels. The quays of Dordrecht have been lined with sandbags; quays in Hoek van Holland, Schiedam, Vlaardingen and Maassluis have been closed as a precaution. The barrier in the Hollandsche IJssel has been closed, and the inflatable bladder

dam at Ramspol, which is meant to stop the wind blowing water from the IJsselmeer into the province of Overijssel has also been deployed, for the first time since 2007. In the southern province of Limburg, pumps are also being installed and dikes are being hurriedly raised along the Maas. A disaster like the Disaster this certainly is not—but suddenly water is the lead story on the evening news and in all the newspapers.

Meanwhile, the winter of 2011/2012 is the mildest since 1901. And as soldiers, dike wardens and police work through the night to hold back the water, probably no one will think about the fact that November was the driest November in the 105 years since such records have been kept.

In 1993 and 1995 the water also rose so high that large-scale evacuations took place, involving 250,000 people but also 1 million animals, both pets and livestock. During the arid summer of 2003, a section of a dried-out peat dike shifted in the middle of the night in the village of Wilnis, causing a flood—the damage was local, but the shock was nationwide. In November 2007, the seawater stood so high that the moveable Maeslantkering barrier at Rotterdam was closed for the first time—the sea had not risen to such a level since the 1953 disaster.

And in the media, news about storms and flooding elsewhere in the world crop up in quick succession: Hurricane Katrina, which devastated New Orleans in 2005 and—like the 1953 disaster—claimed around

1,800 victims; the storm Xynthia, which ravaged western France in 2010, killed 53 people and caused 1.5 billion euros in damage; and then the floods in places like Bangkok, Pakistan, the south of Italy, Rio de Janeiro, Australia's Queensland, the Philippines and China. And nearly New Orleans again in 2011, when a second flood was prevented only by opening emergency sluices for the first time in 38 years and flooding thousands of hectares further upstream.

The Dutch Delta is the safest in the world, says the Dutch government. At the same time, it is generally accepted that this low-lying delta is vulnerable to climate change, which involves not only rising sea levels but also more frequent and more intense rainfall. How safe are the Dutch? How safe do they feel? Do they know what to do if the worst happens? The dikes and the Delta Works that were built after 1953 have magnified the sense of security, so the concentration of people, property, housing and industry behind the dikes has increased astronomically. The chances of a disaster are small, but the consequences would be enormous. Should the Dutch have flood insurance? Is there anything they can do as individuals to protect themselves? Is there still a trace of self-sufficiency, or do they lay their heads in the lap of Rijkswaterstaat and the water districts?

Sense of Security
To the surprise of many foreign visitors, Dutch people do not keep a suitcase packed at home, ready to flee from the water at a moment's notice. They feel quite safe and secure. A 2010 survey by Alterra, in association with Wageningen University, on the Dutch level of satisfaction with high-water defenses showed that no less than 87% of respondents feel safe. That sounds nice—but that sense of security is often based on ignorance. Of the people whose homes would be standing in 2 m or more of water as a result of a flood, 68% underestimate how high the water would reach. Even worse, 26% of the people in these areas are convinced that their homes cannot be flooded. On the other hand, 56% do honestly admit they are not properly prepared for a flood. And this while 76% say they believe floods and other water problems are going to take place more often.

But let's be honest: the Dutch population as a whole doesn't do much about it; the people feel safe. They rely on the government: Article 21 of the Dutch Constitution states, after all, that it is the "concern of the authorities to keep the country habitable and to protect and improve the environment." A while ago,

various adverts were broadcast on radio and television about the emergency kit that should be in every Dutch home—for however safe things are, one should always be prepared for trouble. I looked up online what was in this emergency kit, which you could buy as a shoulder bag or in a convenient water-tight plastic barrel for 40, 65 or 85 euros: a whistle, a box of matches, tea-lights, first-aid sticking plasters, an emergency foil blanket. You can buy all that yourself, I thought, and for less money. I confess: even though I live here too, and should be water aware, I never got beyond a water bag from the camping supplies shop and some toilet paper. I take comfort in a line from an article by Bas Kolen, disaster management consultant with the HKV agency: "Because floods are so rare it is improbable that citizens are prepared for them."

It might seem silly, so little account taken of the fact that the Dutch live significantly below sea level. The Alterra researchers, however, put this attitude into perspective. Safety is invisible to the citizen, they write: until, essentially, water is pouring out of or over the dike, it is difficult for lay people to gauge whether safety has increased or in fact decreased. The image of the water manager plays a more significant role in the sense of satisfaction about the government's handling of water defenses, according to Alterra—at least to the extent that we have any image of that manager at all. Respondents turned out to know little about the water board districts, for example, except that they collect taxes. And in 2008, when elections were held nationwide to select water district administrators in a democratic process for the first time, the turnout was only 23%. The water, or more accurately everything involved in keeping our feet dry, is set up so well, and is so much the private domain of professionals, that we know nothing about it anymore.

The Once-in-Ten-Thousand-Years Storm
The United Nations University's *World Risk Report* puts the Netherlands in 69th place in terms of the risk for natural disasters. In this index—the first, according to the UN, to take into account social and economic as well as physical conditions—the Netherlands comes off well because the social, economic and ecological factors compensate for low elevations and the susceptibility to climate change. In other words, the country is well defended and wealthy. On the other hand, the Netherlands is near the top of the index among the 27 countries of the European Union, immediately followed by Greece, Romania and Hungary. But is Holland safe?

Yes: Thanks to the centuries of experience the

Netherlands has in living with and keeping back water, the entire country is based on this, especially, of course the two-thirds of it that lie below sea level.

No: <u>The safety standards for the Delta Works and the dikes date from the years immediately following the Second World War.</u> There were far fewer people living behind the dikes at the time, and the economic worth was a fraction of what it is now, particularly in the low-lying Randstad, the urban conglomeration in the west of the country. <u>What's more, not all dikes meet even these outdated standards.</u> As a member of the second Delta Commission, Marcel Stive, professor of coastal engineering at Delft University of Technology, has been campaigning to raise these safety standards by a factor of ten. "The government sets immensely high safety standards for business, but far less for water safety," he says. <u>"Even worse, one-fourth of the dikes do not even meet the outdated standards.</u> In spite of the fact that the chance of a flood is greater for the Netherlands than the chances of all other disasters put together."

Yes: Dutch safety standards, as outdated as they may be, are very high, the highest in the world. As a comparison: in many places in the world it is normal for dikes to be built to withstand the type of storm that might occur once every 100 years, but the calculations for the Dike Ring 14, which encircles the Randstad, were based on one that might occur every 10,000 years.

No: <u>Dike Ring 14, the largest in the country, is so large, without any compartments, that if—if, if—this dike should breach, the damage here would be immeasurable.</u>

Yes: After the high water levels in the river region in 1993 and 1995, new legislation on water defenses (the Wet op de Waterkeringen, or Water Defenses Act) was adopted, requiring primary defenses—dikes, dunes, dams—to be tested every five years. The Netherlands, after all, has a high-maintenance landscape. The water districts have been in place since the Middle Ages, forming not only the foundation of our democracy, it is said, but also guaranteeing an intricate system of maintenance for the waterways.

No: <u>In 2011, after the third test, junior minister for Infrastructure and the Environment Joop Atsma informed the Dutch parliament that one-third of the defenses are not up to standards, especially in the river region.</u> According to the minister, this is not because they have deteriorated, but because more defenses are being tested (a number of dikes and structures in Limburg were added) and the testing has become more rigorous. This testing has problems of its own, an evaluation by Twynstra Gudde commissioned by the ministry has revealed. Since mid-2011 the Dutch

state no longer funds all improvements to primary defenses; the water districts now have to pay half. They are not happy about this. "It is not likely that the new funding system will lead to a better embedding of water defense testing," Twynstra Gudde dryly observes. In addition, <u>there is too little capacity and expertise available to carry out the testing, making Rijkswaterstaat, the ministry's operational agency, and the water districts too dependent on engineering firms.</u> This dependency "threatens the quality of the testing," in Twynstra Gudde's view.

Yes: The Netherlands spends money on maintenance and water safety in a systematic way. Today, this represents 0.14% of the gross national product. Proposed by the Delta Commission, there will now be a separate Delta Fund, which will receive over 10 billion euros between 2020 and 2028. This will raise the figure to 0.22% of GNP.

No: This fund will only come into existence in 2020. What's more, the current High-Water Protection Program is costing billions more than anticipated. Instead of the 1.6 billion euros planned for work on 367 km of dikes and 18 structures, costs have reached 3.8 billion, almost 2.5 times as much. This is why the water districts have to pay for half of the maintenance on the primary defenses they control.

Yes: Not only is the Netherlands the best-protected delta in the world, it also has a finely meshed and well-maintained network of roads and motorways. Furthermore, the Dutch are also digitally connected, with a high percentage of Internet connections and smart phones. It should be possible to organize emergency situations.

No: <u>The most densely populated country in Europe, which generates more than 60% of its economy below sea level, has no national evacuation plan.</u> Even though floods are a far more likely threat to the Netherlands than terrorism or a nuclear disaster, <u>"A preventive evacuation of the coast in the event of a flooding threat is not feasible within the realistic forecast period of 48 hours,"</u> writes Dutch disaster management consultant Bas Kolen in his 2010 article "Self-Sufficiency in Floods and Large-Scale Evacuations":

<u>For the Randstad, we expect only a small percentage to be able to evacuate preventively.</u> It is unclear to citizens what they can and cannot expect from the government and emergency services and what the government can actually do.

Yes, no, yes, no, yes, no. Safety in the Netherlands has been thought about extensively and in depth—that much is clear. On paper, everything has been set out in details in covenants, accords, guidelines and ordinances. What about reality?

Evacuation

After the threat of floods and evacuations in the mid-1990s in the river region of the Netherlands and the devastation of New Orleans by Katrina, the Flood Management Taskforce (TMO) was set up here in 2006. It was supposed to improve organizational preparations for flooding within two years. The taskforce's final act was a massive national disaster drill, christened "Waterproef" (Water test). At the end of the two years, the taskforce itself was forced to conclude that the preparations were not complete, especially at the national level:

> National operational planning is still urgently needed and requires high administrative priority. As an example we cite the necessity for a national evacuation plan, including an approach to traffic management during floods. This evacuation planning is indispensable, offers our inhabitants and businesses security and creates order in the chaos caused by floods.

How difficult collaboration can be had already been demonstrated by the "National Security Project" in 2007. This broad-based report on security in all kinds of areas had been intended for internal use and only became public when the NOS Journaal, the public broadcaster's news program, filed a claim under the Government Information (Public Access) Act. In the unsettling report, to which every ministry in The Hague had contributed, 400 civil servants complained about collaboration with other organizations. With regard to the risk of flooding, according to an article in the NRC Handelsblad newspaper, they say there is a worrying "gap" between the concrete threat of floods and the response to it. "Due in part to population growth, the chance of numerous victims from floods is greater by a factor of ten than all other external risks put together." This threat is insufficiently recognized, they feel. Conclusion: an unacceptable risk over the long term.

Climate Game

The young company Tygron also feels that collaboration on water issues has to improve. This can be done through "serious gaming," a digital game intended to provide insight, using policy simulations, into how diverse interests affect one another—and when collaboration produces better results. "We started this four years ago as a spin-off of Delft University of Technology," explains co-founder Jeroen Warmerdam, originally a computer engineer:

> We saw that you could use the techniques and concepts from the world of games for uses other than pure entertainment. In one game, you can show the physical as well as the social consequences of all sorts of decisions— before the decisions have been taken.

It started with a game that—as student assistants— they developed in cooperation with Delft University of Technology's Faculty of Technology, Policy, and Management for the Port of Rotterdam Authority. "We, the university, and the port authority each became one-third stockholders. This is how our company was formed, and we now focus specifically on the domain of the physical living environment, water, and climate." Tygron has developed a number of digital simulation games: a single-player game for Rijkswaterstaat about water management, a game based on the water problems in Tiel-Oost for the educational program Leven met Water (Living with water), and games on shrinkage in Groningen and Heerlen. With the Vietnam Game, they were also on hand when Vietnam and the Netherlands signed an agreement in March 2011 to jointly draw up a Delta Plan for Vietnam. The players were Crown Prince Willem-Alexander, junior minister Atsma, and several Vietnamese guests. The company has also been given an innovation subsidy from the Ministry of Infrastructure and the Environment.

Today I'm the city of Delft, and I'm going to do my best in the Climate Game to outmaneuver—uh, no, cooperate with—the housing corporation, the university's real estate department, and the water district. There are several issues we need to resolve together: there is flooding in the city when it rains, there is heat stress during droughts, and housing has to be built. We "fly" over the Schie River, passing over the familiar buildings of the university, like the auditorium and the library. No sooner have I decided where I want to set aside space for water storage than a building application flashes up from the housing corporation, which wants to build 300 homes in that exact spot. Perhaps I should build a water plaza?

I click on the "weather simulation" button and it starts pouring with rain onscreen. It looks like half of Delft is turning blue—how am I supposed to lay out green space and have money left over for housing improvements? All the while, the light for the corporation's building application is flashing impatiently: I am—already?! —at risk of exceeding the legal response deadline. I could also cheerfully strangle those people from the water district: they're only looking at flooding risk while as a city administrator I have to keep all these interests in the air like spinning plates. No wonder I can scarcely come to a decision. The afternoon is almost over already, the applications are piling up, I'm running out of money and space, and I am totally worn out.

But I do now understand what Tygron is saying, that you develop an understanding of one another's problems and start to think more in terms of possible solutions. In that process, the social aspect is at least as important as the technical issue. Where is the button that will let me ask the mayor to go have a drink with the dike warden?

Campaigns

"We are safe," says the government, but citizens should still be prepared in the event a flood comes. This is realistic, since no one would really believe a 100% guarantee from the government. To quote a famous Dutch comedian: "The flood of 1953 was the last natural disaster. Everything that's happened since is the government's fault." And yet it is confusing to be at times reassured and at times urged to prepare for danger. "The national discourse about water and safety runs along two lines," say cultural sociologists Baukje Kothuis and Trudes Heems, who work and study together under the name Waterworks. "The government thinks in terms of threat, but the citizen still thinks that the story of the Netherlands and water is one of conquest. So they talk at cross purposes," says Kothuis. "This gap leads to a great deal of indignation, because the citizen's trust in government is undermined."

Waterworks has compared a number of government campaigns in order to examine the implicit assumptions behind them. Heems explains:

The campaign "Nederland leeft met water" [The Netherlands lives with water] from the Ministry of Transport and Water Management ran concurrently with "Denk vooruit" [Think ahead] from the Ministry of the Interior. The first is the discourse of the conquest, the second

that of the threat. The government had hoped that this communication trajectory would lead to greater awareness of the risk and therefore more acceptance of measures against flooding. What we're seeing is that it does lead to greater awareness, but also to indignation, loss of trust, and an even greater need for a guarantee of security.

Dordrecht alderman Piet Sleeking has a similarly dual message: he wants to show the inhabitants of his city that the government is keeping the city safe, and on the other hand he wants them to know what to do if a flood happens. "Every year, everyone in Dordrecht gets a brochure from the city about where the sandbags are, and we hold a drill with the flood bulkheads," he says, "but we have to do more to get residents involved in safety." He has on old gym shoes and zip-off trousers for the occasion, because he's about to go into the Flood Room, a shipping container in which a room has been partitioned off with glass. It's a project of the Technische Universität Hamburg intended to raise flood awareness. Natasa Manojlovic, who has come along from Hamburg, says: "We call it the Panic Room, but they didn't like that name." The alderman is ready in his glass cubicle, with a laptop on the table, folders of documents on shelves on the wall, a table lamp, a TV on the wall— an average office or living room, except for the yellow rubber duck that starts bobbing up and down as soon as Natasa turns on the tap. "Where are my documents? Is the lamp going to short out? Where am I going to put the laptop?" she yells through a microphone in order to create the right panicky atmosphere. "Where is my money? If I have to evacuate, I will need money—where did I put it?—I don't know when I'll be coming back! What do I take—I can't take everything!" She turns off the tap when the water is knee-high—not a disaster, but it is cold, says the alderman. "And remember," says Natasa, "this is clean water. In real life it's dirty and muddy."

Johan van Nieuwenhuizen of the Hollandse Delta water district has come to enjoy watching the alderman get wet. "The water district's greatest concern is that attention to safety is diminishing, and that there is no willingness to allocate money for this." A year and a half ago a risk map was distributed to every home in the whole region, he says. "But the awareness among people is very vague."

Beer for Awareness!

In Krimpen aan den IJssel colors are being used to stimulate this awareness and make the water visible and tangible. In the Hollandsche IJssel, as the first Delta Works after the 1953 flood, a barrier was built that is closed only when the water reaches dangerously high levels. It is a fine piece of infrastructure, with tall control portals and two enormous steel slides 80 m wide that can go up and down. According to the Rijkswaterstaat website:

> Many of the current inhabitants of this area did not experience the North Sea Flood. They probably know they live in one of the lowest-lying areas of the Netherlands, but they are often not aware that they are protected from high water by a movable storm-surge barrier.

This led to the idea of lighting the control portals at night: bluish-purple when the barrier is open, red if it has to be closed because of high water. Artist Henriette Waal uses beer rather than colors as a means of getting people to think about the water we take so much for granted:

> Brewing beer and purifying water are things people used to do themselves. In fact brewing beer was a way of purifying water that was too dirty to drink. Today, water purification is done by the state and breweries are worldwide corporations. And the water in the city is a kind of black hole, across which you might sail in your pleasure boat, if at all, but any other kind of connection has been lost.

For the public event "Edible Landscape" on the grounds of a former water-treatment plant in Tilburg, Henriette Waal came up with a mobile beer brewery on a trailer. For a whole summer she organized an "open-air brewery" in which she demonstrated the culture of amateur brewing:

> There turned out to be a whole community of amateur brewers, but they mostly wanted to imitate the big beer varieties. I thought, on the contrary, as an amateur you should want to break free, even make something new? We took rainwater, wild hops and all kinds of other herbs that were growing there in the field: clover, yarrow, elder, linden, even just grass. Unhindered by knowledge of any kind I threw all sorts in. And I have to say, some very special beers came out.

> Waal was then invited to make beer using water from the Sloterplas lake in Amsterdam; to make it

drinkable, she added the water-filtration tower. "I added it later to be able to purify even the filthiest water right on location. The filter consists of sand, ceramic, and charcoal; the brewing process does the rest." This was followed by a brewery on the Erasmusveld, a project of the Stroom art centre in The Hague. "There was no amateur brewers' club in The Hague; I had to find them by word of mouth and online. They all made their own beer in sheds, kitchens, and garages." For the Erasmusveld, Waal founded the "Beer School," to get local people involved in the idea that their neighborhood was going to produce its own beer—using water from the ditch:

> It's not so easy to get people involved in purifying their own wastewater—unless the purification plant is a brewery! There's something magical about that brewing and distilling, a kind of alchemy. I think it's great if we can set up a collective amateur brewery in The Hague in which amateur brewers share one installation. Just like a food cooperative, but for beer—a water-purification association with beer as the final product. This home brewing draws a short line between the Dutchman, his beer, and the water that surrounds him.

Build Your Own Dike
If you ask an American farmer what the best way is to protect your property from the water, you might get the following answer, born of his philosophy of ultimate self-sufficiency: build your own dike. In the Netherlands this is inconceivable: water safety is still a collective good, even if this is less and less coordinated at the national level and increasingly delegated to provinces, water board districts and municipalities.

What is the best way to make and keep the Netherlands safe? For centuries the Dutch have done this by resisting the water with dikes and pumps—and they will never be able to do without them. Yet the first National Water Plan and the Water Security Policy Paper, both for the period 2009–2015, feature a turnaround in policy. There is now a three-stage rocket called "multilayer security" and primarily aimed at limiting damage in the event of a flood. The first layer, which will always remain the most important, is prevention. The other two are aimed at limiting the effects of a flood: intelligent use of the space behind the dikes (for example, the selection of building sites, or the adaptation of vulnerable facilities like hospitals) and effective disaster response and evacuation plans.

Resisting where necessary, moving with the water where we can. In late 2011 the Oranjewoud engineering firm produced a report on six pilot projects using multilayer security. The key conclusion: very promising, but difficult to measure the precise contribution of spatial planning and disaster management to security. (And therefore to allocate budgets as well.)

In any case, Han Vrijling, professor of hydraulic engineering at Delft University of Technology, is sticking to Layer 1: prevention through proper dike maintenance. He defended his viewpoint with fervor during a debate organized by the engineers' professional association KIVI NIRIA between the "Delft engineers" (the classical approach) and those who feel a fundamental change in direction is needed, "the new water builders," as Vrijling calls them. Terps, floating homes, and more room for water—a tad trendy, he feels:

> It must remain clear what is needed for safety and what is needed for beauty and nature. Let's not, under the pretext of security, build dikes that are merely beautiful. It's remarkable to note that the new water builders are suddenly and selectively losing faith in polders.

A much bigger problem than rising sea levels, he feels, is the maintenance backlog of the existing dikes and barriers. Almost a billion euros is needed every year, he says, to get water management in the Netherlands in order. "Every euro we spend on anything other than the dikes is a wasted euro!"

In that gathering of hardcore engineers, his opponent, Jeroen Aerts, professor of risk management at the VU University, was in the minority, but he courageously fought his corner. Our vulnerability to floods is increasing, Aerts said, due to climate change but also due to urbanization behind the dikes:

> There are other solutions besides just dikes. You can also limit the damage by using the land differently, by building the houses differently. Let's start by agreeing that electrical outlets and boilers have to be installed at least 1 m above the skirting-board.

Insurance

Protection from the water is a national concern, and until now coverage of the damage has been as well. Since the 1953 North Sea Flood, the Dutch government, through the Disasters and Serious Accidents Compensation Act, covers the effects of a disaster—but not all damage is necessarily a national disaster, says Jeroen Aerts. It is therefore high time to introduce flood insurance, he feels:

> The discussion on climate change has made us much more aware of uncertainties, including in the protection we can provide. The value of everything that takes place behind the dikes, in terms of human lives and economic activity, has increased astronomically. Your premium increases if you live in a dangerous area. This automatically raises your awareness of the impact of water on the place where you live.

It is just like car insurance, he says: the car has to leave the factory in good order, but you could also have an accident—so you get insurance against that remaining risk. The increasing interest in building outside the dike—in the former docklands of Rotterdam and Dordrecht, for example—is also generating discussion about insurance, because the government does not cover damage outside the dikes. There is already a change afoot in agriculture: since 2010, crop farmers, arboriculturists, horticulturists, and fruit growers have been taking out insurance against extreme rainfall and extreme drought, and the government contributes up to 50% of the cost of the premiums.

> Here too Aerts finds an opponent in Han Vrijling: If your prevention is up to par, you don't have to take out insurance. And if a flood does occur, it's quite conceivable that the insurer will not pay, because it will argue that the government, as the entity responsible for water defenses, failed to meet its obligations and is therefore liable. Moreover, the potential damage is very great. The insurer has to have reserves equal to this damage available when the coverage starts. Therefore it will seek to limit its risk. From a macroeconomic standpoint, private insurance leads to much higher costs than prevention with dikes.

Who pays for the damage? Eight years later, that is still the question in Wilnis, a village between Utrecht and Amsterdam that suddenly became nationally famous on 26 August 2003. The summer of 2003 was hot and dry, and the peat dike in Wilnis had dried from the inside without anyone noticing. In the middle of the night, a 50- meter- long section of the ring dike of the Groot-Mijdrecht polder suddenly shifted. The waterway emptied into the village. "I think one resident saw it happen, someone who couldn't sleep and was smoking a cigarette on his balcony," says Johan de Bondt, dike warden for the Amstel, Gooi, en Vecht water district—at the time just four weeks into the job.

On a summer afternoon I walk with De Bondt along the ring waterway in Wilnis. It was decided to leave the section of damaged dike as it was, as a monument and a reminder. There's a sign, and cows are lying on it, placidly chewing their cud. There is also a metal bench along the water, bearing texts by the village children: "We live and work on pudding." "Our house has been shifted back." "Now the water is in the right place." And of course: "There was no school that day."

The water board district did not rebuild the dike layer by layer—that would have taken too long—but built instead a new bank of steel retaining walls—an expensive but quick and sturdy solution. "This must be one of the strongest dikes in the Netherlands now," the dike warden chuckles. He points to the other side. "Those people were of course looking at the flooded neighborhood and happy they'd been spared. Until their gardens started sinking into the mud too. The damage there was from the water draining from the waterway!" It was in fact on that side that damage was greatest: the soil subsided between 1 and 2 m and there were cracks in all the infrastructure: the roads, the pavement, the utilities. De Bondt points to a bungalow on the opposite side. It stands on a foundation of piles, and the dike warden was able to walk right under it, because so much of the soil had been eroded. Its occupants happen to be using sand right now to raise their garden another 30 cm. "The water damage was not the worst part," says De Bondt:

> It was mostly damage to parquet floors and removing sludge from the gardens, a total of about 1.5 million euros. The neighborhood where the water flooded in was evacuated for half a day and after that it was a matter of cleaning up. But the repairs to the infrastructure cost between 10 and 15 million euros.

Yet the damage has still not been settled; against its will Wilnis has become a test case. De Bondt sighs:

> An independent commission was formed that came to the conclusion that the dike met safety standards and this was a unique event that could not be foreseen. So the water district is not to blame. The damage fell under the National Disaster Fund, which

covers 90% of the damage, but the residents have to pay for the remaining 10%. By private initiative a fund was formed to compensate them. The city and the water district also contributed to this; that was settled rapidly and correctly. But then the city wanted to recoup its losses from the water district. They lost in the lower court, but they appealed. The Court of Amsterdam sided with the city with the argument that the dike is a structure, as though a dike were a building. That's completely wrong, of course! Because according to the higher court the owner is liable. Half of our dikes are owned by farmers; we simply manage them. You could never expect an individual farmer to pay for the failure of a dike like this; it would destroy the whole system of liability. So with the support of the Union of Water Board Districts, we appealed to the Supreme Court, which has since referred the case back to another court. We're still waiting for the decision.

The name of Wilnis will forever be linked to this bizarre and unsettling event. But Wilnis is not unique, says Johan de Bondt. "A week later the same thing happened near Rotterdam, but there the water just flowed into a field—so no one there remembers it."

Back to the Terp

Like the comedian said, everything that's happened since the North Sea Flood of 1953 is the government's fault. Emotions are certainly running high in the places where the government wants to give the land back to the water. Zeelanders, in spite of European Union agreements, were recently able to block the depoldering of the Hedwigepolder on the Westerschelde. In another part of the country, in North Holland, the inhabitants of the Horstermeerpolder staged a "coup" on 17 February 2010, declaring their polder a republic, with its own president and constitution; at the boundaries of the village they stopped cars that were not flying the flag of the brand-new republic. The depoldering plan has been shelved for now.

Things went differently in the Overdiepse Polder in Brabant. There, the 17 farmers took matters into their own hands when they heard their land was to be flooded. "It was at an informational evening about the Ruimte voor de Rivier [Room for the River] program," farmer Nol Hooijmaijers remembers distinctly:

> The province said they couldn't go on raising the dikes forever; more room had to be created for

the river so that at high water Den Bosch would not be flooded. The province had decided that the dike along the north side of our 550-hectare polder had to be excavated, to allow the river the Bergsche Maas to flow over it. To our amazement, the whole Overdiepse Polder was colored in blue, and we knew nothing about it!

Hooijmaijers has just finished his morning rounds to tend to his 75 cows. 'There'd be 100 if we did have to move.' In the scullery he takes off his overalls and rubber boots, washes his hand and goes to take a comfortable seat at the kitchen table:

My neighbor Sjaak Broekmans and I contacted everyone through the homeowners association to come here, and we sat under the chestnut tree. We figured, we can just dig our heels in the sand, but wouldn't it be better if we come up with a plan ourselves?

That's how they devised the idea of creating several terps, or mounds, along the south side. Outside Hooijmaijers's window a dragline excavator is preparing the first. "Yeah, I know," he laughs, "with terps you think of the earliest inhabitants of Friesland and Groningen. But when the water rises, everyone moves to higher ground."

The terps had to be big enough to really be farmed. Each will be over 2 hectares in area and 6 m high. That means there is only room for eight terps—and there were 17 farmers in the polder:

Our neighborhood association carried out a survey of who wanted to stay and who didn't, because some people wanted to give up their farm anyway or preferred to move out entirely. We opened the 17 envelopes with the Rabobank director as a witness. Eight wanted to stay, six wanted to leave, two weren't sure. But then one of the neighbors who wanted to leave found out the plans made his house unsellable—there was only one buyer, and that was the government.

Tension also arose between those who already owned land in the south part of the polder and those in the north part who would have liked to buy a terp but knew they wouldn't get a spot. And the Hooijmaijers family? "My son is interested in taking over the farm." Still, some neighbors in the Overdiepse Polder now

no longer greet each other. This pains Hooijmaijers. "This polder was a village; we'd do anything for each other. But the community has suffered a real blow because of this change. But you can't treat everyone equally." The social workers of the Zuidelijke Land- en Tuinbouw Organisatie said that from their experience, it would take three or four years before the atmosphere stabilized. "Sometimes I feel like a social worker. But those cows have to be milked twice a day."

The land in the polder has dropped in value now that it can occasionally be flooded. But the government is paying the 111-million-euro cost of the project, including the building of the terps, which has been underway for ten years already:

This is a model project, so it had to be just right, but it went on and on . . . The Netherlands is drowning in laws and regulations. Once the plan was done, it took another year because of a bat survey.

The Overdiepse Polder is now a phenomenon, even at the international level. Crown Prince Willem-Alexander has been by, junior minister Melanie Schulz, minister Tineke Huizinga, delegations from China, Vietnam, South Africa, CNN, and the French newspaper *Le Monde*. And at the official launch of the project Hooijmaijers and Broekmans each received a ribbon from the mayor. They are now knights of the Order of Orange-Nassau. "Still, I wouldn't want to do it again."

New Orleans after Katrina
With a flamboyant swerve—New Orleans is, after all, a flamboyant city, nicknamed the Big Easy—David Waggonner turns the car into the car park. From there we walk across another paved expanse to the mid-height levee wall. He points to the deep, wide, fast-moving river below us—the mighty Mississippi. "This is the only place in the city where you can still see the river," he says. "And you have to go to quite a bit of trouble to see it. But why would you want to come here?" He gestures broadly to the surroundings. "Nothing, no bar, no restaurant, nothing attractive, a public space where ten blind horses wouldn't cause any damage. New Orleans was built here because of the river, but the connection with the water has really been lost completely."

New Orleans will forever be associated with Hurricane Katrina. On 29 August 2005, it tore across the city and the region, killing over 1,800 people—as many as during the North Sea Flood in the southwest of the Netherlands in 1953. It caused over $80 billion of damage. <u>This destruction took place primarily in the poorest neighborhoods of New Orleans, which—not entirely by coincidence—are situated in the lowest-lying areas of the city.</u> But there was also spectacular damage further down the coast, such as on the low-lying peninsula of Biloxi, Mississippi, where enormous houses were reduced to piles of sticks.

New Orleans managed to set a large-scale evacuation of 1 million people in motion, but many chose to remain behind—because they had no car, or did not want to leave behind their homes or relatives in need of help, or simply because they did not believe it could be so bad. In the days and weeks that followed, as large parts of the city remained under water, the world watched mesmerized as the Superdome stadium turned into a hell of filth, theft, rape, and administrative impotence. In the wake of the natural disaster, another kind of nature emerged here. And in May 2011 the city was almost flooded again, this time because the river was so swollen by the rains. <u>A new disaster was averted only by opening sluices further upstream and preventively flooding thousands of hectares of farmland—along with farms and villages.</u> The great flood of 1927 was evoked again, when the Mississippi overflowed its banks and set an area of 80 by 160 km under as much as 10 m of water.

I'm walking around the city with Waggonner, an architect who was born and raised here, and I can't believe my eyes. It's 2009; Katrina is now four years behind us, but in many neighborhoods you still see whole streets of houses hanging sideways over their foundations, or foundations without houses. The National Guard has left its tag of colored spray paint, like a grisly kind of graffiti: a circle with four quadrants, showing the date they were there and what they found, living or dead. Waggonner manages to put my stupefaction into some perspective. "Many of these houses were already vacant long before Katrina. The city's population has been declining since the 1970s." A little while later we drive through the stately old districts of the city (on higher ground), with enormous Southern mansions on either side of Charles Avenue and their Greek columns and sumptuous verandas. In the historic city center and now entertainment district of the French Quarter, also on higher ground, you can buy a T-shirt on Bourbon Street that reads "Make Levees Not War," and a local bar is offering the Katrina Cocktail, "Voted #1 Cocktail to Blow You Away!!!"

It is no secret that the French decision to build Nouvelle-Orléans here in 1718 was unwise. That decision was driven purely by economics because of the strategic location on the river, but the low-lying terrain that makes it difficult for water to drain away, combined with staggering amounts of rain, mean the city has always been engaged in a struggle with the water. <u>"An improbable city," Craig Colten calls it in his ecological history of the city, *An Unnatural Metropolis*. "Keeping the city dry, or separating the human-made environment from its natural endowment, has been the perpetual battle for New Orleans."</u>

Things have not improved since. <u>To accommodate shipping, the Mississippi, notorious for its power and its whims, has been squeezed into an ever tighter and deeper strait-jacket.</u> The storms and hurricanes are becoming more intense and more frequent; the dikes are officially calculated to withstand storms that can occur once a century, but such storms are taking place increasingly often. <u>The wetlands that protect the coast are disappearing at a rapid pace</u>, according to many due to the activities of the oil industry in the Gulf of Mexico—a theory that gained even more credence with the oil leak of British Petroleum's Deepwater Horizon in 2010.

In the wake of Katrina, the inhabitants of New Orleans cursed the authorities for the slowness of their response, but money was eventually provided to improve the infrastructure. In 2010 I'm in the Big Easy again for the annual conference of the American Planning Association (APA)—only a couple of months before the BP oil disaster, it will turn out. The APA has organized a visit to the new storm-surge barrier in the Mississippi River–Gulf Outlet Canal. This canal was supposed to provide a shorter route for shipping between the Gulf and the port—but it also proved to be a shorter route for a wall of water that Katrina swept from the gulf into the city. A storm-surge barrier has now been installed, with the participation of the American-Dutch firm Arcadis. It is a colossal concrete structure with sliding panels 10 m tall, which are always open, except when the water reaches dangerous levels.

Colonel Robert Sinkler of the US Army Corps of Engineers stands ready in camouflage fatigues, at attention, feet apart and hands behind his back, to tell us about it: "There is so much concrete in here that you could fill the Dallas Rangers football stadium four and a half times, and metal reinforcements equal to 50 times the Eiffel Tower. We completed this job, which would normally take two years, in 14 months. That was possible because so much funding, around $1 billion, was made available."

Sinkler is glowing with pride. It's no accident that the US Army Corps, the equivalent of the Netherlands' Rijkswaterstaat, is still a military organization that takes on the water as though it were the enemy.

Waggonner, on the contrary, is convinced that the salvation of his city lies not in a hard infrastructure of football stadiums of concrete, or in concrete walls that cut the water off from its surroundings, but in a restoration of the city's connection with the water. He draws inspiration for this from the Netherlands. In cooperation with the American Planning Association and Dale Morris of the Dutch Embassy, he initiated the "Dutch Dialogues," a series of three workshops in 2008 and 2010 in which designers, landscape architects, urban planners, engineers, and water experts from the US and the Netherlands worked out plans to make the water visible and tangible in New Orleans once again. I'm at the third session in 2010, where Dutch and American experts are working on plans for four different neighborhoods in a kind of pressure cooker in a room at Tulane University. "South Louisiana, like the Netherlands, must adapt to the threats inherent to living in a subsiding delta," Waggonner says:

'Living with the water' has recently become an ordering, corollary principle of Dutch policy. Dutch Dialogues participants believe that adapting a Living with the Water principle is necessary in post-Katrina New Orleans; they likewise reject the false choice posited by those who see only a choice between safety or amenity from water in the Louisiana delta. Indeed, Dutch Dialogues posits that both safety and amenity from water are crucial to a future in which New Orleans is robust, vibrant, and secure.

As in the Netherlands, he says, they will always have to pump in New Orleans, but natural systems like rain gardens and more surface water in waterways and canals can also reduce the pressure on the mechanical systems. If New Orleans fails to adapt, the city will be gone in a hundred years—this is Waggonner's somber conviction. "The military solution is not a solution." With several Dutch Dialogues participants and in cooperation with Rotterdam and Amsterdam, his architecture practice, Waggonner & Ball, has been commissioned by the Regional Economic Alliance to design a comprehensive, sustainable, integrated water-management strategy. The master plan is to be ready, and political decision making is set to start, in the autumn of 2012.

Bringing water back into the city is perhaps more a mental and emotional matter than a design issue. For the city's inhabitants, many of them poor and with little education, water, certainly since Katrina, remains associated with fear—not least because four out of ten of them cannot swim.

"That $3 billion investment by the Army Corps in all kinds of defense works—we don't think it's all that smart," says Han Meyer, a professor at Delft University of Technology and a Dutch Dialogues participant:

The walls along the canals are now impenetrable barriers between the neighborhoods on either side. We propose taking out these walls and lining the canals with broad, floodable banks. In normal dry conditions, these banks can serve as new public space from which the adjacent neighborhoods will benefit and which will connect the neighborhoods with one another.

Mary Landrieu, who represents Louisiana in the US Senate, also believes in the Dutch approach and has even visited the Netherlands on a couple of occasions. "The Dutch have learned over the centuries to keep their feet dry through a combination of water management and proper urban planning," she told visitors from the American Planning Association conference. "We've spent $150 billion on clean-up since Katrina. We could have avoided that through prevention."

Han Meyer emphasizes that the aim of Dutch Dialogues is not to impose the Dutch approach on New Orleans:

The whole context is different there. It's a different kind of delta, with a different soil and a different climate, and it is also very different from the Netherlands socially, societally and politically. All the same, it is also worthwhile for New Orleans to combine solutions for water management with solutions for improving the urban planning structure.

Dutch Dialogues is primarily focused not on safety from storm surges but on a fundamentally different system of water management:

The system had to be able to process the huge quantities of rainwater. Three times more rain falls in New Orleans every year than in the Netherlands; in fact sometimes the same amount of rain falls in one week in New Orleans as in a whole year in the

Netherlands! In spite of their enormous pumping capacity, they can't get rid of it, with regular flooding as a result. Our proposal is to create more space for the temporary collection of rainwater. You might call this the "neo-Dutch approach," because this approach—first hold rainwater, then store it and only discharge it afterwards —has also only recently been applied in the Netherlands as well.

Not everyone shares the belief that the Dutch approach can be applied in the US. "What might the United States learn from all this?" Cornelis Disco of the University of Twente wonders in his essay "Delta Blues" for the journal *Technology and Culture*:

A new attitude towards risk and water? Perhaps, but it is hard to see how such a new attitude could result in concrete flood-control policies given the ingrained, constitutionally buttressed and currently hyped-up American distrust of big government.

In the Netherlands the whole country deals with water directly, but not in the US: "New Orleans's tragedy is that the Mississippi Delta is a small tail attached to a great big dog." The Netherlands has learned from New Orleans: as a result of Katrina, a start was made on evacuation plans here.

There is new construction in New Orleans, sometimes with exciting architecture and often on stilts. In the Lower Ninth Ward, one of the poorest districts, actor Brad Pitt and his Make It Right foundation have had several unique houses built, including by the Dutch practice MVRDV. In May 2011, 80 of the planned 150 houses in Brad Pitt's project had been completed. But even Pitt cannot bring back the density and energy that existed before Katrina. And the houses on stilts seem to prefer to avoid the water than to engage with it. They are living above, rather than with, the water.

TOOLS OF THE RISING [73, 81, 85, 99] ANCHORED OBJECTS [77] ARK [24] ARTIFICIAL ECOLOGY [67] BARRIER ISLAND [32] BASINS [99, 101, 102, 103, 104] BIODYNAMIC PRODUCTION [85] CALIBRATED ECOLOGY [24] CANALS [24, 49, 62, 69, 93] COASTAL BUILDING CODES [35] COASTAL ENGINEERING [35, 36, 41] COMMUNICATION SYSTEM [69] CONCRETE CHANNEL [93] CONCRETE SHELTERS HIGH ATOP PILINGS [35] CONTROLLED CANALS [24] CONTROLLED LAKES [53] CONTROL STRUCTURES [93] CROP EXPLOITATION [85, 87] DAMS [34, 41, 53] DETAIL [81] DIKES [24, 35, 36, 39, 40, 41, 44, 45, 46, 48, 53, 69] DIKES AND PUMPS [44] DISASSEMBLE [25] DRAINAGE [34, 36, 53, 73, 93] DYKE [69] ELEVATION [25, 32, 34, 35, 36, 73, 74, 75, 77] EMERGENCY KIT [40] ENGINEERING [35, 60] EVACUATE [41, 43, 53] EVACUATION [23, 24, 35, 41, 42, 44, 48, 50] FILL MATERIAL [25, 75] FLOATING HOMES [45] FLOODABLE FOREST [85, 87] FLOODGATES [24, 69] FLOOD INSURANCE [40, 45] FLOOD-PROOFING [24, 77] FLOOD RHYTHM IN ACCORD WITH THE AGRICULTURAL RHYTHM [69] FLOOD ROOM [43] FLOODWAY [63] FLOOD ZONE [25, 75] GIANT TIDE GATES TO CONTROL THE IN-AND-OUT FLOW OF WATER FROM THE LAND [35] GROINS [35] GUTTER [99] GUTTERS [99, 101, 103] HEROIC IMAGERY [65] HEROIC TECHNICAL, JOURNALISTIC, AND GRAPHIC LANGUAGES [61] HIGHER GROUND [47, 48] HOUSES ON STILTS [50] HYBRID INFRASTRUCTURE [93] HYDRAULIC LOGIC [85] IDEALIZED GEOMETRY [24] IDEOLOGIES OF RATIONALIZATION AND CONTROL [24, 61] IDYLLIC WATERFRONT DESTINATION [73, 77] IMPORTED FILL [25, 73] INFORMATION GRAPHICS AND MAPS [65] INTRICATE SYSTEM OF MAINTENANCE FOR THE WATERWAYS [41] JETTIES [35] LARGE-SCALE EVACUATIONS [39] LARGE-SCALE EVACUATIONS [41] LARGE STAINLESS-STEEL GUTTERS [99, 103] LEVEE [24, 47, 59, 63, 64] LEVEES [35, 36, 63, 93] MACHINIC ECOLOGY [59, 67] MACHINIC LANDSCAPE [59, 65] MACHINIC PHANTASMAGORIA [63] MIGRATE [32] MODULAR [24, 69, 81] MULTI-FUNCTIONAL SYSTEM [93] NATURAL MEANDER [85, 87] NATURE-QUELLING MACHINE [61] NEW DATUM FOR INHABITATION [25] NEW ERA OF WATERFRONT RECREATION [73] NEW TOPOGRAPHY [25, 73, 75] NEW TYPOLOGY OF WATER MANAGEMENT [93] OPEN AIR BAPTISTERY [99] OPERABLE BAYS [59] PERMEABLE PAVING [75] POLDERS [45] PREVENTIVE EVACUATION [41] PRIMARY BARRIERS [53] PUBLIC ASSURANCE [59, 61] PUBLIC BASINS [99] PUBLIC EDUCATION [65] PUMPS [39, 44, 53] PUMP STATIONS [93] RAIN WELL [99] RECREATIONAL SPACE [99] RELOCATION [36, 81] RESERVOIR LAKE [69] RESILIENT PUBLIC SPACE [77] RING [45, 46, 69] RIVER LANDSCAPE [85, 87] SALT-TOLERANT TREES [73] SALT-TOLERANT TREES [75] SCIENTIFIC RATIONALISM [65] SEAWALLS [24, 31, 32, 33, 34, 35, 36] SELF-SUFFICIENCY [40, 44] SHARED SPACE [25, 85] SHIFTING DUNES [81] SLOPED RIP-RAP REVETMENT [75] SLUICES [40, 48, 53] SMART GRID [25, 96, 97] SMART NETWORK [93] SMART SWAMPS [93] SOFTSCAPE [73, 75] SOFTSCAPE PERIMETER [73] SPECTACLE [24, 59, 61, 63, 65, 66, 67] SPECTACULARIZED IMAGERY [59, 63] SPILLWAY [24, 58, 59, 60, 61, 63, 65, 67] STORM GATES [53] STORM-SURGE BARRIER [44, 48] STORM WALLS [53] SWAMPS [92, 93] SYSTEM OF SEAWALLS [24] TECHNOLOGICAL PHANTASMAGORIA [63] TEMPORARY RIVER [85, 87] TERPEN [53] TOPOGRAPHY [25, 69, 73, 75, 76, 77, 78, 85, 87, 90] TRAFFIC MANAGEMENT DURING FLOODS [42] UNDERGROUND INFILTRATION DEVICE [99] URBAN FLOOD PROTECTION [65] VERSATILE TWO-WAY GRID [81] WARNING SYSTEMS [35] WATER DEFENSE TESTING [41] WATER SQUARE [99, 101] WATER WALL [99] WEATHER SIMULATION [42] WEIR [24, 59, 60, 61, 65] WELL-AERATED TOPSOIL [77] WELL-PUBLICIZED EVACUATION PLANS [35]

ACT ONE:
THE RISING
DEFEND

To construct a mechanism to keep rising water levels away from people, buildings and cities.

--- ---

Index of tools | primary barriers, controlled lakes, dikes, dams, sluices, pumps, storm walls, storm gates.

THE DELTA WORKS PROJECT

Netherlands. 1953-present

Benjamin Gregory

The Dutch are often cited as the quintessential Western "water" culture, and for good reason. Due to their position at the delta of three major European rivers (the Rhine, the Maas, and the Scheldt), vast tracts of land have existed in flood plains and below the level of the North Sea for much of their modern existence. Their history is one intertwined with maintaining "dry" land. Early formation of the land into mounds, called terpen, responded to the ever-fluid boundaries between wet and dry, providing an escape from the intermittent rising waters.[1] Agricultural practices involving land drainage, along with peat harvesting, over time led to compaction of the water table, and severe land subsidence, pushing "dry" land further and further below the level of the North Sea (Hoeksema, p. 10, fig. 1-8). Economic affluence and advances in technology in the 16th century allowed vast infrastructures of agriculture and settlement to take root in these vulnerable areas (p. 20). And the tides and storms of the tidal waters of the Netherlands have ever been a threat, with significant events often breaking the threshold between wet and dry, resulting in dire consequences, yet eliciting an even greater resolve from the Dutch to maintain their dry land. Perhaps the greatest of these events, and the one which elicited the greatest of responses, was the flood of 1953.

The aging water infrastructure of the Netherlands, a decentralized collection of dikes, dams, sluices, and pumps, was no match for a storm that attacked the southern tidal areas of the country, the most vulnerable of the nation's coasts, in 1953. The storm damaged 500 linear miles of dikes, puncturing 67 flow gaps, and inundating 770 square miles of land. About 1,800 people perished, and another 72,000 were forced to evacuate populated areas (p. 106). The Dutch response was swift. They could either increase the height of all 435 miles of dike along the coast, or they could dam the tidal inlets, greatly shortening the coastline by essentially turning the tidal estuaries into controlled lakes (p. 108, Fig 8-8). Preferring the scale and control the latter option gave them, they quickly began plans to create a series of massive infrastructural projects to insulate the coast from the temperamental North Sea. The Delta Works project was born.

The Delta Works project slowly evolved over time, from a simple series of dams to a complex system of dams, sluices, storm walls, and storm gates. To maintain the greatest degree of control, the Dutch not only had to separate the tidal estuaries from the sea, with a series of primary barriers, but also had to separate each estuary from the other, such that the flows within them from the rivers would remain relatively unchanged (p. 109). The order of projects was based on importance and simplicity; protecting those areas of economic significance occurred first, and the simplest projects were completed first, in order to provide experience for more complex ones. On the following page is a map showing the extent of the project, and highlighting the most significant interventions.

The effect of this project on the region has been enormous. As Tracy Metz outlines in "Catastrophe," reprinted here from her book *Sweet and Salt*, the safety promised by the dike system historically, and by the Delta Works system most recently, has resulted in a very high population density in the southwest region— despite the inherent danger of living and investing in land below sea level. In addition, there has been a great degree of negative environmental and economic impact from the Delta Works project. Isolated estuaries, whose tidal variations were eliminated when they were cut off from the sea, lost much of the flora and fauna that depended upon those daily variations. The Zandkreek and Veerse Gat Dams caused a complete loss of mussel and shrimp harvesting from that stretch of southern estuary (Hoeksema, p. 111). Similar issues occurred with the Brouwers Dam, whose design was retrofitted to allow tidal variations, and to somewhat restore the damaged estuarian ecosystem (p. 114). Later projects, such as the Oosterschelde Barrier, were designed to allow a maximum degree of daily tidal flow, and close only in the event of a threatening storm (p. 115).

Notes ---

1. R. J. Hoeksema, *Designed for Dry Feet: Flood Protection and Land Reclamation in the Netherlands.* Reston, VA: American Society of Civil Engineers, 2006. All further citations are to this source and are given in the text.

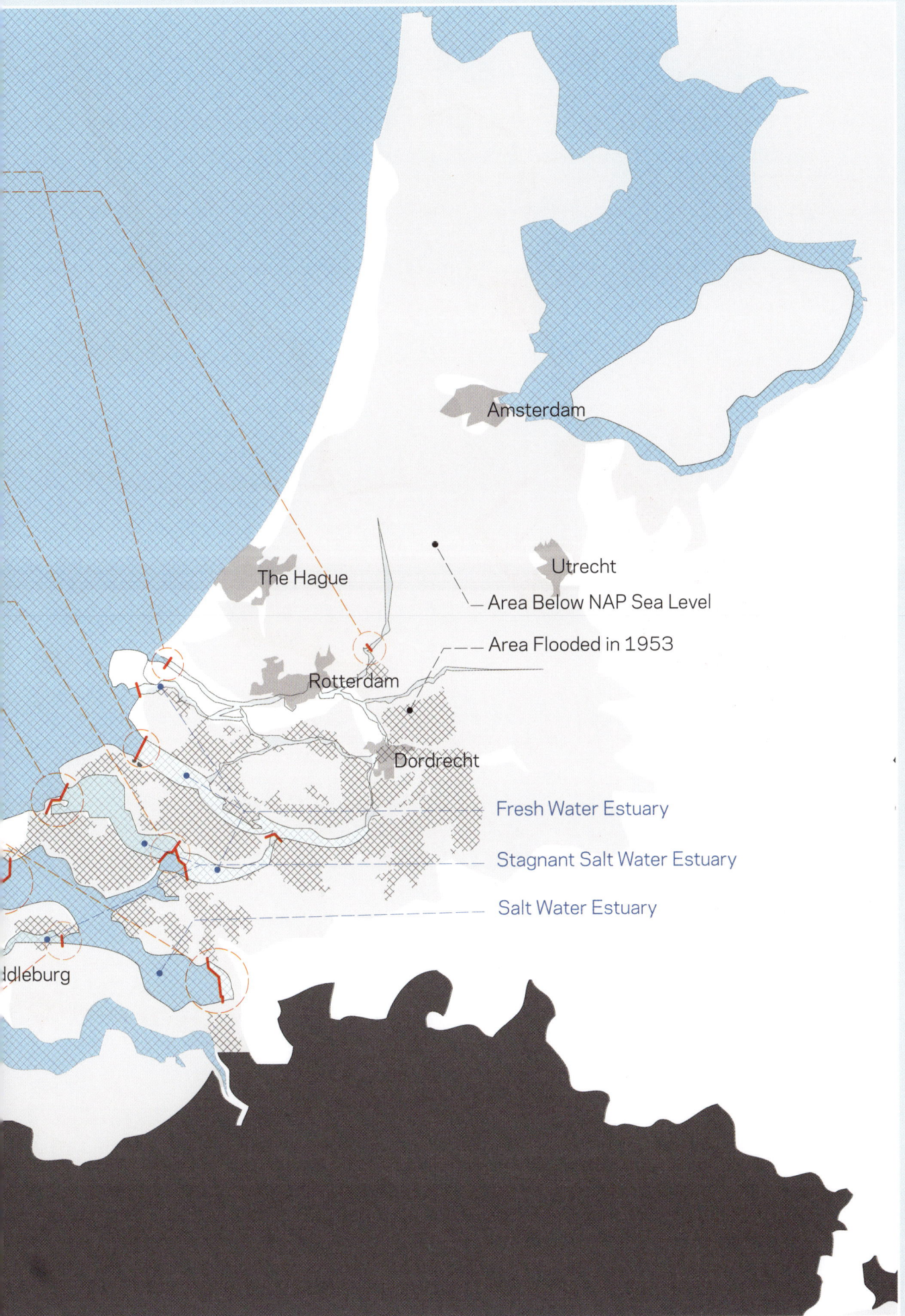

Amsterdam

Utrecht

The Hague

Area Below NAP Sea Level

Area Flooded in 1953

Rotterdam

Dordrecht

Fresh Water Estuary

Stagnant Salt Water Estuary

Salt Water Estuary

dleburg

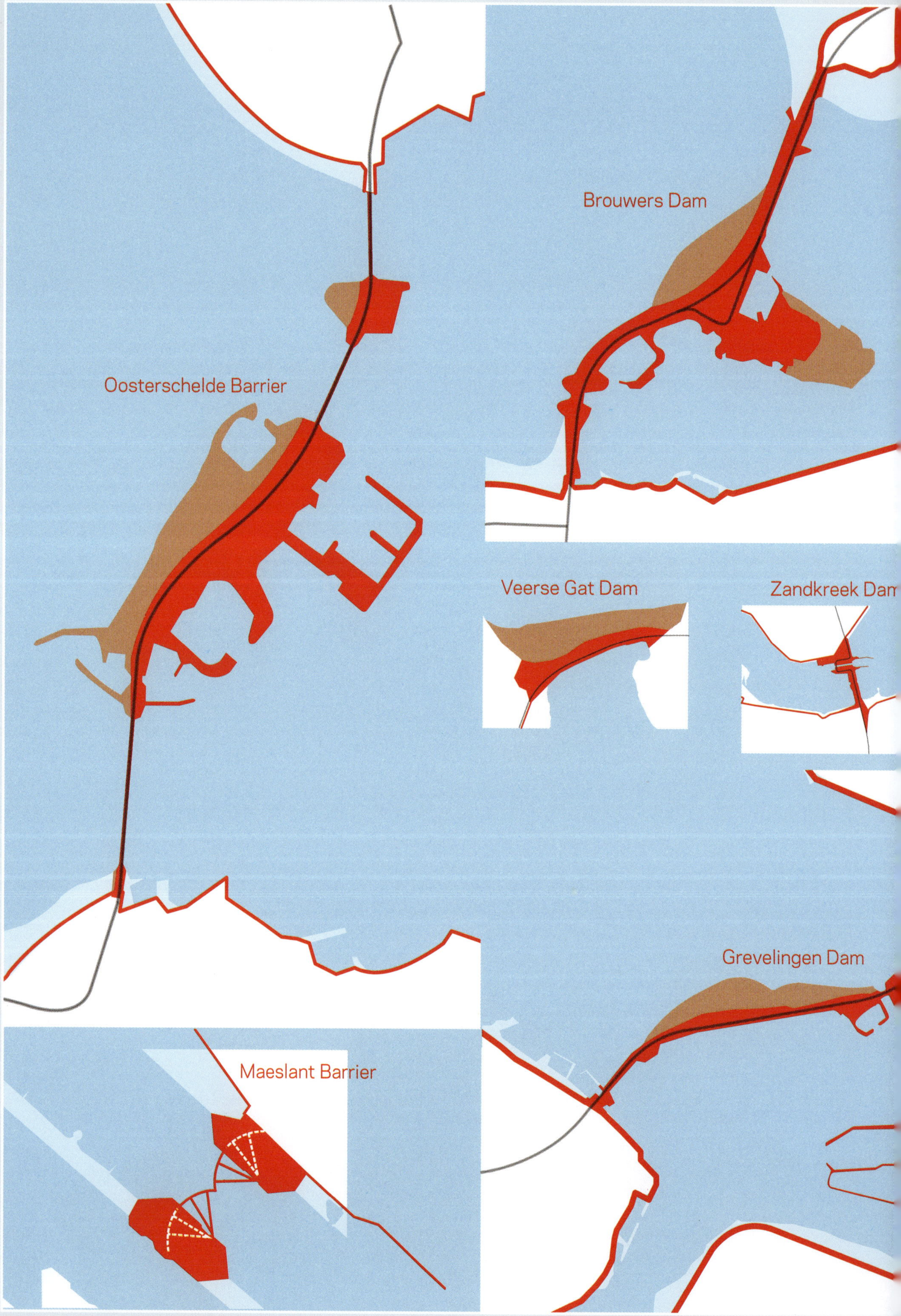

Brouwers Dam

Oosterschelde Barrier

Veerse Gat Dam

Zandkreek Dam

Grevelingen Dam

Maeslant Barrier

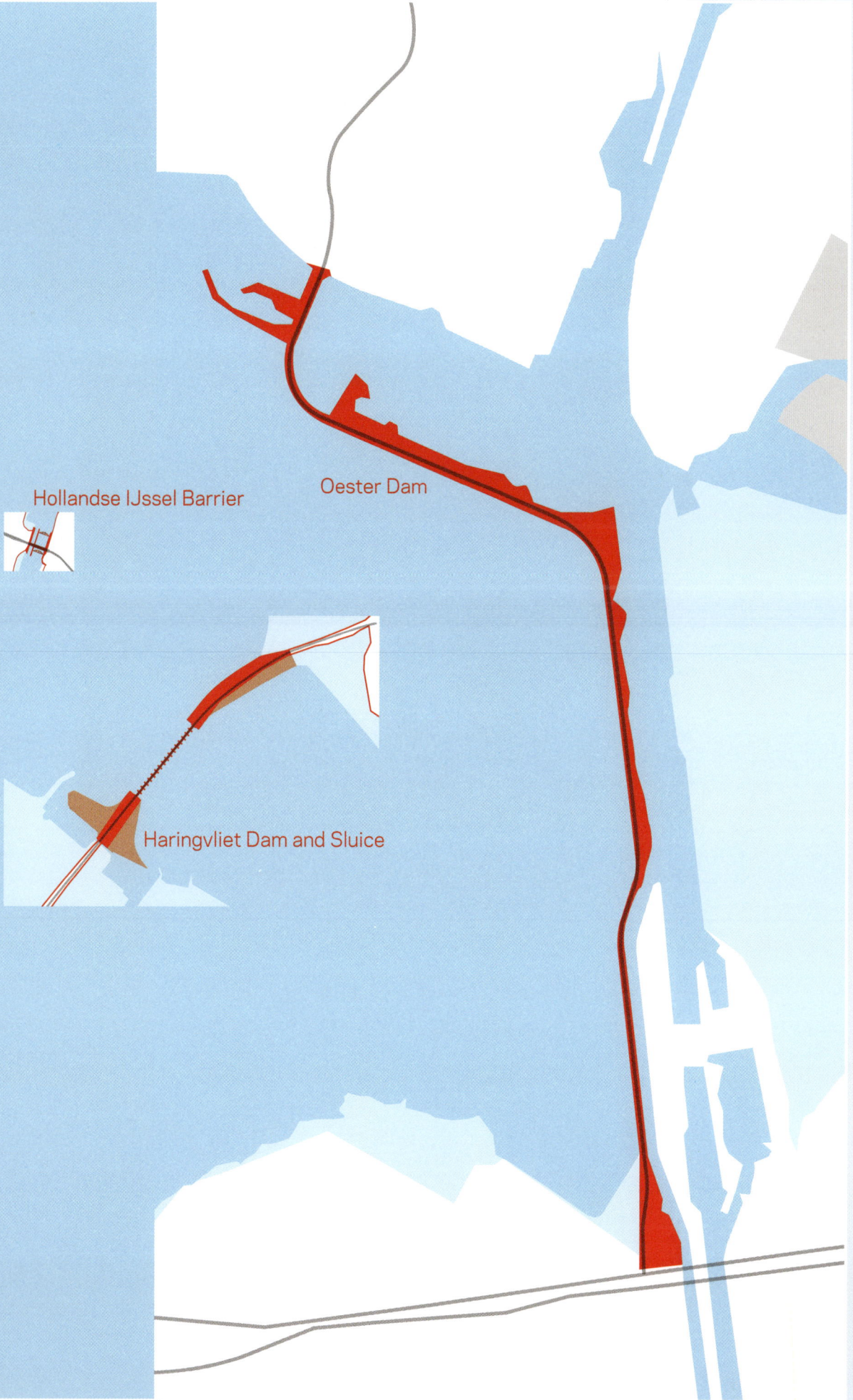

Hollandse IJssel Barrier

Oester Dam

Haringvliet Dam and Sluice

Above:
Spillway Opening 2008
The spillway releases fresh water
heavily saturated with excess
nutrients, oxygen, and sediment
into the brackish Lake Pontchartrain
to aleviate spring flooding on the
Mississippi in 2008.
(Source: NOAA & NASA Earth
Observatory)

--- ---

Index of tools | levee, weir, operable bays, spillway, public assurance, spectacularized imagery, scientific rationalism, machinic ecology

THE SPECTACLE OF WATER AND MACHINE: THE IDEOLOGIES AT WORK IN THE BONNET CARRE SPILLWAY OPENINGS OF 1937 AND 2011

Travis K. Bost

Stretching 7,000 feet in length atop a section of levee 32 miles north New Orleans is a concrete weir structure of 350 relentlessly severe, individually operable bays. Completed in 1935, the Bonnet Carré Spillway, a US Army Corps structure, is opened in times of periodic high water, allowing as much as 250,000 cubic feet per second to be siphoned off the river and drained to the lower Lake Pontchartrain, to quell raging floods. Every bit the heroic modernist ideal, the precision-calibrated device and its machinic landscape would tame a threatening river beast. Matthew Gandy writes, "the American modernist ideal was founded on a close synthesis between the transformation of nature and the iconography of machine technology."[1] Such transformation is not only physical but ideological, together producing an entirely new "urban nature," the production of which is "a microcosm of wider tensions in urban society."[2]

The hallmark of modernist infrastructural production is so often that nature is seen as threatening "outside" or "other." Margaret Fitzsimmons implores us to "recognize that externalized, abstracted, Nature-made-primordial provides a source of authority to a whole language of domination."[3] Of course, there is no shortage of critical research on the domination of nature by modernist

design, development, and engineering, but all too often such criticisms examine only the most immediately obvious tools and manifestations of domination, usually hubris-driven environmental destruction.[4] Less tangible are the ideological tools which constructed and routinized such vilified nature into modern urbanism. Examining the first and the latest opening events of the spillway as laboratories, this essay takes up the foremost of these modernizing tools: the spectacle, as it celebrates, reconfigures, and popularizes altered images and imaginaries of nature and machine.

Opening 1937:
The flooding of spring 1937 was never estimated to be as severe as that of 1927. Of the main tributaries to the Lower Mississippi trunk, only the Ohio River was in flood, as opposed to nearly all in 1927. However, this flood would be the first opportunity for the Army Corps to put the spillway to the test, having completed construction only two years before. The Corps gained a publicity boost fighting the flooding upstream in January with the inaugural detonation of the fuse-plug spillway at Birds Point, Missouri, to equally explosive fanfare. Following that, the Corps' technocrats were eager to showcase another success,

Above:

Inaugural Spillway Opening 1937

The concrete spillway weir with several initial bays opened drains river water into back swamplands for the first time, 7 February 1937.

(Source: United States Engineer Office Second District)

Left:

Declaring Victory for Engineering

Heroic images are circulated, widely declaring "victory" for rational engineering, in this case of the spillway structure with reassuring images of men of reason pasted on top out of perspective, Sunday, 21 February 1937.

(Source: The Sunday Item-Tribune)

Right:

Heroic Engineering Aesthetic

This Army Corps photo poses the seemingly infinite, orderly bays of the weir as heroic soldiers against a wild and raging river, 7 February 1937.

(Source: United States Engineer Office Second District 1937)

this time with a mechanical solution. Additionally, the country's recovery from the depression was still in progress, and given that the spillway's physical construction had been undertaken by the CCC and WPA work programs created by President Franklin D. Roosevelt to aid the recovery, there was a clear political-economic imperative to showcase accomplishment and success, and to garner public confidence. And the river—portrayed in a snarling, savage caricature—became an all too easy target for "progress."

Locally, similar pressures were at work. In 1927 there had been great losses of investments and national trade, and ten years later nervousness circulated widely among investors, especially East Coast financiers. Worried by "national misunderstanding of our situation … calculated to do a great deal of harm" expressed in several national newspapers[5]—"some of them deliberately fomented for ulterior reasons"[6]—city officials and industry leaders saw publicity quickly becoming a liability in addition to being their own promotional tool. Despite the assurances of the engineers' work, *The Times-Picayune* decried: "It is only the flood of unfounded rumors that the scientists and the engineers have been unable to curb."[7] Countless personal public attestations were made, including from senior editors of newspapers and the mayor himself. Finally, *The Times* claimed that the "premature opening of this artificial canal was in response to pressure of opinion created from outside, and not because of any local alarm." Those outside concerns—distant investors—manifested into local ones to the degree that their investments in the city's traders, heads of exchange, and port facilities were threatened.[8] Concerns, local and global, therefore governed the shaping of ideologies of the river and the spillway machine. As high water rolled downriver, speculation grew on rumors and suspicions over the safety of fields, people, and investments in the river's delta. Seeking to avoid destructive frenzy, social (locally) and economic (globally), a cabal of local newspapers, engineers, politicians, business interests, and scientists conspired around a mission of public assurance. Each with the specialized tools of their trade worked in the collective effort to "educate" by taming a wild nature with heroic technical, journalistic, and graphic languages. Circulated nationally were reassuring quotations from leading engineers, from engineers' matter-of-fact warranty:

The world should understand that the defenses of New Orleans against flood danger are properly regarded as safe against all conceivable floods. [9] to the frustrated and accusatory:

There need never have been fear or hysteria on the lower Mississippi; if persons who knew nothing about the situation had kept their mouths shut there never would have been any hysteria.[10]

In photos, the stark modernist geometries of the structure stand out against a "raging" river, while simplified maps and cartoons summarized lengthy newspaper listings of gauge readings and mechanical minutia, calming spectacular speculation in a wide audience. One such graphic was simply titled "Map Shows Why New Orleans Is Safe from Flood."[11] The impetus for such marketing was an alloy of public fascination with a "grotesque" nature, technophilic fetishism of a new nature-quelling machine, civic duty to maintain calm, and calculated economic effort to maintain investments. There are myriad forms and motivations (often conflicting) behind the equally numerous examples of spectacle played out in newspapers of the time. But while each group used spectacle to their own advantage, wielding social, political, or economic control, all ultimately relied on the spectacularly estranged and violent nature personified by the Mississippi in flood. Again we are reminded of Fitzsimmons' linking of nature-as-other to domination, as is done in one headline, "Men and Cutoffs Transform Snarling Wildcat into Purring Kitten."[12] The spectacularizing of the river as dangerous beast proved useful for pushing through political power while subjugating an externalized nature in the process. Spectacle, along with ideologies of rationalization and control, is a fundamental tool of modernity.

After a month of speculation as the river continued to rise throughout January, the first of the 350 bays began to be opened on 30 January 1937. The two cranes riding the top of the weir on rails continued to remove needles with the rising river stage for forty-six more days, finally closing on 16 March. Fortuitously, the flood arrived during the culmination of carnival season, Mardi Gras being on 10 February, when the city was packed with visitors and locals alike ready for a show. Since the buildup of waters in the delta and rumors in the national press, authorities were all the more glad to have an audience they could overtly capitalize upon. The New Orleans mayor himself promoted the event in an editorial in *The Times*,[13] while other officials arranged for a public display of the Corps maps in a city museum downtown.[14] In addition to the fact that

Above:
Interior Spillway Lands
The cleared tract of land at the spillways center
is engineered to allow the speeded flow of
floodwaters, but it also allows for the development
of a new synthetic wetland ecology, August 2011.
(Source: Travis K. Bost)

Below:
Routinizing the Industrial Natural Landscape
Overlaid borrow canals, power lines, railroads, and
pipelines allow a correlate system of ecological
and recreational programs, August 2011.
(Source: Travis K. Bost)

Mardi Gras provided a crowd, a palpable parallel runs through the novelty and spectacle implicit in both carnival and the spillway opening: a questioning of the normative function of the urban landscape. In downtown street parades or on the city's periphery, normalized power structures are subject to both prideful display and vicious parody, whether between social classes or between machine-equipped modern society and primordial nature. In each, the "uncanny"[15] becomes a tool for questioning the "ordinary" of the urban. Carnivals, Eliza Darling claims, are one of many sites allowing urbanites to gaze upon nature-as-spectacle, "to struggle with and work out their often contradictory and conflicting relationships with it."[16]

The initial opening, a Sunday, brought thousands out of the city—"Traffic Swamps Airline Road at Spillway 'Show'"[17]—to line the guideway levees that flank the spillway path to the lake and jamming to a standstill Airline Highway, which crosses its breadth on piers, all eager to witness the wild river beast tamed by the prowess of modern machine engineering. More than the spectacle of the machine or the "battle," there is the spectacle of "wild" nature itself on display, as it is "tamed" by the impressive might of the engineers. In press photographs, the positioning of spectators is also significant, perched always on high ground, levee, or causeway, inevitably looking down as though at the lion's den at the zoo. Spectators look on, in their "Sunday best" for the show, a mixture of fascination and contemptuous control, as water pours through, further enabling a masterful tone. The spectacle is enhanced by yet other machines, namely the airplane which heightens the angle of masterly perspective over the water channeled by strict orthogonal machine geometries. It is truly a machinic phantasmagoria on display.

But lest the public fall to fearing the tempestuous river beast on display, they are made consistently aware—in the run-up to, during, and after the performance—of the assurances from the men of reason positioned in the corner of every news clipping who oversee the spillway's operation, dressed in tie and hat to ensure not only control but professionalism, which speaks to the regularity and calm with which they conduct themselves in the process. Despite the glorification of the machine, it is still important to witness the spectacle of human labor in the process of control over the beast. To this end, the weirs eschew mechanized tainter gates, and are instead operated by teams of men raising the 7,000 creosote timbers by manually operated cranes

as other teams of spotters guide the work, taking no less than 38 hours.[18] The relations of laboring on and with nature have long been the basis for forming identity; in this case it is one of control.[19]

The excessive promotion of the city's safety in local and national press along with the spectacularized imagery of the spillway's operation make clear that the mechanical functioning of the flood control structure was only part of its intended function. There was a clear need to propagandize the floodway in order to popularize the USACE's engineering success, but more importantly to guard against the looming greater potential disaster caused by the flood: a panic of investment and capital circulation.[20] The carefully calculated scapegoat in all this was, of course, the spectacle of a primeval, threatening river.

Opening: 2011
In the run-up to the most recent opening of the spillway, just shy of 75 years later, the climate of relations to water—social, economic, and political—had gone through great upheavals, due largely to three key factors: the urbanization of the lower delta, the economic transformation from intermodal shipping to petrochemicals to tourism, and the federally made crisis of Hurricane Katrina. All of these issues served to isolate residents and financial interests (what were left) from direct concern with the workings of the river.

With multiple generations coming of age after the original construction of the spillway, its structure and landscape have changed dramatically in their ideological association. These subsequent generations of the early moderns who were drawn by technological phantasmagoria are drawn to the site still, but for new reasons. The sons and daughter of Prometheus, as Maria Kaïka calls them, come "[n]o longer in search of the awesome beauty of god-given, pristine nature," but in the conscious pursuit of 'second nature.'"[21] For this new generation, the site is still fascinating for its exceptionalism, but as a new, separate but routinized, "natural." Indeed, the spillway grounds themselves nearly look the part of untouched wilderness, but the many channels, ponds, and cypress stands are almost without exception properties of human infrastructures, whether the spillway itself or the many utility, petrochemical, or industrial activities operating within it. Regardless, nearby residents, in the 75-year interlude, have come to regard the spillway as a natural swampland yielding fishing, hunting, boating, and other outdoor activities that serve to render

THE TRIBUNE, NEW ORLEANS, MONDAY, FEBRUARY 1, 1937
CARNIVAL THRONGS WATCH SPILLWAY WORK

Above:
Spillway Spectacle
Thousands drove from metropolitan New Orleans, forsaking carnival activities, on Monday, 1 February 1937.
(Source: The New Orleans Tribune)

Below:
Spillway Opening 2011
Hundreds of visitors gathered along the guideway levee banks for the 2011 opening, May 2011.
(Source: USACE)

banal any still-lurking animosity, but stop short of any true familiarization with the makings of the place.

This new relationship goes on to play out clearly in the spectacle of the May 2011 opening. Whereas Katrina had largely been <u>sensationalized as a sudden and urban catastrophe</u>, the slow rise of the river brought eminent concern only to those <u>outside urban flood protection</u>, a dichotomy the press would further promote. In the run-up to the opening, national newspapers were quick to drum up drama pitting <u>a battle between city and country</u> with headlines such as "Army Corps Blows Up Missouri Levee,"[22] "Areas Will Be Flooded to Protect Louisiana Cities,"[23] and "In Louisiana, a Choice between Two Floods."[24] Among local urban residents, however, views of the river and the spillway turned away from seeing them as a threatening beast and <u>a heroic machine to a nonchalant routinizing of the machinic landscape</u>, <u>a new "natural,"</u> seemingly always having been there.

But being rendered routine does not mean that spectacle is not present. <u>Spectacle is, after all, a tool for leveraging power by way of nature imagery, and is still a key tool of military, political, and economic interests.</u> Whereas exemplars of 1937 spoke to scientific rationalism and heroics, more plainly read politicking bleeds through in the latest opening. Images published by the Army Corps—largely on their new quick-distribution-friendly Facebook page—continue a reassuring theme that all is under control: always a uniformed official is speaking or a machine is operating. Much the same occurs in all published images of the governor and other state officials, arriving inexplicably by helicopter or surrounded by military uniforms. It is perhaps not surprising that each would harness the spectacle of military reassurance as an opportunity, especially for the governor, to gain back public trust lost during Katrina. But on the whole, <u>the heroic imagery</u> seems to strive most at justifying its own news-worthiness.

<u>Public "education"</u> remained a goal for newspapers, whose extensive use of <u>information graphics and maps</u> continued apace with that of 1937. But instead of foregrounding the fascination of the spillway machine, graphics tended to portray the larger river system, furthering an "us" versus "them" dynamic among urban centers and distant lowland areas and the Army Corps, and were generally unafraid to make less-than-subtle threats. In one *Times-Picayune* story, "Mississippi River Flooding in New Orleans Area Could Be massive If Morganza Spillway Stays Closed,"[25] the accompanying

graphic, labeled <u>"Worst Case Prediction,"</u> shows huge swaths of territory, urban and rural, bathed in saturated purple and red indicating potential flood extent if certain areas are not inundated. The implication to the public is that <u>the city is being held hostage by the Army Corps</u>, and putting pressure on the Corps to open the spillways, <u>never mind the costs to rural residents.</u>

But as the day finally came to open the first bays at Bonnet Carré, 9 May 2011, headlines again promoted <u>the river-machine spectacle.</u> This time, however, they were more likely to be found in back pages such as the music and entertainment section, "Spillway Opening and the Travel Channel's Mark DeCarlo's Book Signing on Monday in New Orleans,"[26] or under politics and government: "Bonnet Carré Spillway Opening Serious Matter, but Residents Enjoy the Spectacle."[27] The headlines cater to the new nonchalant local attitudes, having to remind us, "<u>[i]t's not just for recreation."</u>

Complementing many of the headlines, the images of the opening show quite tangibly the calm public consciousness and comfort with the now routine or <u>banal synthetic nature under operation</u>. Not only does the spillway's operation mimic the historical action of the fresh-water delta marshes, but the machine is somehow better than the original ecosystem as <u>it showcases the ecology as almost cartoonishly verdant wetlands</u>, while also <u>offering opportunities for recreation and sport</u> to be taken advantage of during the short operation. In some way, the banal becomes a tool for enhancing the engagement of urban residents in the dynamic, if synthetic, landscape.

Despite its being cast as banal, <u>the fascination of spectacle has, if anything, grown in popularity.</u> But it is a different, more <u>quotidian spectacle</u>, as if it were the latest rendition of a well-known play—which, in a way, it is. However, whereas the spectacle of 1937 promoted marveling at the machine's triumph over nature's ill will, today there is a gawking—beer in hand—at the caricatured, unthreatening nature that is produced as it pours through the weir. The casual images of residents relaxing or fishing in the flood waters—"Bonnet Carré Spillway Opening a Mixed Bag for Anglers,"[28] we are told—are the height of such sentiment. Like some of the imagery of 1937, there is a great deal of similarity to carnival crowds in the groups of spectators that gather for the spillway's opening. But while visitors of 1937 came in their cars and were well-dressed, not knowing exactly what was to be witnessed but ready to be seen as part of the spectacle, visitors here have come prepared with lawn chairs and ice chests as though for

Above:
Taking in the Spillway
The 2011 opening was widely publicized by local media and the Army Corps themselves to assure accomplishment, while those attending come ready for a entertaining but familiar show, May 2011.. (Source: USACE)

Below:
Permanent Spectacle
The hybrid landscape of industrial activity leaves a surprisingly bucolic landscape that fulfills a permanent, synthetic spectacle.
(Source: Travis K. Bost)

a parade or fireworks display mandated by tradition. Not only are they gathered along the high ground behind the largely ceremonial fence, but some also scatter low in the water as well. This is plainly evidence of a changed consciousness of comfort with the machinic ecology in operation, as opposed to those spectators of 1937 who huddled close to the cranes, and the faith in technology protecting them from the wilds being unleashed beyond. Complementing the naturalization of an artificial ecology, even a "zoo-ification" of the machine appears. In one image in *The Times*, the diesel-powered cranes resemble wild animals, behind bars, going about the best approximation of their "natural" instincts in the approximation of their "natural" habitat. The spillway becomes a zoo for the spectacle of an artificial ecology, a simulacrum even, of a wetland, of which its periodic opening is a special exhibit.

Spectacle, whether of water or machine, is evidence that under modernism, "[n]ature and the world never come to us unmediated,"[29] but we ourselves and our cities are also being synthetically transformed as well, heightening ties between the urban and the landscape. And yet, beneath relationship or event, the spectacle is image. Summing up the 2011 celebration, *The Times* laments, "Spillway Opening Draws Crowd, but Fog Spoils the View."[30]

Notes ---

1. Matthew Gandy, *Concrete and Clay: Reworking nature in New York City* (Cambridge, London: The MIT Press, 2002), 148.

2. Ibid.

3. Margaret Fitzsimmons, "The matter of nature," *Antipode 21* (1989): 109.

4. However, the Flood Control Act of 1928 which called for the spillway's construction is often lauded for its return to a more "natural" operation for the river—perhaps questionably, given that it is a thoroughly mechanically managed operation.

5. James M. Thomson, "Floods and flood fears analyzed," *The Item-Tribune*, 31 January 1937.

6. "River Psychology," *The Times-Picayune*, 25 February 1937.

7. Ibid.

8. "Spillway Opened," *The Times-Picayune*, 2 February 1937.

9. James M. Thomson, "Floods and flood fears analyzed," *The Item-Tribune*, 21 February 1937.

10. Citing General Markham of the United States Army Corps of Engineers, in Thomson, "Floods and flood fears analyzed."

11. "Weather Bureau, United States Engineers Chart Capacity of Lower River Reservoirs," *The Sunday Item-Tribune*, 31 January 1937.

12. This headline is from *The Times Picayune-New Orleans States*, 14 March 1937.

13. Robert S. Maestri, "Gen. Jackson and Lafitte Saved New Orleans in 1815; Today's Heroes are Uncle's Sam's Army Engineers," *The Times-Picayune*, 24 February 1937.

14. John M. Barry, *Rising Tide: The Great Mississippi Flood of 1927 and how it changed America* (New York: Simon & Schuster, 1997).

15. Maria Kaïka uses the Freudian term "uncanny" in discussing moments in the urban fabric when modernizing systems of the city are temporarily undermined, revealing the underlying power structures that weigh over nature, which is subdued as outside or as resource. See Maria Kaïka, *City of Flows: Modernity, nature, and the city* (London: Routledge, 2005).

16. Eliza Darling, "Nature's carnival: The ecology of pleasure at Coney Island," in Nik Heynen, Maria Kaïka, and Erik Swyngedouw (eds.), *In the Nature of Cities: Urban Political Ecology and the Politics of Urban Metabolism* (London: Routledge, 2006), 84.

17. *The Times-Picayune*, 1 February 1937.

18. Emory L. Kemp and Michael C. Robinson, "Stemming the tide: Design and operations of the Bonnet Carré Spillway and Floodway," *Essays in Public Works History* 17 (Chicago: Public Works Historical Society, 1990).

19. White, Richard White, "Are you an environmentalist or do you work for a living?" in William Cronon (ed.), *Uncommon Ground: Toward reinventing nature* (New York: W. W. Norton, 1995).

20. Barry, *Rising Tide*.

21. Kaïka, *City of Flows*, 139.

22. *The New York Times*, 2 May 2011.

23. *The New York Times*, 13 May 2011.

24. *The Washington Post*, 11 May 2011.

25. *The Times-Picayune*, 10 May 2011.

26. *The Times-Picayune*, 9 May 2011.

27. Ibid.

28. Bob Marshall, "Bonnet Carre Spillway Opening a Mixed Bag for Anglers," *The Times-Picayune*, 14 May 2011.

29. Noel Castree, "The nature of produced nature," *Antipode* 27(1995): 83.

30. *The Times-Picayune*, 9 May 2011.

References --

Barry, John. 1997. *Rising Tide: The Great Mississippi Flood of 1927 and how it changed America*. New York: Simon & Schuster.

Castree, Noel. 1995. The nature of produced nature: Materiality and knowledge construction in Marxism. *Antipode* 27.

Darling, Eliza. 2006. Nature's carnival: The ecology of pleasure at Coney Island, in Nik Heynen, Maria Kaïka, and Erik Swyngedouw (eds.), *In the Nature of Cities: Urban Political Ecology and the Politics of Urban Metabolism*. London: Routledge, 75-92.

Fitzsimmons, Margaret. 1989. The Matter of Nature. *Antipode* 21: 106-120.

Gandy, Matthew. 2002. *Concrete and Clay: Reworking nature in New York City*. Cambridge, London: The MIT Press.

Kaika, Maria. 2005. *City of Flows: Modernity, nature, and the city*. New York: Routledge.

Kemp, Emory L., and Michael C. Robinson. 1990. Stemming the tide: Design and operations of the Bonnet Carré Spillway and Floodway. In Essays in Public Works History 17. Chicago: Public Works Historical Society.

Smith, Neil. 1984. *Uneven Development: Nature, capital, and the production of space*. Oxford, New York: Basil Blackwell.

White, Richard. 1995. Are you an environmentalist or do you work for a living? In William Cronon (ed.), *Uncommon Ground: Toward reinventing nature*. New York: W. W. Norton.

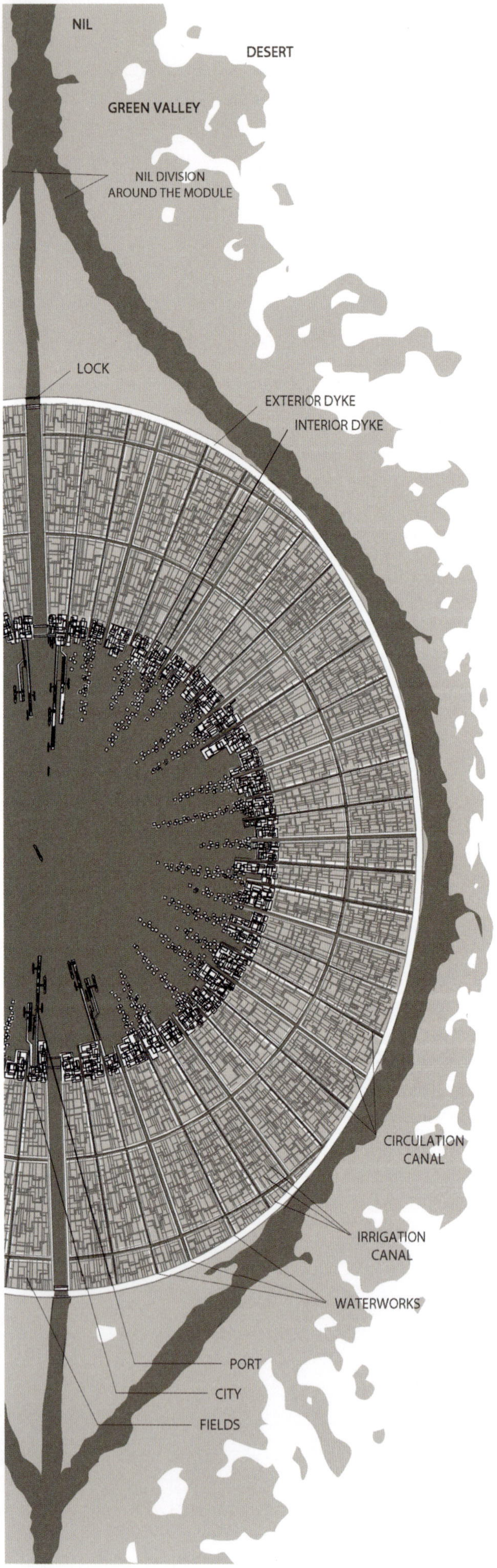

NIL

DESERT

GREEN VALLEY

NIL DIVISION
AROUND THE MODULE

LOCK

EXTERIOR DYKE

INTERIOR DYKE

CIRCULATION
CANAL

IRRIGATION
CANAL

WATERWORKS

PORT

CITY

FIELDS

MEDITERRANEAN SEA

ALEXANDRIE

DAMIETTA

PORT SAID

DELTA DU NIL

TANTA

ZIFTA

LE CAIRE

NIL

GREEN VALLEY

MOUNTAIN

"HYDROPOLIS"

FAIYUM

BENI SUEF

MINYA

مصر
(EGYPT)

ASYUT

DESERT PLATEAU

TAHTA

FARSHUT

LUXOR

NAGAA
EL-SHAIKH

AL MADIQ

ASSOUAN

ASWAN
DAM

LAC NASSER

ABU SIMBIL

Index of tools | sea wall, dyke, floodgates, canals, ring

HYDROPOLIS
Nile Valley, Egypt. Speculative. 2012.

Marion Ottmann, Margaux Leycuras, and Anne-Hina Mallette

Problem

In the past, the Nile Valley lived to the rhythm of the rise of the water level, taking the advantage offered by silt to fertilize its farmlands. However, these water variations are irregular, periods of flooding often followed by periods of drought. Thus, the Egyptian government in 1902 began the construction of a dam seven kilometers upstream from Aswan to tame the river, creating a huge lake, Lake Nasser, which floods part of the territory. This dam was consolidated many times and rebuilt completely in 1950.

This dam has a negative impact on the ecosystem. Indeed, the absence of silt in the valley no longer compensates for marine erosion along the coast, causes a decline in the fishing, and no longer fertilizes farmlands. In addition, saline intrusion in the Nile Delta is causing a precipitous coastline erosion. The rapid filling of the reservoir by the deposition of silt will saturate the capacity of the dam in less than a century, hence its inefficiency in the long term.

Concept

In order to solve these problems and restore the ecosystem in the Nile Valley, this project develops an alternative project to the Aswan Dam which takes advantages of the contributions from the flood while controlling the river. It creates a modular system of cities along the river Nile, turning a technical problem into an ideal city.

The main idea of the project is to divide up Lake Nasser into each of the city modules in a reservoir lake 200 meters deep. This lake, combined with a hydrological system, succeeds at getting a more natural flood level control. Each city is surrounded by the Nile, connecting all city modules together by Egypt's first communication system. These cities are formed of an enveloping seawall, adapting the topography of the mountain bordering the valley. From this dike emerges a traffic flow (roads, bridges) linking the city with the outside (surrounding villages).

At the urban scale, the module is structured by a complex hydrological system. Indeed, the main fluvial axis, controlled by locks, crosses the city to reach the internal port area situated on the reservoir lake. In addition, irrigation of the fields is provided by the combination of dikes, canals, and valves. The whole converges towards the central reservoir lake, and on its circumference is located a ring forming the city. It sets up on the bank surrounding the lake and extends on the water. Thus the city straddles between the land (agriculture) and the lake (exchanges).

This hydrological system puts the flood rhythm in accord with the agricultural rhythm. From July to September, the Nile is in flood. The valves of the exterior dam are opened, then the water and the precious silt is deposited in the fields. The valves of the inside dam are also open in order to fill in the reservoir lake. Then the Nile returns to its riverbed, the lake reservoir is full, and the water is retained by the interior dam, the valves of which are closed. The valves of the exterior dam are also closed to irrigate and leave 50 centimeters of water in the fields. This is the time of plowing, planting, and growing rice shoots.

For five months, the rice will grow its roots in the water. May is the month of harvest. Three weeks before, the floodgates will open to the external dam, and the rice will finish its maturation out of the water. From June to July is the drought, drying, and winnowing of the rice. Finally, at this time, the level of the Nile is at its lowest level, but the city benefits from the water of the lake reservoir.

Opposite Left:
Hydropolis module

Opposite Right:
Hydropolis city modules along
the Nile

JULY

SEPTEMBER

OCTOBER

DECEMBER

MAY

Left:
Flood cycle and rice culture
adaptation by a hydraulic system

Above Top:
An oval Hydropolis module adapting
according the geographical situation

Above Bottom:
Cultivating Rice.

ACT ONE:
THE RISING
RETREAT

To move people, buildings and
cities away from rising water levels.

Index of tools | lifted island, new topography, imported fill, softscape perimeter, elevation, idyllic waterfront destination, drainage, salt-tolerant trees, hills

GOVERNORS ISLAND PARK

New York City. Phase 1 Opened 2014.

West 8 and Mathews Nielsen Landscape Architects

Governors Island offers a world apart from New York City and an extraordinary vantage point on New York Harbor. West 8 understood from the outset that the 87 acres of new park and public spaces would need to be resilient in the face of rising waters. In order to build a park that is sustainable across the next century, West 8 had to plan for the continuing long-term increase in mean sea level and for the more frequent and violent storms that are expected to accompany climate change along the eastern seaboard.

Developed from the 2006 international design competition winning entry, and the Park and Public Spaces Master Plan, Phase 1 represents a shift in the character of the city's park system towards a new era of waterfront recreation. Rather than withdrawing in the face of the advancing waters, the park designed by West 8 allows people to enjoy the connection to the salt winds, swirling waters, and expansive views of the harbor while ensuring that the trees planted in 2013 will grow into a great forest over the next few generations.

Phase 1 of Governors Island Park and Public Spaces pays homage to the vast water and skies of the harbor, and to a democratic notion of an Island accessible to all. The 30 acres of new park which were opened to the public in spring 2014 include:

- Liggett Terrace, a sunny, 6-acre plaza with seasonal plantings, seating, water features, and public art
- Hammock Grove, a verdant 10-acre space that is home to 2,000 new trees, play areas, and hammocks
- The Play Lawn, 14 acres of play and relaxation on the water's edge that includes two turf ball fields.

It also encompasses the rejuvenation of the historic landscapes, two new arrival areas and buildings at Soissons and Yankee Landings, a re-graded Parade Ground, the completion of the northern half of the Great Promenade, and key visitor amenities in the Historic North Island, including lighting, seating, and signage.

These parks and public spaces have always been conceived within the framework of rising waters. Hurricane Sandy brought the future sooner than expected, and the power and height of the storm surge on the island proved the importance of integrating resistance to the rising waters into the DNA of the park. WEST 8's design strategy for the island specifically focuses on the following elements.

Above:
Governors Island Master Plan ©
West 8

Right:
By raising the grade, trees planted
will have their roots above future
flood elevation for generations of
growth. Elevations shown for Sandy
Hook datum © West 8

Transformation through Topography
We lifted the majority of the island out of the flood zone.

Unlike the naturally raised topography of the historic North Island, the South Island was a pancake-flat landfill. Without intervention, the majority of the South Island would be inundated in a projected 100-year flood. By importing hundreds of thousands of cubic yards of fill material, West 8 has created a new island elevation that lifts the root zones of the new planting away from brackish water and above projected flood levels. This has the added benefit of raising "ground level" in the adjacent development zones.

These raised areas, and the enrichment of soil conditions, provide sustainable habitat for trees and other plantings to thrive. The new topography will reuse on-site materials from demolished buildings and parking lots, keeping those materials out of existing landfills. The second phase of the island's development, the hills are the culminating feature of the Governors Island Master Plan and will rise 30 to 80 feet above sea level.

The design diminishes the impact of a wave attack. During intense storms, wave action can erode and scour the land. The construction of a new sloped rip-rap revetment made of large boulders will dissipate the wave energy. A seat edge designed by West 8 will run along the western park edge and act as the second barrier against wave attack. With this strategy in place, water from a particularly high storm surge will still flow into the park spaces but with reduced energy, minimizing erosion.

Resilience in the Face of Flooding
Use a Lawn at the Perimeter

West 8's design places lawn areas, which will easily survive repeated inundation by brackish waters, around the perimeter and reserves special plantings for the interior. Irrigation installed in all new planting areas will allow salts from flooded lawn and plant areas to be leached quickly. The replacement of acres of impervious asphalt with softscape, plantings, and permeable paving reduces stormwater runoff, maximizes infiltration, and decreases the urban island heat effect.

Plant Salt-Tolerant Trees at Low Elevations
West 8 has located less tolerant species in the raised areas of the park, like Hammock Grove, and more salt-tolerant species like London plane trees around the perimeter.

A Perimeter that Can Withstand Flooding
The future Great Promenade will be at an elevation that receives flooding during extreme storms. Park amenities at the perimeter—balustrade, lighting, benches, etc.—are specifically designed to withstand flooding.

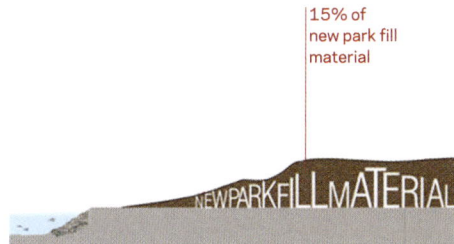

15% of new park fill material

NEW PARK FILL MATERIAL

ROOT ZONE ABOVE PROJECTED FLOOD LEVELS

2FT SEA LEVEL RISE BY THE YEAR 2100 (99% CONFIDENCE)
PROJECTED FEMA 100 YR FLOOD LINE
EL. + 12.0' (DATUM NGVD 1929)

1

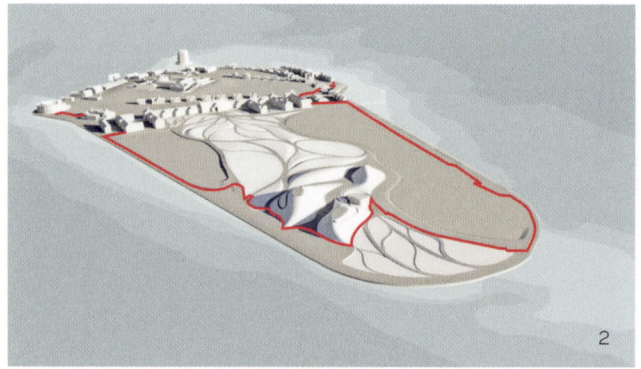

2

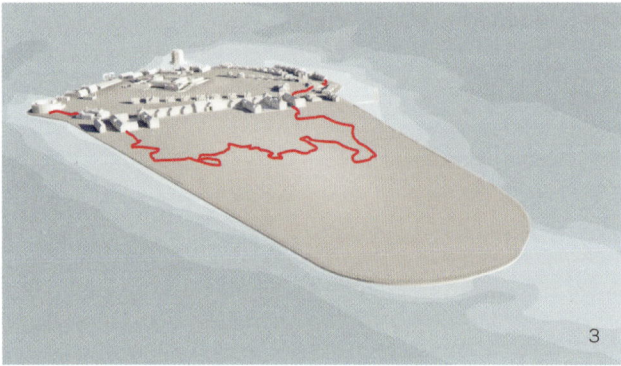

3

4

Existing Island

Topography

Edging

Circulation and Paving

Planting

These individual systems comprise the core anatomy of the plan.
Facing page: The park and public spaces are an integrated system of layered components.

1 - Original size and shape of the island. © West 8

2 - Governors Island topography in 2006, with projected 100-year flood line. © West 8

3 - Proposed topography moves the projected 100-year flood line. © West 8

4 - Proposed topography raises part of the landscape for long-term planting above projected flood elevations. © West 8

Below: Park Anatomy © West 8

Resilience in the Face of Violent Storms

Design for High Winds
Most fixed park objects (lamp posts, benches) on the island will be anchored to withstand a typical storm event, but trees cannot be. In order to combat tree loss due to high winds, West 8 devised a unique planting strategy. Smaller caliper trees have been planted to allow trees to adapt to the unique island environment and windy conditions. Provision of ample root space (large, contiguous areas of topsoil and plentiful depths) also accommodate a broad and deep root zone.

Design for Heavy Rain
Large volumes of rain can saturate soils, cause damage to pavement areas, and weaken the resiliency for tree roots to keep trees upright during wind gusts. The park topography is sculpted so that stormwater will always gather in the lower areas, and is available for reuse on- site. West 8 has designed a park that will drain quickly and will not be swampy and unusable for days or weeks. Well-draining and well-aerated topsoil and fill materials will also make the new trees more resistant to high winds.

Changes to the Limits of Projected Flood Zone
Hurricane Sandy put some of these already-implemented design elements and site-specific strategies to the test. Despite the fact that Sandy's storm surge exceeded the projected 100-year flood in the year 2100 by about a foot, Governors Island made it through the storm relatively unscathed. Shipping containers, flotsam, jetsam, and other debris washed over the seawall and could be found throughout the South Island—everywhere except where we were building the new 30-acre park. Superstorm Sandy re-emphasized the importance that elevation and flood-proofing have on the durability of the project.

This natural bay where the Hudson and East Rivers meet is without comparison. Phase 1 of Governors Island accentuates the experience of the harbor, weather, and water, and establishes a strong foundation for future growth. It transforms this abandoned military base and forgotten, windblown island into a resilient public space, a green broccoli in the water, and an idyllic waterfront destination for the people of New York.

The island offers a world apart from New York City, an extraordinary vantage point on New York Harbor, and the chance to experience the sensations of a green island surrounded by water and sky. Governors Island is re-emerging as an extraordinary new public park that embraces all New York Harbor, its ecology, its history, its culture, and its magnificent beauty: an icon for the city, a beacon in the harbor.

Phase 1 of Governors Island Park and Public Space opened to the public in May 2014, and Phase 2 will be completed in 2015.

Left:
Transformation through topography
conceptual sketch © West 8

Right:
Governors Island Liggett Hedges
2014 © Timothy Schenck
Photography

Above:
Governors Island: transformation through topography. © Timothy Schenck Photography
Below:

Phase 1 of Governors Island Park and Public Space opened to the public in May 2014 © Jim Navarro

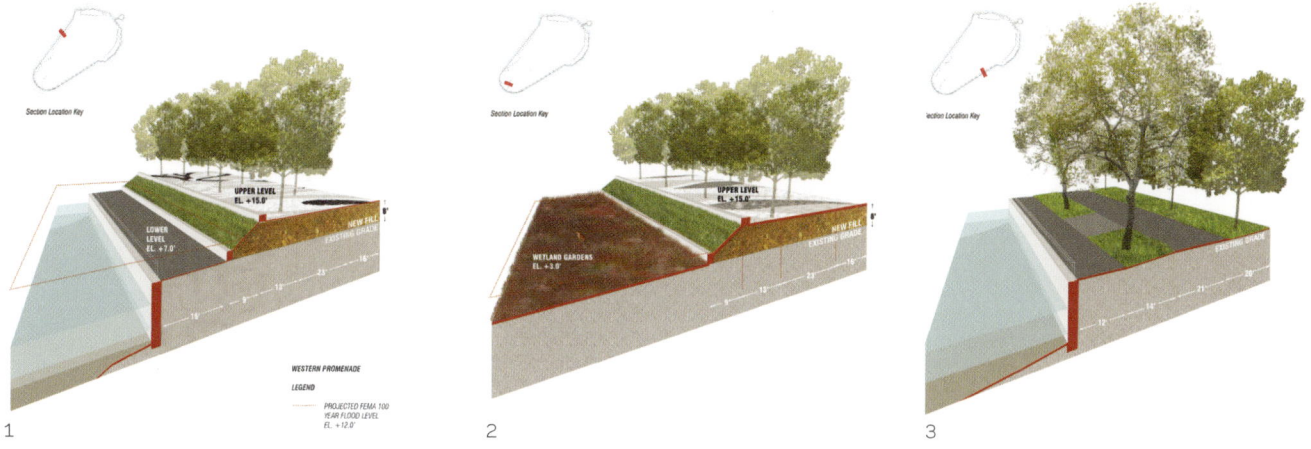

1

2

3

Above:
1. Western Promenade: A section showing the Western Promenade when elevated 7ft above existing grade with new trees © West 8
2. South Prow Overlook: A section showing the South Prow Overlook

when elevated 7ft above existing grade with new trees © West 8
3. Eastern Promenade: A section showing the Eastern Promenade at the existing grade, which preserves existing London plane trees © West 8

Medium:
Phase 2: Framing views of the Statue of Liberty from between the Hills © West 8

Below:
Governors Island under construction 2014 ©The Trust for Governors Island

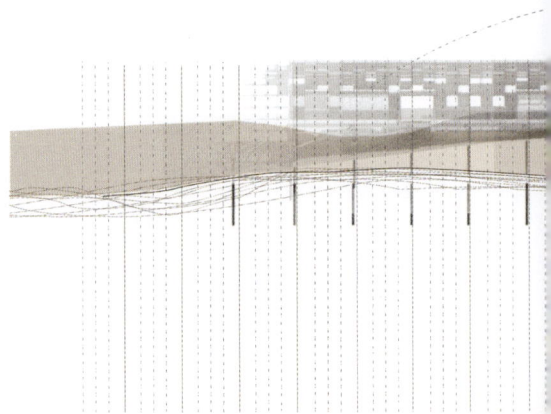

--- ---

Index of tools | detail, joint, shifting dunes, undermined foundations, relocation, modular, versatile two-way grid.

CONNECTIONS ON UNCERTAIN GROUND

Cape Cod, Massachusetts. Speculative 2012.

Benjamin Gregory and Ed Ford

Cape Cod is a landscape in flux. Formed by geologic processes over the past 10,000 years, and continually shaped by the oceanic tides and coastal erosion, no ground in Cape Cod is permanent. The National Seashore must always stay aware of these processes, and adjust by moving or completely demolishing and reconstructing the numerous cultural and institutional built resources along the shore: the lighthouses, bathhouses, and visitor centers that greet the thousands of visitors each year. What unique approaches can be taken to address these concerns?

By turning the design process on its head, starting with a detail and working to the larger picture, this studio sought to investigate what a building could be under these unique conditions. The design attempts to confront this ever-changing landscape. By perching the series of bathhouses and changing-rooms atop a cliff, the scheme allows for the closest and most convenient relationship to the beach below. To respond to the 3-foot-per-year receding of the cliff, the canopy and series of volumes beneath are detailed at the most granular scale practicable. The design, which started with a 4-foot by 4-foot bent wood box, connects to other such boxes by steel connectors forming a versatile two-way grid, and is able to be disassembled, bay-by-bay, from the front of the structure and added to the back end, matching the erosion of the cliff on which the building sits. This causes a series of formal and spatial changes which provide a vehicle to express the rate of change of the landscape they rest upon.

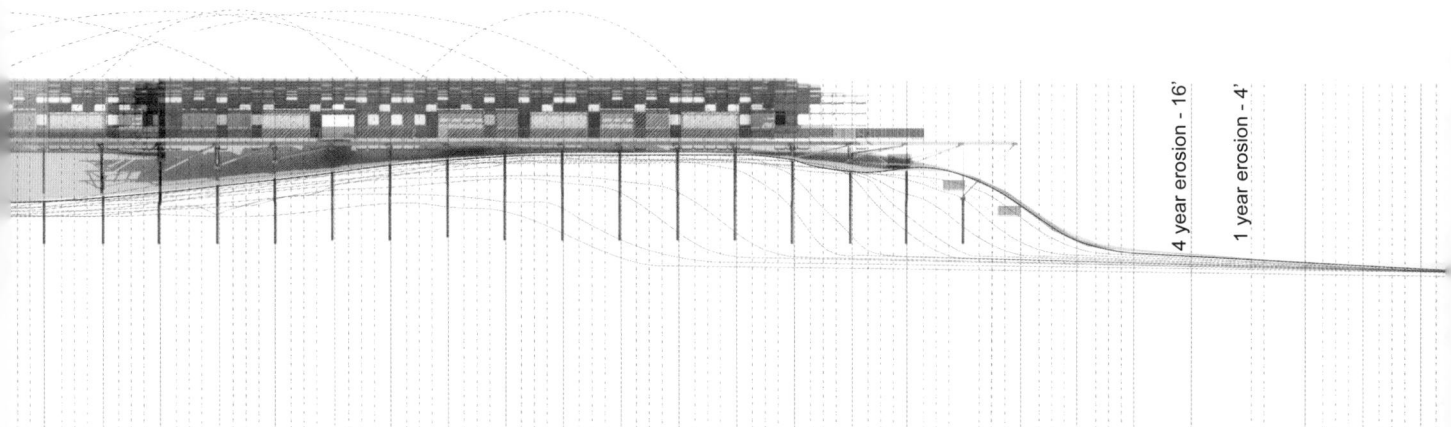

4 year erosion - 16'

1 year erosion - 4'

15 YEARS   20 YEARS   27 YEARS   36 YEAR

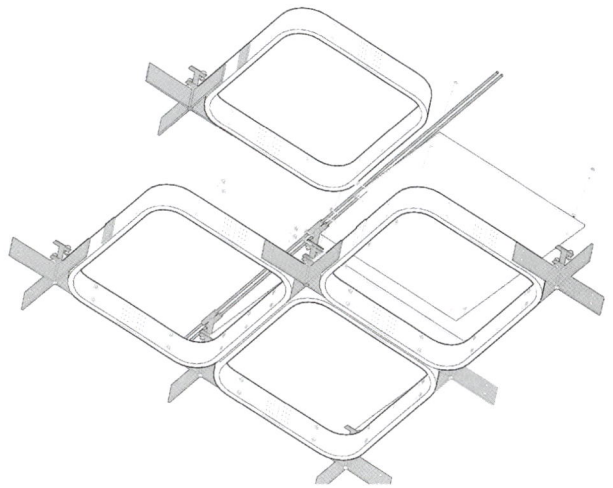

MODULAR CANOPY SYSTEM
SCALE 1/2" = 1'-0"

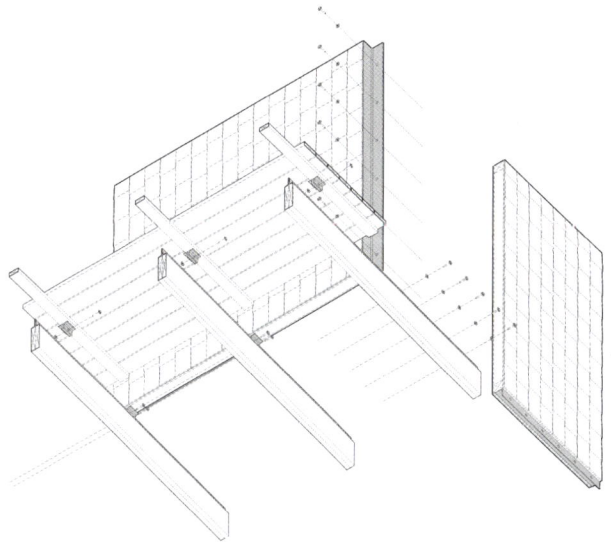

MODULAR WALL and FLOOR SYSTEM
SCALE 1/2" = 1'-0"

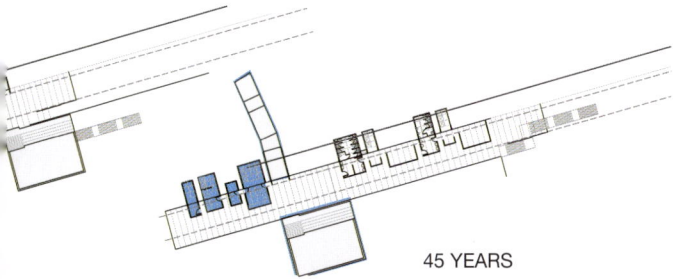

45 YEARS

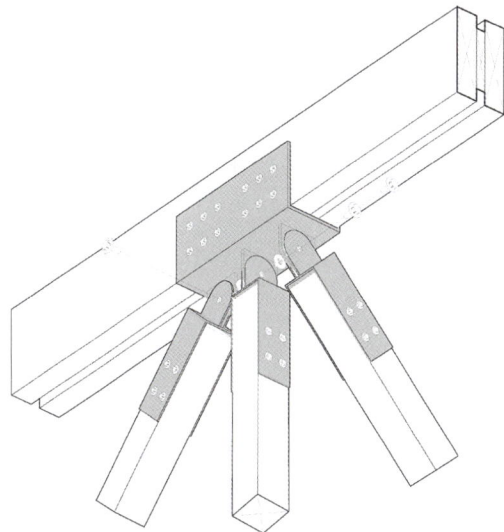

COLUMN and BRACING SYSTEM
SCALE 3/4" = 1'-0"

ACT ONE:
THE RISING
ADAPT

To allow rising water levels to enter the spaces of cities and communities, prompting buildings, landscapes, and people to transform in an effort to acclimate to the presence of water.

--- ---

index of tools | biodynamic production, natural meander, crop exploitation, floodable forest, temporary river, shared space

ARANZADI PARK

Pamplona, Spain. Completed 2013.

aldayjover architecture and landscape

Aranzadi Meander, A Privileged Park
This is an unusual place, a beautiful meandering between Pamplona and the old neighborhoods of Rochapea and Chantrea with a landscape of gardens and vegetation of beautiful specimens, yet settled, in a suitable climate for the development of a splendid park for the public. To rebalance the position of strength with which man has been linked with the environment, looking for a balance and a covenant, is the ultimate goal of this proposal.

Environmental Public Park
This is a unique and sensitive site in terms of the different requirements that must be addressed. We propose to restore the dynamism of a natural meander, in terms of its environmental role with the river corridor, which means working the vegetation, wildlife, and hydropower. We seek a balance between the needs of high-quality space for citizens, river dynamics, crop exploitation of local varieties recovered by organic farming, and the environmental role of the park in the Arga River corridor.

A River Landscape
Based on the pre-existing Aranzadi meander and doing a thorough analysis of its micro-topography, we deduce and understand its hydraulic logic in flooding, suggesting that the park should work with the functionality of the Arga River when the water flow increases, decreasing slightly the frequency of flooding of the gardens while simultaneously creating a river landscape within the park, a natural occurrence subject to the seasonal dynamics of flood water. Water is landscape in the park.

Matters of Memory and the Future.
We start with the certainty that the landscape's existence is defined by its behavior and its history. We propose a strategy for enhancing the potential and the traces of the place. The territory has in its settings and behavior a past which is the basis for its future.

Previous State
This is an unusual place, a beautiful meander hidden between Pamplona and the neighborhoods of Rochapea and Chantrea where the local farmers have grown vegetable gardens within the oldest biodynamic production in Spain. Nowadays, some of the plots are abandoned and the river has been forced inside a tight channel that floods suddenly. It has been kept

as a reserve, surrounded by the river and blocked by several buildings. Only narrow service roads among the plots, shared by occasional walkers and the farmer's vehicles, allowed for a fragmented visit to the place. The secular "soft" intelligence of the farmers, filtering the flood and its solids (stones, tree trunks, trash…) through bushes in between plots, is combined with "hard" urgent solutions. Dikes have tried to deny the river dynamics and the fertility of the new soils sedimented by the floods. Topography has been changing progressively for centuries and the river beach has now become higher than the interior, encroaching the riverbed—a channel whose bed is dug deeper and deeper by the river's energy.

Aim of the Intervention

This is an opportunity to rebalance the position of strength with which human has been linked with the environment. A new relationship needs to be established between the river and the city, incorporating the natural dynamics and production into the urban realm. The paradigm of protection has to be switched. Protection, against what? Are not rivers the reason for almost all human settlements? The conflictive relation between dependence and defense is a constant in agricultural and urban environments. The concept of "catastrophe" reveals a conflict that new approaches to nature should change. In the urban context, river dynamics and floods need to become part of a public space that is given life and made meaningful by them, with positive instead of catastrophic consequences. We propose to restore the dynamism of a natural meander within the river corridor. We seek a balance between citizens' needs for space, river dynamics, and crop exploitation of local varieties recovered by organic farming. We started with the certainty that the landscape exists both in its behavior and its history. We propose a strategy for enhancing the potential and the traces of the place. The territory has in its settings and behavior a past which is the basis for its future.

Description of the Intervention

An agricultural landscape. The edge of the meander, along the river, keeps its vegetable gardens, structured for different kinds of social production and integration of handicapped people, and research/conservancy of endemic varieties.

The subdivisions are low, filtering bushes that allow views, following the tradition for dealing with floods.

A river landscape. The micro-topography of the meander suggests a new temporary river course for high waters. Part of the park is a space shared by the river and the citizens. The floodable forest slightly decreases the frequency of flooding of the orchards and gardens while simultaneously creating a river landscape within the park, subject to the seasonal dynamics of the flood. Water is landscape in the park. On the riverbank, the riparian forest is widened and the paths are kept high and without lighting, far from the delicate fringe of contact between water and bank, the richest in natural life.

A citizen's landscape. The meander becomes visible and accessible, in all its variety of spaces. In the second phase, the inner part will be transformed into gardens with open space to play, gather and celebrate, and stay and contemplate. New pedestrian bridges over the river will link the park with the adjacent neighborhoods.

Evaluation

The park has been in use constantly during the construction. Agricultural plots were protected and kept their activity. Along with farmers and the people involved in the activities of integration through agriculture, citizens had regular access during construction. The sense of ownership has been increased with the process of discovering a space regarded as the hidden treasure of the city. Some restaurants serve the local "crispilla" lettuce and other specific varieties of the meander. The third user, the river, has occupied the park several times as well. The most notable has been the September 2013 flood, the largest since the city began keeping records of water volumes. At that time, the work was in the last phase of planting and finishing details, and the only damage happened in the slope in front of the water entrance, which was not yet reinforced with the benches that overlook the river (now placed). Besides this, some stones were captured with the concrete wall deflectors under the pedestrian bridge, and some small trees were torn down by a big tree trunk carried by the current until the walls of the first low bridge over the forest stopped it.

Opposite Above: Aranzadi Park, unfooded.

Opposite Below: Aranzadi Park, flooded meander.

Above:
Plan of the meander and the
surrounding urban context.

Below:
 Water Entrance view of the end of
the floodable forest and the slope
towards the river, low bridges over
the picnic and work tables.

CURRENT FLOODING CONDITION — NATURAL CONDITION — 2.33 YEAR — 10 YEAR — 25 YEAR

PROJECTED FLOODING — NATURAL CONDITION — 2.33 YEAR — 10 YEAR — 25 YEAR

WATER   Avenue Q5 Actual   Avenue Q5 Projected

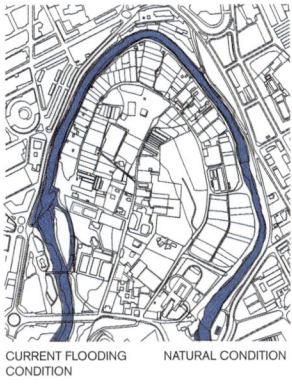

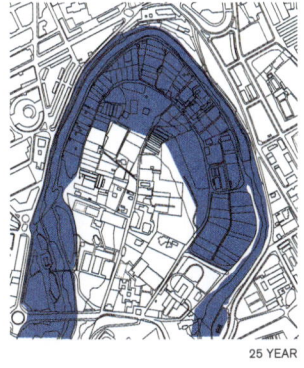

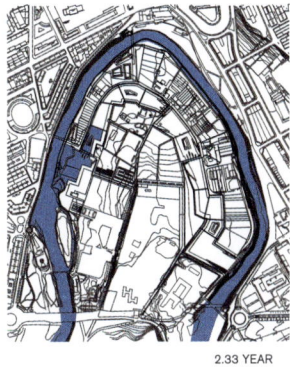

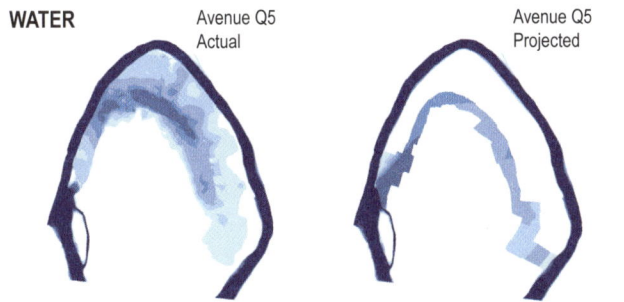

The parks river landscape reduces the frecuency of flooding from the garden

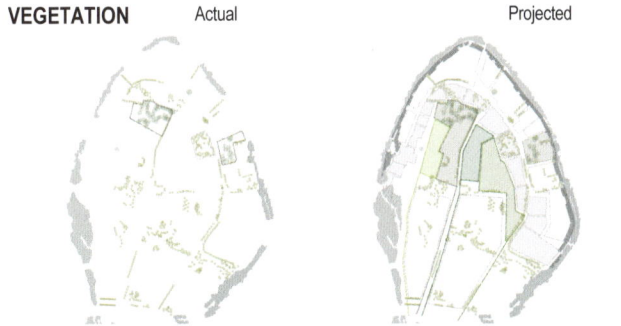

VEGETATION   Actual   Projected

The support structure for the land is a poetic, environmental landscape

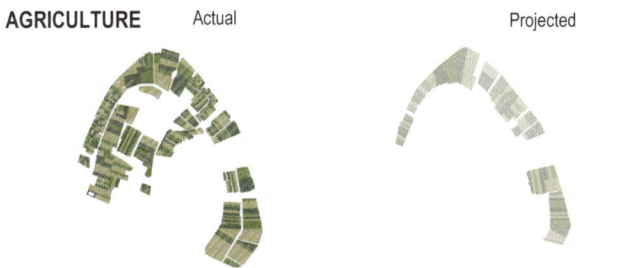

AGRICULTURE   Actual   Projected

The most fertile and varied garden is consolidated at the border, creating a magical space

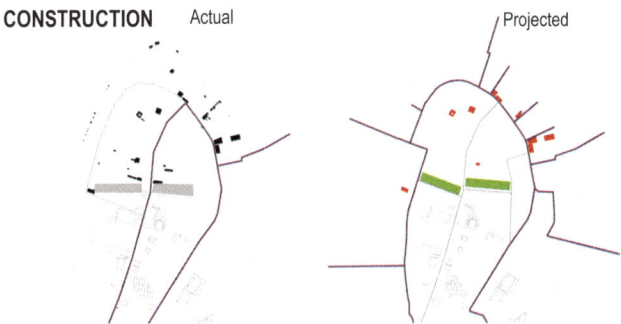

CONSTRUCTION   Actual   Projected

Aranzadi is enhanced with more connections and more civic uses

Above:
Diagrams demonstrating the change in the way the river floods, from simply spilling over its inclined banks as is the current condition, to a controlled meander which helps manage the river around the various programs on the site.

Below: Plan Diagrams of water, vegetation, agriculture and construction layers of site.

Above:
Sections showing the natural
topography of the river (above), its
modified form as a channel (middle)
and the proposed form which
provides a controlled outlet, or
meander, for the rivers flood waters,
around programs in the park.

Below:
Landscape meander section

Opposite Above:
Floodable meander and bosque.

Opposite Below:
A footbridge over the floodable
portion of the landscape helps
maintain paths of circulation through
the park at all times.

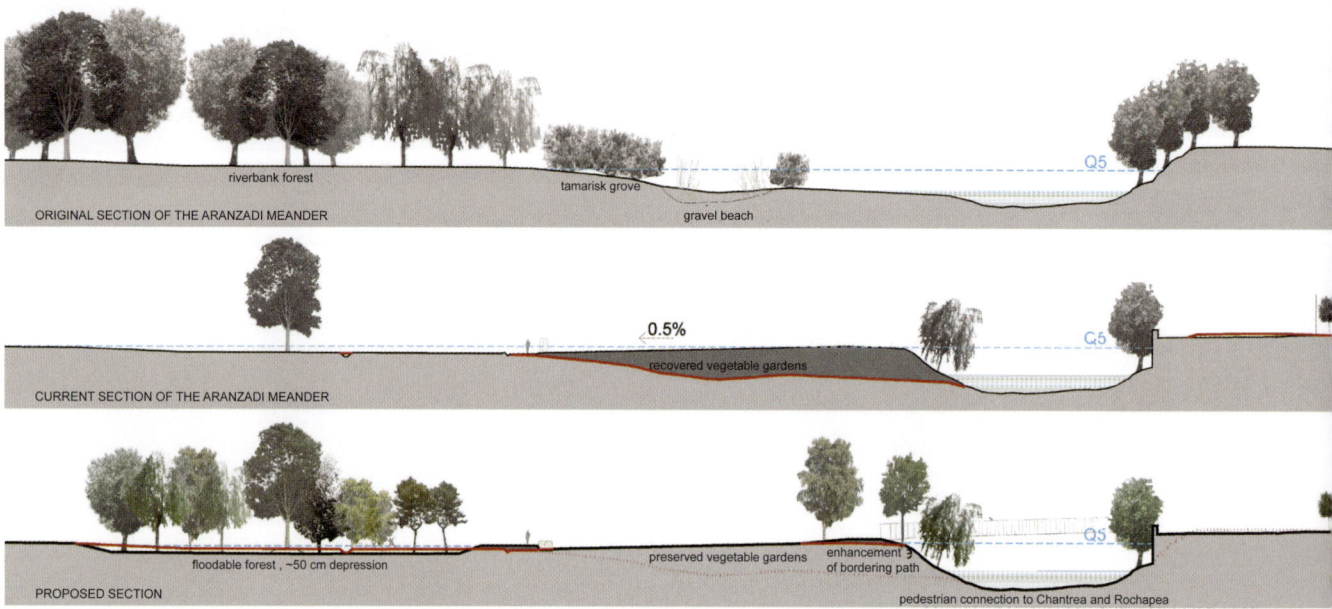

riverbank forest

tamarisk grove

gravel beach

Q5

ORIGINAL SECTION OF THE ARANZADI MEANDER

0.5%

recovered vegetable gardens

Q5

CURRENT SECTION OF THE ARANZADI MEANDER

floodable forest , ~50 cm depression

preserved vegetable gardens

enhancement
of bordering path

Q5

PROPOSED SECTION

pedestrian connection to Chantrea and Rochapea

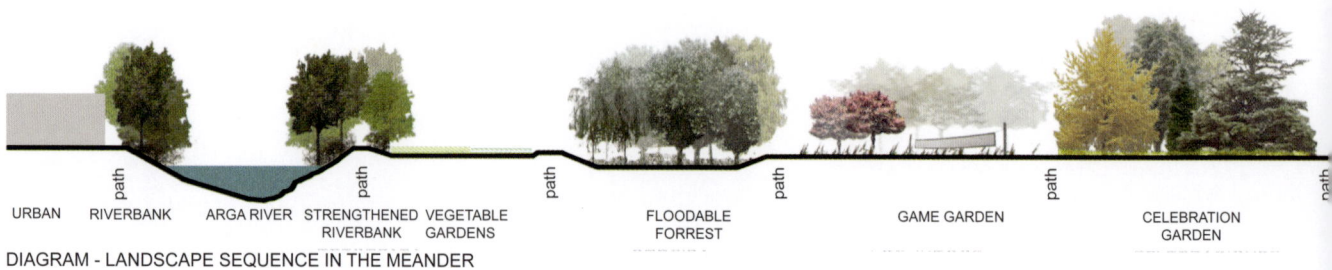

| URBAN | RIVERBANK | ARGA RIVER | STRENGTHENED RIVERBANK | VEGETABLE GARDENS | | FLOODABLE FORREST | | GAME GARDEN | | CELEBRATION GARDEN | |

path   path   path   path   path   path

DIAGRAM - LANDSCAPE SEQUENCE IN THE MEANDER

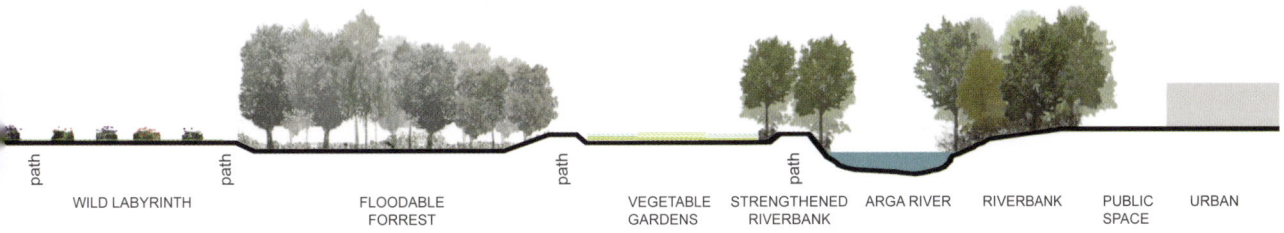

path

path

path

path

WILD LABYRINTH

FLOODABLE
FORREST

VEGETABLE
GARDENS

STRENGTHENED
RIVERBANK

ARGA RIVER

RIVERBANK

PUBLIC
SPACE

URBAN

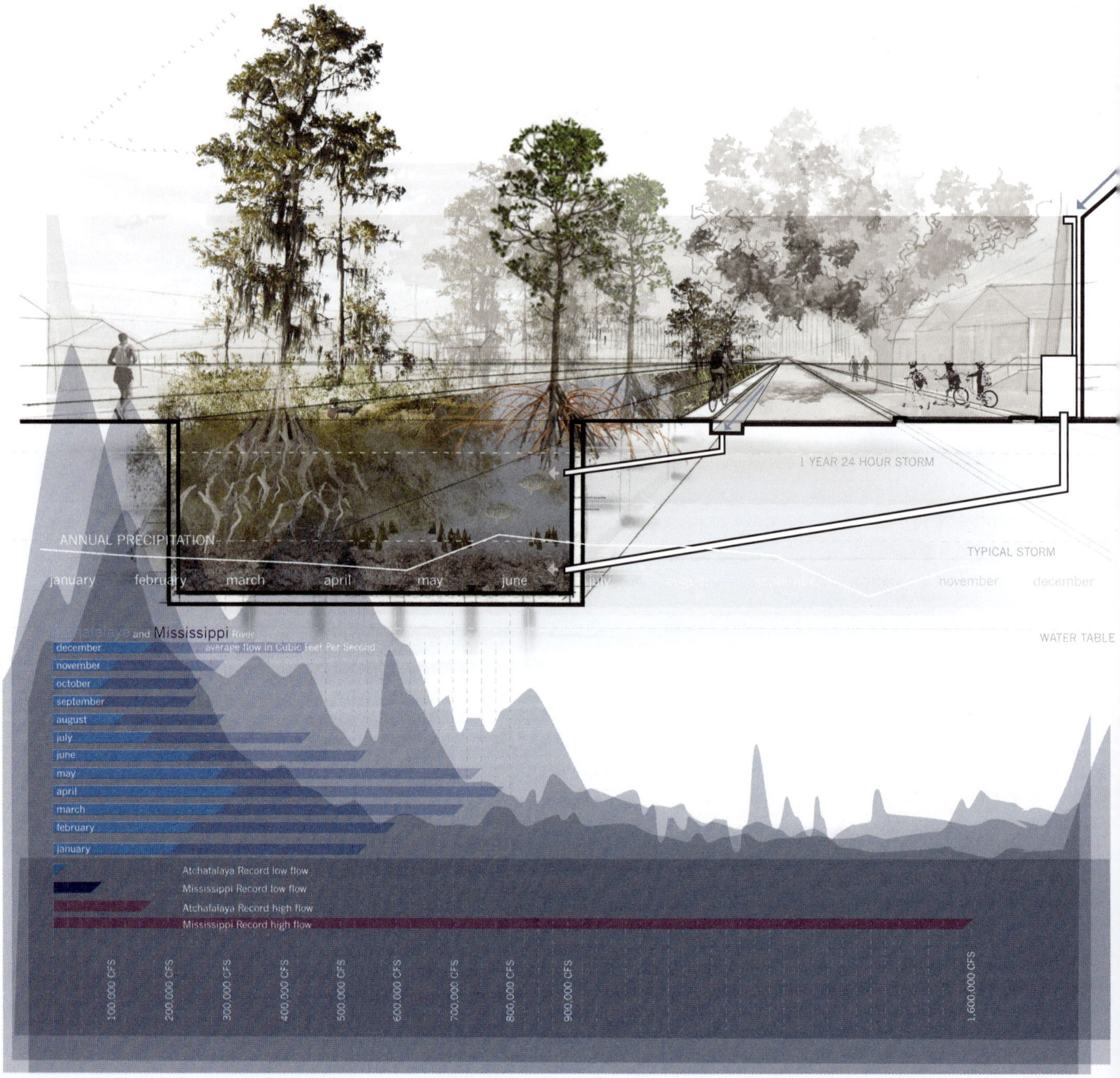

ANNUAL PRECIPITATION

january   february   march   april   may   june   july   august   september   october   november   december

1 YEAR 24 HOUR STORM

TYPICAL STORM

WATER TABLE

Atchafalaya and Mississippi River
average flow in Cubic Feet Per Second

december
november
october
september
august
july
june
may
april
march
february
january

Atchafalaya Record low flow
Mississippi Record low flow
Atchafalaya Record high flow
Mississippi Record high flow

100,000 CFS   200,000 CFS   300,000 CFS   400,000 CFS   500,000 CFS   600,000 CFS   700,000 CFS   800,000 CFS   900,000 CFS   1,600,000 CFS

Mississippi Flux
Analyzing the flows of the Mississippi
River and exploring their expression in
the city at the body scale. How does
reintroducing swamps into the city
reconnect inhabitants to the dynamic
processes of a delta ecology?

--- ---

Index of tools | smart swamps, hybrid infrastructure, multi-functional system

SWAMP THING: SMART GRID, SMARTER WATER MANAGEMENT IN NEW ORLEANS

New Orleans, Louisiana. Speculative, 2012.

Isaac Cohen, Kate Hayes, and Jorg Sieweke

Throughout history, the Mississippi River has jumped its channel every 500 to 1,000 years in order to follow the shortest and steepest route to the Gulf of Mexico. In the past century and a half, the Army Corps of Engineers has ground this evolution to a halt and guided the river's fate, turning the mighty Mississippi into a concrete channel of revetments, levees, and control structures, and leaving cities and towns along its course vulnerable to flooding and worse. This project imagined the "nightmare" scenario of the Mississippi forging a new course, away from New Orleans. This scenario has the potential to leave New Orleans low, dry, and without a fresh water source. In this proposal, we consider how long static controls can hold up against impending forces, and alternative methods of control that are more progressive and adaptive.

To better manage water in a zone deprived of its fresh water supply, we reintroduce swamps into the public realm as a vital part of a smart water grid. This system acts in contrast to the current monofunctional system whose only purpose is to pump water out of the city. It transforms the city's infrastructure from a system that views water as a burden to a system that values water as an indispensable resource in a new situation.

We live in an age of "smart technology," two-way communication that enables infrastructure to serve demand in a dynamic way. This proposal introduces a smart network of swamps into a system of existing pump stations, drainage canals, control structures, and locks that will effectively distribute rainwater based on demand.

The project imagines the reintroduction of swamps as a vital public space and as part of a hybrid infrastructure that dynamically tunes the input and output of existing infrastructure, modulating and allocating water throughout the city. Water is redirected, cleansed, and repurposed to serve the city's water needs—drinking, municipal, and industrial. This flexible water system works with the existing civic infrastructures to create a new typology of water management, one that is productive, adaptive, and multifunctional. While this smart grid is fitted to New Orleans, the idea is transferrable to other cities with similar water infrastructure issues. It is an idea that can easily be nested within other, larger urban systems.

Citywide System Section
A typical section of New Orleans
showing the stormwater
infrastructure and the movement of
water across the city in the existing
and proposed conditions.

Below:
Citywide System Section
A typical section of New Orleans
showing the stormwater
infrastructure and the movement of
water across the city in the existing
and proposed conditions.ç

Right:
Narrative Section of a Changing
System (Opposite) Exploring
ecological and infrastructural
changes—from the premodern,
to the contemporary, and on to
the proposed condition—as a first
generative exploration.

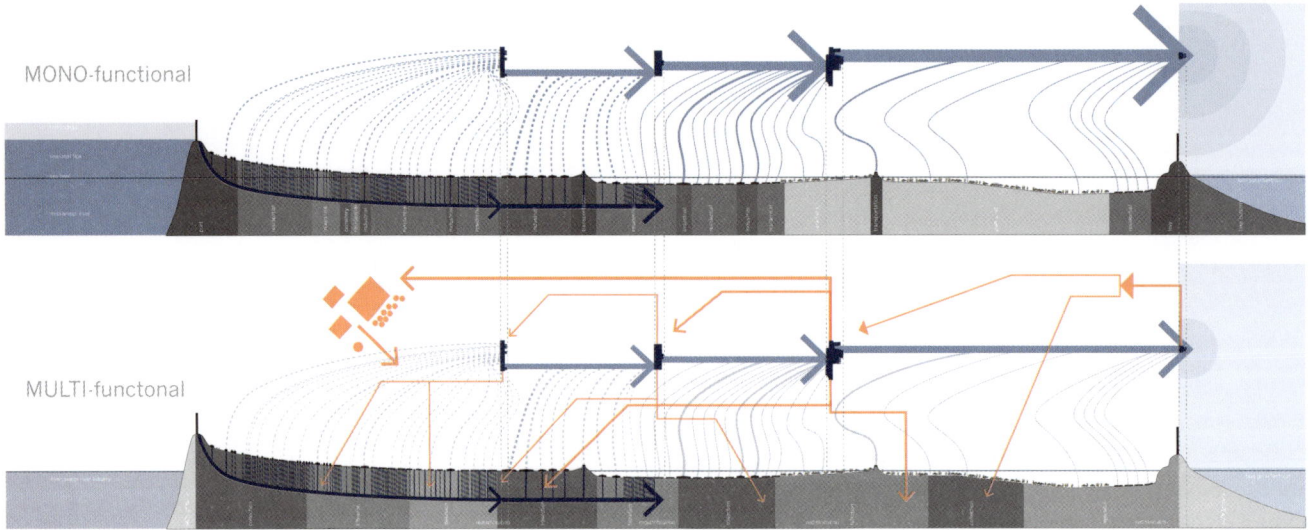

MONO-functional

MULTI-functonal

TAXODIUM
DISTICUM

pump station:
NEW PUBLIC INTERFACE

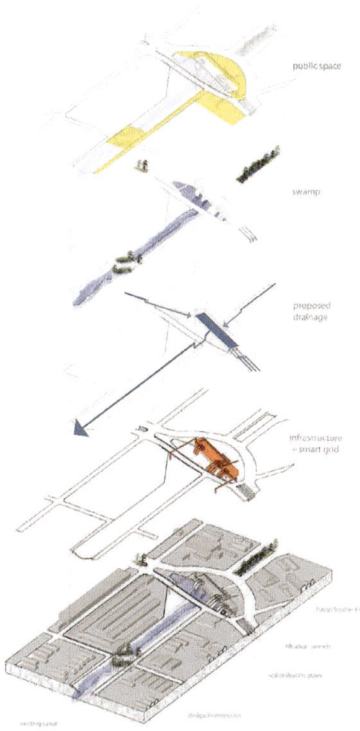

mixed-use cluster:
BRIDGING SCALES

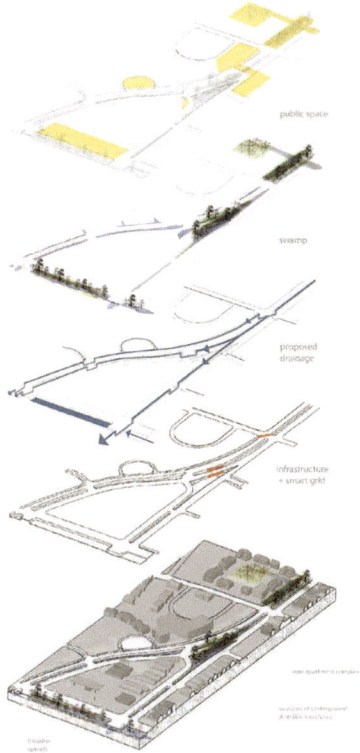

port:
HEADWATERS

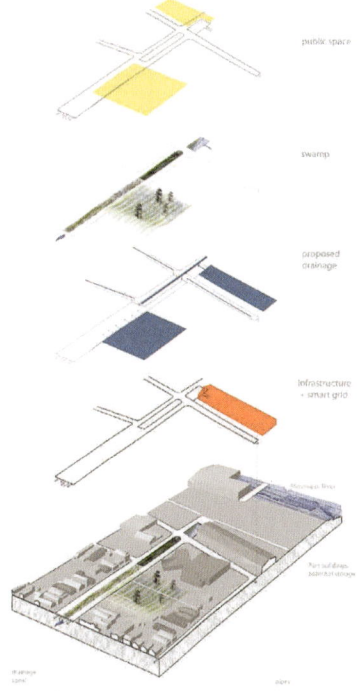

public space

swamp

proposed drainage

infrastructure + smart grid

public space

swamp

proposed drainage

infrastructure + smart grid

public space

swamp

proposed drainage

infrastructure + smart grid

PUMP STATION

collect

pump

expel

mono-functional infrastructure

PUMP STATION

collect

redistribute

react to the systems needs

multi-functional infrastructure

functions of the swamp return to the system and

serve multiple needs with multiple pumps

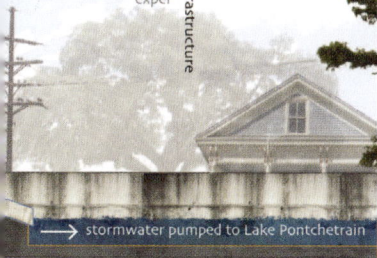

→ stormwater pumped to Lake Pontchetrain

stormwater cleansed and slowed

→ stormwater reutilized within the city

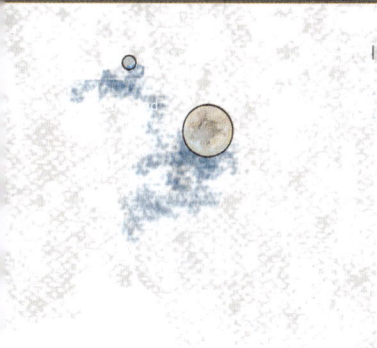

Above:
Swamp Corridor
The Swamp Thing | Smart Grid
creates a new public space to
allow for habitation of municipal
infrastructure and a new form of the
swamp in the city.

Below:
Long Lot Water Storage (Below)
The new multifunctional system
allows for the storage and reuse of
water, as well as physical interaction
with it. The system engages
individuals and communities with
water as a resource, not as a
destructive force.

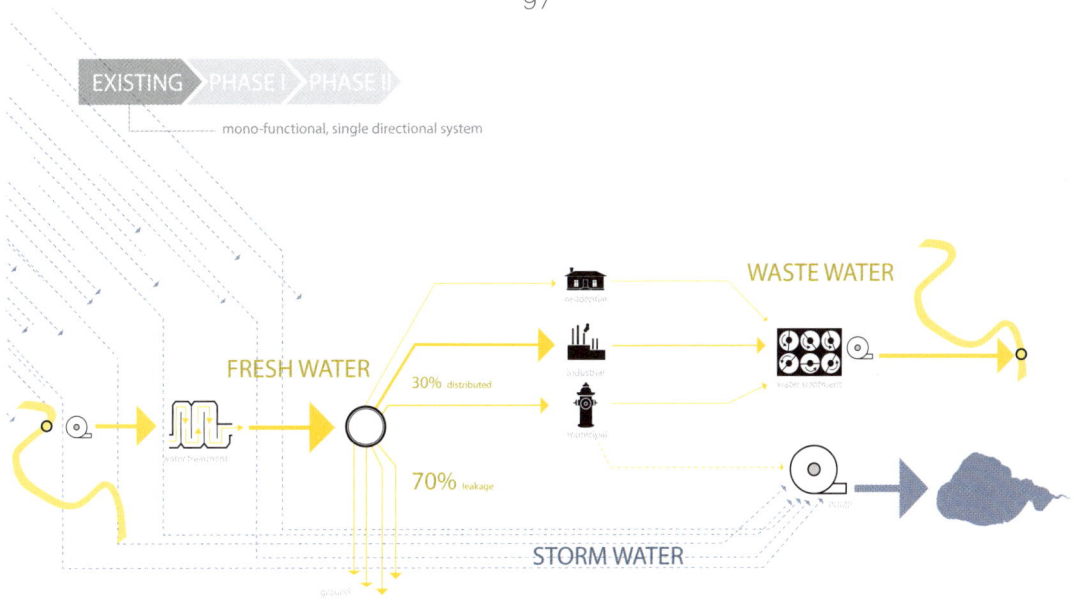

EXISTING PHASE I PHASE II

mono-functional, single directional system

WASTE WATER

FRESH WATER

30% distributed

70% leakage

STORM WATER

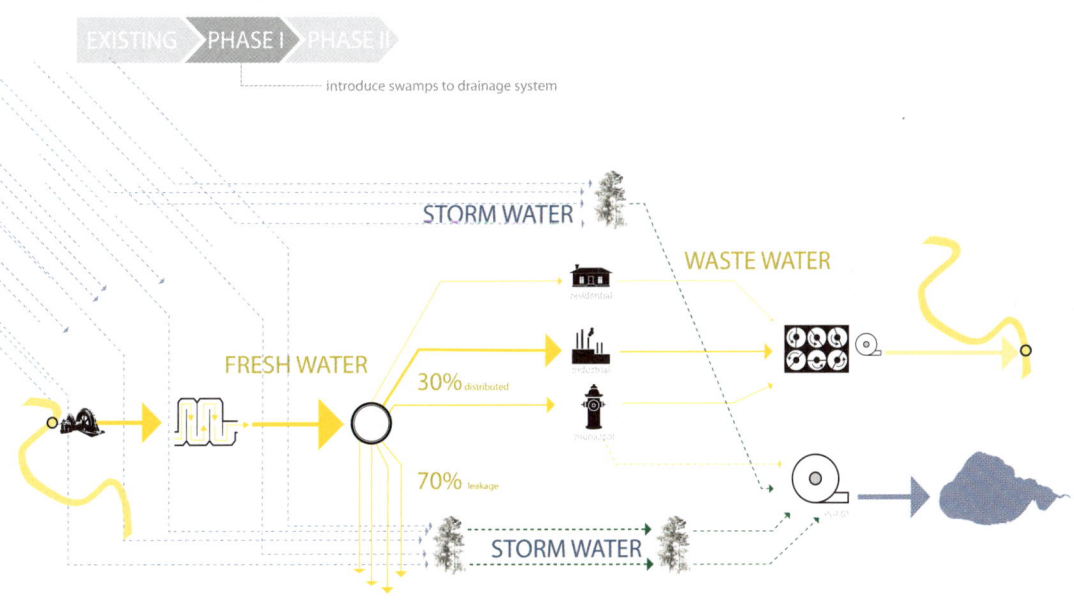

EXISTING PHASE I PHASE II

introduce swamps to drainage system

STORM WATER

WASTE WATER

FRESH WATER

30% distributed

70% leakage

STORM WATER

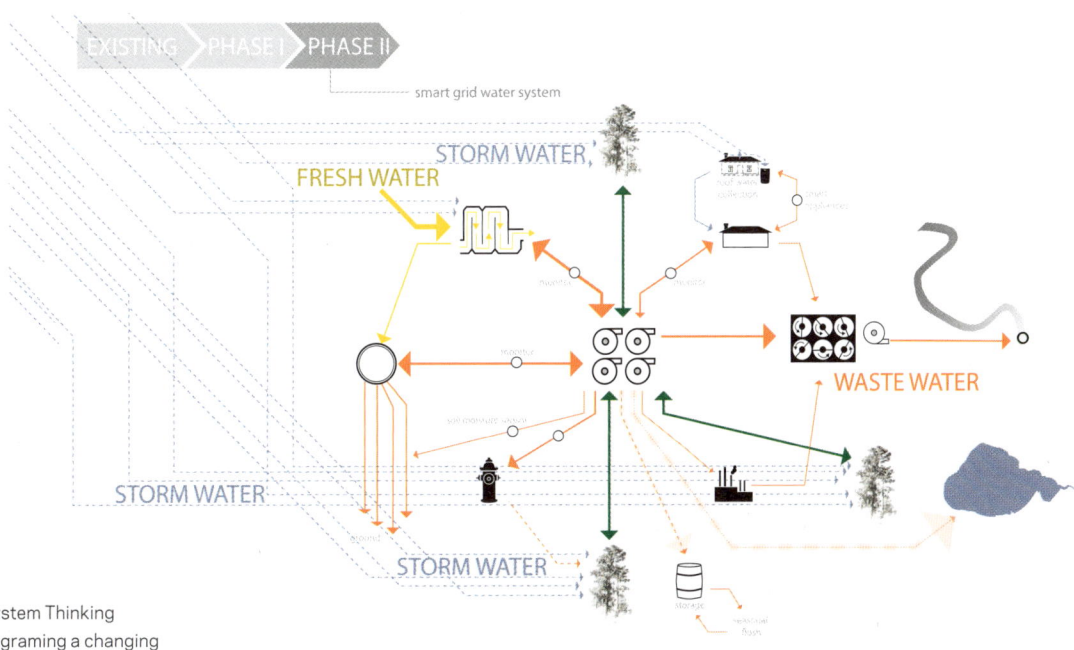

EXISTING PHASE I PHASE II

smart grid water system

STORM WATER

FRESH WATER

WASTE WATER

STORM WATER

STORM WATER

Smart Grid System Thinking
(Opposite) Diagraming a changing
system—from overlapping
monofunctional water systems to an
integrated multifunctional network
of water, ecology, and use.

WATERSQUARE BENTHEMPLEIN

Rotterdam, Netherlands. Constructed 2012.

De Urbanisten

The water square combines water storage with the improvement of the quality of urban public space. The water square can be understood as a twofold strategy. It makes money invested in water storage facilities visible and enjoyable. It also generates opportunities to create environmental quality and to give an identity to central spaces in neighborhoods. Most of the time the water square will be dry and in use as a recreational space.

In Benthemsquare the first water square has been realized. In an intense participatory trajectory with the local community, De Urbanisten jointly conceived ideas about the square that included students and teachers of the Zadkine College and the Graphic Lyceum; members of the adjacent church, youth theatre, and David Lloyd gym; and inhabitants of the Agniese neighborhood. In three workshops they discussed possible uses, desired atmospheres, and how the stormwater could influence the square. All agreed: the water square should be a dynamic place for young people, with lots of space for play and lingering, but also nice, green intimate places. And what about the water? This had to be excitingly visible while running over the square: detours were obligatory! The enthusiasm of the participants helped to make a very positive design.

Three basins collect rainwater: two shallow basins for the immediate surroundings will receive water whenever it rains; one deeper basin receives water only when it consistently keeps raining. Here the water is collected from the larger area around the square. Rainwater is transported via large stainless-steel gutters into the basins. The gutters are special features; they are oversized steel elements fit for skaters. Two other special features bring stormwater onto the square: a water wall and a rain well. Both dramatically gush the rainwater visibly onto the square. The rain well is designed as a special beginning to the stainless-steel gutter lifting itself from the ground. This well brings the water from the adjacent building into the gutter. The water wall brings the water from further away into the deep basin. Here a rhythm of waterfalls is being directed in relation to the amount of water falling from the sky. Two more water extras complete the picture. An open air baptistery is placed next to the church that is situated on the square. Here a small fountain starts from which the water meanders over the square into one of the shallow basins. And in the deep basin we "join the pipe" and plant a drinking fountain for all thirsty athletes to enjoy.

After the rain, the water of the two shallow basins flows into an underground infiltration device and

Opposite:
Three basins collect rain water: two shallow basins for the immediate surroundings will receive water whenever it rains, one deeper basin receives water only when it consistently keeps raining.

Above:
View of the deep basin during a rain.

Below:
The deep (third) basin is a true sports pit fit for football, volleyball and basketball, and is set up like a grand theatre to sit, see and be seen. The water of the deep basin flows back into the open water system of the city after a maximum of 36 hours to ensure public health.

from here gradually seeps back into ground water. Thereby the ground water balance is kept at level and can also cope with dry periods. This helps to keep the city trees and plants in good condition which helps to reduce urban heat island effect. The water of the deep basin flows back into the open water system of the city after a maximum of 36 hours to ensure public health. All the storm water that has been buffered does not flow into the mixed sewage system anymore. Like this the conventional mixed sewage system is relieved and lowers the frequency of the relatively dirty water to overflow in the open water whenever it reaches its buffering capacity. By separating storm water gradually from the black water system with each intervention, the entire system step by step moves towards an improvement of the overall quality of the open water in the city.

When its dry, the square is a feast for active youth to sport, play and linger. The first shallow basin is fit for everybody on wheels and whoever wants to watch them doing their thing. The second shallow basin will contain an island with a smooth "so you think you can dance" floor. The deep (third) basin is a true sports pit fit for football, volleyball and basketball, and is set up like a grand theatre to sit, see and be seen. On each entrance we create more intimate places to sit and linger. The planting plan emphasizes the beautiful existing trees. We plant high grasses and wild flowers surrounding

the trees framed by a concrete border at seating height to offer many informal places to relax here.

The color scheme emphasizes the function of the water square: all that can flood is painted in shades of blue, and all that transports water is shiny stainless steel. This means gutters receive extra attention and are made beautiful. And the floors of the three basins are painted in blue colors that match the colors of the surroundings. The space is gently defined and subdivided by a green structure that makes a difference in planting colors between the entrances and the center of the square. The water square creates a new context for the great modern building of the architect Maaskant and allows the fantastic artwork of Karel Appel to receive more attention.

De Urbanisten invented the typology of the water square in 2005 for the International Architecture Biennale Rotterdam (IABR), "The Flood." Typological research into the design of water squares was carried out in 2006-2007. The water square became official policy on an urban scale in the "Rotterdam Waterplan 2" in 2007. A pilot study was carried out in 2008-2009. In 2010 their graphic novel *De Urbanisten and the Wondrous Water Square* was published by 010, Rotterdam. In 2011 the preliminary design for the Benthemsquare was made. In 2012 they finished the final design and construction started. On 4 December 2013 the water square was officially opened.

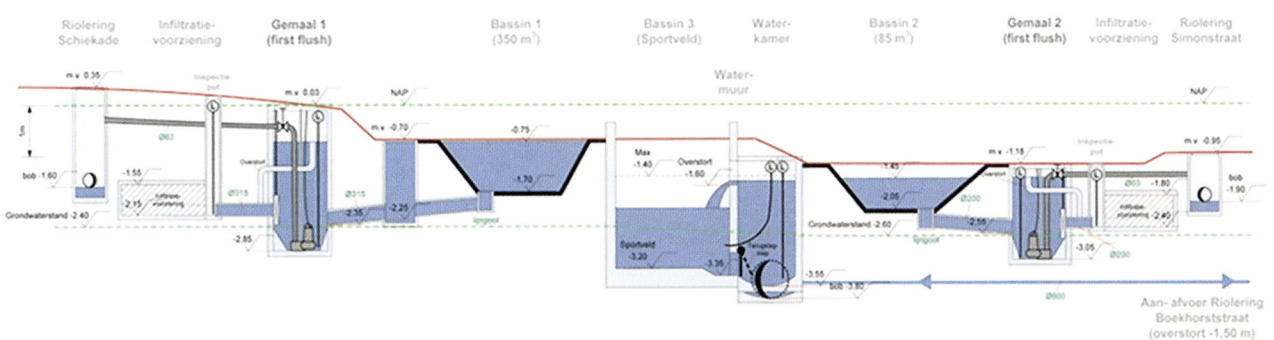

A section illustrating the three basins and the sources of water as they flow into and out of them.

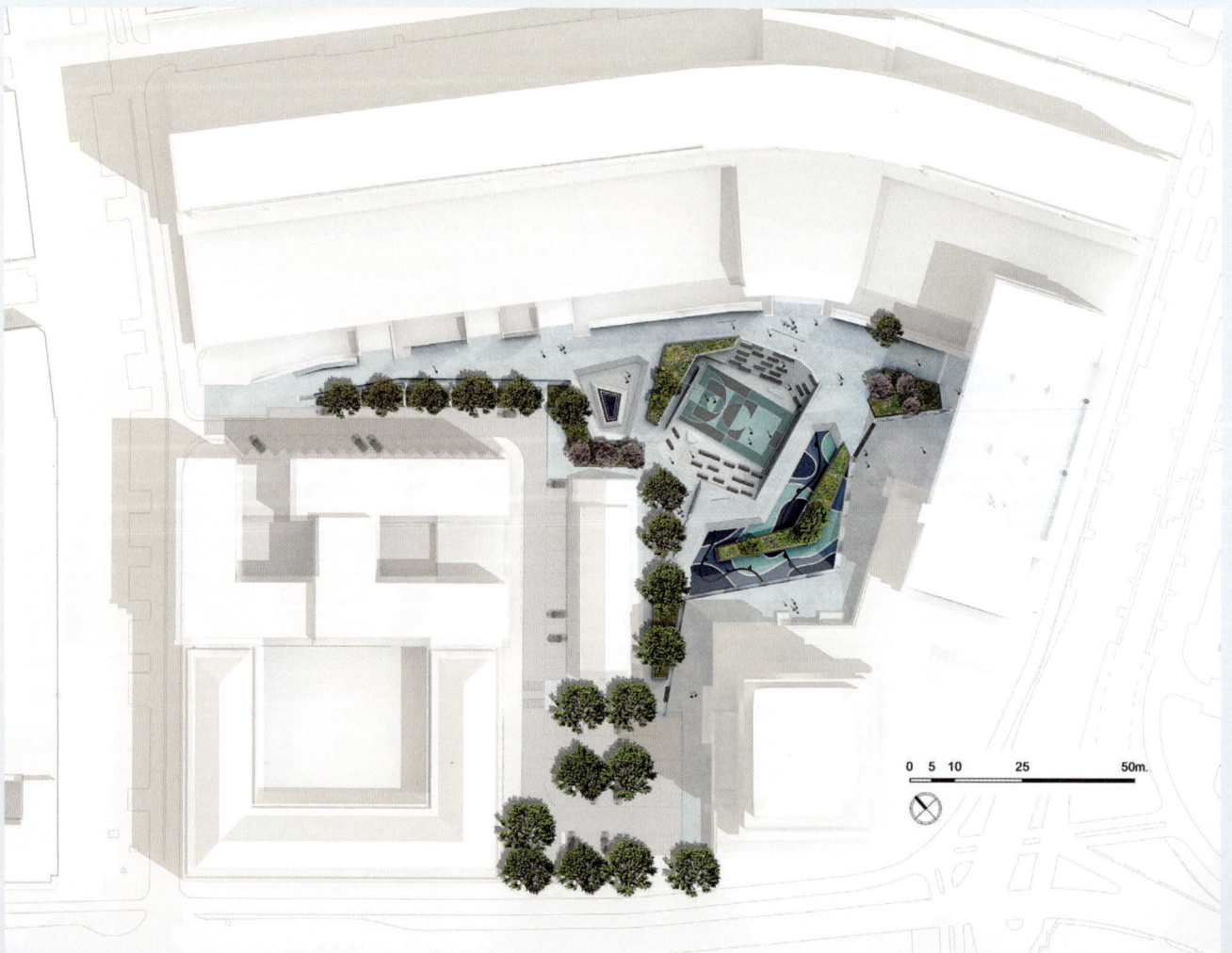

Above:
The water of the deep basin flows
back into the open water system
of the city after a maximum of 36
hours to ensure public health.

Below:
A plan highlighting the three
collection basins of the square.

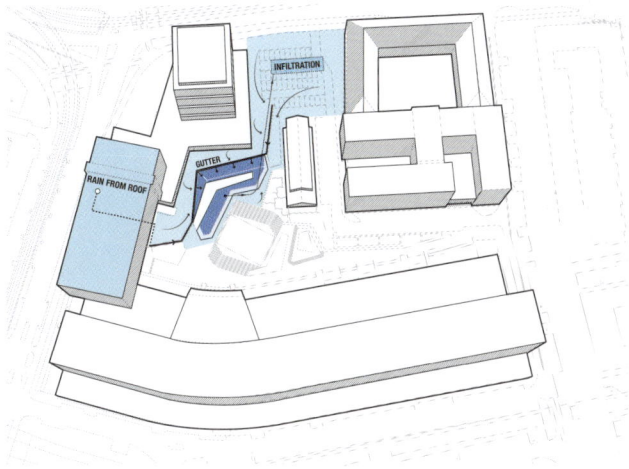

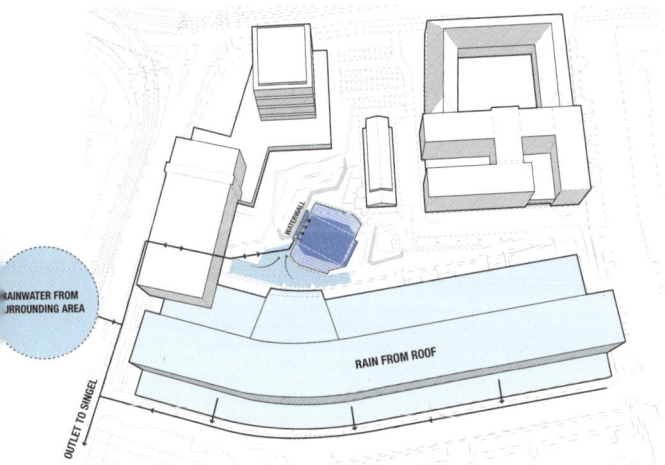

Above:
Three diagrams showing the basins and the various sources of stormwater runoff which it manages.

Above Left:
Rainwater is transported via large stainless-steel gutters into the basins. The gutters are special features; they are oversized steel elements fit for skaters and also mark the path taken by the water that is collected within the square.

Left:
The Rain Well highlights where water from the rooftops of adjacent buildings is brought into the system of stainless-steel gutters.

Above:
A model showing the public square and its relationship to the surrounding buildings.

Below:
The Rain Wall dramatically deposits offsite water into the largest of the three basins.

-1.60 -1.64 -1.72 -1.88 -1.90
BOTTOM OF SLIT

D -1.60
C -1.64
B -1.72
A -1.88

WATERCHAMBER
-0.65

RAINWATER SUPPLY
FROM SURROUNDINGS

OUTLET TO SINGEL

-4.10

STAINLESS STEEL SLIT

WATERWALL

BASIN 3

MAXIMUM WATERLEVEL
-1.40m.

DRAIN

CLOUDBURST

HEAVY RAIN

NORMAL RAIN

RAINFALL
INTENSITY

THE SOURCE

RISING WATER
PRESSURE

600mm
600mm
100mm
50mm
500mm

STAINLESS STEEL GUTTER

1000mm

VISIBLE RUNOFF
TO BASIN 1

UNDERGROUND SUPPLY TO 'SOURCE'

RAIN ON ROOF

RAINFALL

INTERNAL DRAINAGE

The Rain Well highlights where
water from the rooftops of adjacent
buildings is brought into the system
of stainless-steel gutters.

ACT TWO: THE CONTAMINATED

The realization that water quality has a direct relationship to health and disease is a recent phenomenon in the vast history of humankind. This realization arrived in the mid-19th century when "water, like other facets of urban nature, was incorporated into an increasingly rationalized and scientifically managed form."[1] The advent of the modern city was characterized by a transformation in the relationship water and the human body. A new "hydrological order" reflected a growing concern for Edwin Chadwick's "sanitary idea"—the notion that the physical environment "exercised a profound influence over the well-being of the individual" and that health has a direct correlation to sanitation.[2] Nineteenth-century Europe began to clean the contaminated waters of the city through "complex technological networks, changing patterns of everyday life and the establishment of new modes of municipal administration."[3]

This is the situation of "The Contaminated": an investigation of water as a medium that transfers unwanted matter, often at a microscopic level, undetectable to the human eye. This act is about quality, not quantity. The projects in "The Contaminated" work to examine how the city, the building, and the landscape are structured to maintain or improve water quality. Today, water contamination comes from a plethora of sources: the fuels that power our machines, the plastics that package our products, the chemicals that fertilize our plants, or the excrements that escape our bodies. One in eight people in the world lack access to clean water and 3.3 million die from water-related health problems each year.[4] Providing clean water is a design problem. Policy and economics must parallel design, but ultimately it is the physical environment that contributes to the quality of water. There is no one answer, as some situations may call for a centralized, municipal format, while others may demand a decentralized, local operation.

The second act for this index of water propositions investigates how people pollute water and how the built environment can work to minimize this effect of civilization. Two Calls to Action highlight the urgency of tackling water contamination by framing it within a historical context. Matthew Gandy's "The Bacteriological City and Its Discontents," first published in *Historical Geography*, documents how developments in science and technology began to influence urban form in the

19th century. Gandy explains that with the advancement of bacteriology, it became clear that environmental conditions, especially related to water, have a direct impact on the health of citizens. The concept of the "bacteriological city" that emerged in the 19th and early 20th centuries presented a centralized technological model aimed at controlling the flow of pure and impure water within the city. Gandy argues: "The contemporary transition away from the bacteriological city can only be fully appreciated in the context of the innate weaknesses within this centralized technological model."[5]

Following Gandy's survey of 19th-century political and technological formations relating to water contamination, Martin Melosi's chapter "Bringing the Serpent's Tail into the Serpent's Mouth: Edwin Chadwick and the 'Sanitary Idea' in England" from his book *The Sanitary City: Urban Infrastructure in America from Colonial Times to the Present* (2000) is reprinted here in order to delve into the specific development of England's sanitary system. This text charts the components of Chadwick's 'arterial-venous" system as a response to the emergence of bacteriology in the late 19th century when, Melosi writes, "The advent of the sanitary idea offered a clearer rationale and newer strategies for improving sanitary services first in England, and then throughout the world."[6]

These two Call to Action texts present a historic moment in the municipality's control over water quality. They outline a period when fatalism was replaced "by a new faith in the power of scientific control of the physical environment."[7] The urgency described by Chadwickian sanitarianism is ever present today as every city struggles with the many effects of increased density, imbalanced wealth, and outdated or inadequate waste systems. Following these prompts, three strategic responses are presented: a response of defense, a response of retreat, and a response of adaptation.

The strategy of defense is charted through two projects: the cleanup of the Deepwater Horizon oil spill by British Petroleum (BP) in 2010 and Steven Holl and Michael Van Valkenburgh's Whitney Water Purification Facility and Park in Connecticut, United States. The Deepwater Horizon oil spill is more accurately described as an event rather than a project and is included to present a sense of scale. The spill produced a horrifying realization of the enormous impact human error can have on the environment: "Over two months, an estimated 4.9 million barrels of oil billowed from the earth's crust and into the Gulf of Mexico," producing the largest accidental marine oil spill in history. It is this type of event—the instantaneous accident of gigantic proportion—that initiates the mode of defense. To contain the bleeding, sometimes all that can be done is a retroactive installation of booms and berms to trap the contaminating agent.

The Whitney Water Purification Facility is discussed as a defense strategy because it is part of a centralized system for water purification. With the treatment facilities located beneath a public park, the purification process occurs in a controlled compound, largely concealed from public view. The water purification process is used as a design metaphor rather than a direct ordering strategy.

These two entries exemplify the operation of defense, in which mechanisms are constructed to keep contaminated water away from people, civic space, and fragile ecologies.

To retreat from the contaminated involves mobility or separation—moving people and the built environment away from the source of contamination. This strategy is investigated by Suzanne Harris- Brandts in the project "Seeping Boundaries: Informal Infrastructures of Dirt, Demolition, and Sewage in the West Bank," as well as by Kimberly Garza in the project "Dendritic Zoning: Establishing a New Urban Gradient in the Garden State." With "Seeping Boundaries," Harris-Brandts suggests a tactical method for intervention in the politically charged Palestinian West Bank using the political agency of sewage. Liquid waste is a staggering problem in the Middle East, and this project co-opts the destructive force of the Israeli demolition of Palestinian houses as a key ingredient for environmental change. Here, forced retreat, instigated by political turmoil, paradoxically enables the construction of wastewater filtration systems along the West Bank.

Garza's "Dendritic Zoning" takes a less literal approach to the strategy of retreat. This project examines the highway (the US Interstate system) as a mechanism for ecological recovery and water management. A transportation network is hijacked and rezoned as an ecological framework. This project is positioned within the strategy of retreat because it is a solution for managing water quality that emerges in response to expanding urban development. The tool that enables retreat, the highway, creates an ecological corridor, while the uninhabited buffer zone surrounding the transit infrastructure is redefined as a hydrological and urban gradient.

To adapt to the contaminated is to allow contaminated water to enter the inhabited spaces of cities and communities, prompting the built environment to cleanse the water close to its point of use. This is the contemporary strategy, if not an obsession, for control lies in the hands of the designer. Examples of this approach are plentiful, and we have provided a wide range of both realized and speculative projects of fluctuating scales.

Detoxi-city by Zuhal Kol, Rodney Bell, and Julia Gamolina is an urban plan for Barra Do Furado, Brazil, that fuses an offshore drilling industry with an aquaculture industry allowing for scalar urban growth. This hybridization works to counteract the negative environmental effects of oil drilling with an ecologically sensitive aquaculture.

Moving down in scale is Parallel Networks by Op.N (Ali Fard + Ghazal Jafari), a proposal that imagines New York City's waterways as a sixth borough. The project deploys floating wetland modules to create a symbiotic relationship between the shipping industry and marine life. Architectural intervention creates "a network which is as much a performative waterscape as it is a connective urban tissue."

The idea of a performative waterscape for the city can also be seen in a series of realized projects by the landscape architecture office Turenscape. The Qunli Storm Water Park in the Heilongjiang Province of China and the Shanghai Houtan Park both present urban interventions at the scale of a park. These parks perform remediation on polluted water while simultaneously creating a public space. While the two projects explore different formal and technical strategies, they both use living systems to remediate dire environmental situations. Thus, design is a regenerative act, turning industrial brownfields into lush, constructed wetlands.

This approach can also be seen in D.I.R.T. Studio's Vintondale Reclamation Park, where landscape architecture is deployed to reverse the effects of coal mining in the Appalachian Mountain region of Pennsylvania, United States. While this project addresses a condition existing outside of an urban core, it employs calibrated ecologies in an adaptive fashion to enable the decontamination of acid mine drainage.

Adaptation is also examined at the scale of the home. "Water Core Home" by Dan Williamson, David Karle, and Sarah Thomas Karle provides an example of how the single-family home can become a key component of a decentralized micro-watershed management system. This project "challenges existing urban residential blocks, lots, and homes relying on combined sewer systems to be spatially reconfigured and architecturally address water performance." The Water Core Home leverages "internal and external spatial strategies that aimed to reduce the quantity and improve the quality of the water before it reached the combined sewer system." Following a methodology of adaptation, architecture thus becomes the mitigating mechanism for transforming water quality.
In the following pages you will embark on the situations and solutions of "The Contaminated." The projects address contamination as an operation related to industry, bacteria, sewage, resource extraction, and urbanization. They range from fantastic speculations of cities formulated from a desire to manage pollutants to realized constructions of stormwater parks. While some provide very specific, technical solutions, others are included to generate a larger, conceptual discussion. The hope is that this index of possibilities for the future can catalyze a discourse about contaminated waters and design methodologies to deal with them.

Notes ---

1. Matthew Gandy, "The bacteriological city and its discontents," *Historical Geography* 34 (January 1, 2006): 3.

2. Martin V. Melosi, *The Sanitary City: Urban infrastructure in America from colonial times to the present* (Baltimore: Johns Hopkins University Press, 2000), 43.

3. Gandy, 3.

4. ""Water: Our Thirsty World," National Geographic 217, no. 4 (April 2010): 112.

5. See Gandy, 138.

6. See Melosi, 145.

7. Charles-Edward Amory Winslow, *The Conquest of Epidemic Disease* (1943; New York: Hafner, 1967), 243.

ACT TWO: THE CONTAMINATED CALL TO ACTION

The following text is reprinted from the essay "The Bacteriological City and Its Discontents" by Matthew Gandy, first published in *Historical Geography*, volume 34 (January 1, 2006). Reproduced by permission of Matthew Gandy.

THE BACTERIOLOGICAL CITY AND ITS DISCONTENTS

From *Historical Geography*, Volume 34, 2006.

Matthew Gandy

Introduction

In the rapidly gentrifying Oderberger Straße in the Prenzlauerberg district of Berlin lies a curious building that resembles a medium-sized factory. Now a semi-derelict venue for alternative cultural events, it was until 1994 a public baths and swimming pool. The imposing Stadtbad, first opened in 1902, is a remnant of a distinctive phase in urban history when the benefits of regular washing and exercise were promoted as part of a wider attempt to improve the health and well-being of the general population.[1] The changing relationship between water and the human body in the modern city reflects a distinctive "hydrological order" characterized by the extension of complex technological networks, changing patterns of everyday life, and the establishment of new modes of municipal administration. Water, like other facets of urban nature, was incorporated into an increasingly rationalized and scientifically managed urban form.

The history of urban infrastructure is now the focus of a vibrant debate that combines the established insights of urban history with emerging perspectives drawn from other fields such as architecture, critical theory, and urban studies. Emphasis on the administrative, technical, and political dimensions of 19th-century urban reform has been supplemented by a greater emphasis on the micro-spaces of the modern city—in particular, the body and the domestic interior—along with an expanded theoretical discussion of themes such as the ideological rationale for urban governance, the role of public works projects in the construction of a functional public realm, and the social, cultural, and economic implications of technological networks in urban space.[2]

Implicit within this current debate is a sense that a *longue durée* extending from the mid-19th century until the last quarter of the 20th century has been partially supplanted by a new set of socio-technological developments. This essay explores the movement towards a distinctive constellation of space, society, and technology that is referred to here as the "bacteriological city" in order to differentiate this historical phase from the early industrial era and also from a range of developments over the last thirty years associated with the emergence of neo-liberal approaches to public policy.[3] Placing an extended period of urban history under one conceptual frame risks a degree of elision between different developments, but it does help to identify some of the commonalities and anomalies that have characterized processes of capitalist urbanization since the middle decades of the 19th century. This urban epoch has been variously referred to in the literature as the "hydraulic city," the "sanitary city," or the "modern infrastructural ideal," but the term "bacteriological city" is deployed here to denote a distinctive set of interrelated developments ranging from science and technology to new forms of municipal administration.[4]

The term "bacteriological" is especially apposite for an exploration of the relationship between water and cities since technical and political discourses cannot be easily disentangled from advances in disease epidemiology that influenced developments in civil engineering, planning, and public health. At the same time, however, the term "bacteriological" is not intended to give undue weight to the medical or scientific dimensions of urban policy making but will be related to wider themes, such as the role of urban networks in mediating the relationship between the body and the city. In exploring the development of water infrastructure, this essay examines the transformation of the modern city as part of an interrelated set of developments that transcend the interventions of individual engineers, planners, or medical advocates. The relatively stable urban form that emerged out of the chaos of the 19th century is presented as a historical compromise that emerged in order to enable the modern city to function more effectively. Yet in circumstances where the modernization process was never fully completed—most notably, in a colonial context— the underlying weaknesses of the bacteriological city as a universal ideal are sharply revealed.

Delineating the bacteriological city

The 19th-century city, as the political and economic fulcrum for industrialization, posed a complicated set of dilemmas for the scope and effectiveness of modern government. A particular challenge during the first half of the 19th century was the marked deterioration in urban living conditions punctuated by devastating outbreaks of infectious disease. Though the public health crisis affecting rapidly growing cities was readily ascribed to atrocious physical conditions, this masked competing interpretations of the problem and the degree to which public health was conceived as part of a wider set of social and political reforms. Moralistic interpretations of ill health, for example, co-existed with a miasmic emphasis on "mephitic exhalations" associated with the dangers of stagnant air and water. The relationship between poverty, disease, and the physical environment remained a confused arena in the pre-bacteriological era in part because few professional discourses engaged with urban problems in any systematic way that might enable the political, economic, and technical spheres to be considered in relation to one another. In any case, diseases such as cholera and typhoid threatened not just the poor but entire populations, and problems with water supply were generally conceived in terms of taste or convenience rather than outright threat.

With the development of the empirical sciences in the early decades of the 19th century, however, the pattern of mortality and morbidity could be conveyed far more accurately than in the past. From John Snow's classic survey of the incidence of cholera in Soho to Parent-Duchâtelet's olfactory investigations of underground Paris, we find an emerging classificatory impulse towards the *terrae incognitae* of the modern city. The surveys and writings of figures such as Friedrich Engels, Henry Mayhew, Thomas Southwood Smith, and others placed the living conditions of the modern industrial city under unprecedented critical scrutiny. In so doing, the scope of modern governance was widened to include not just the "modern subject"—a new kind of urban citizen amenable to the emerging discourses of hygienism and social control—but also the recognition of governable spaces that had previously not been systematically identified. The issue of public health became an increasingly significant concern for the modern state, such that the health of the population acquired a strategic importance that had previously been neglected. The development of more systematic forms of data collection and the expansion of state activity into hitherto neglected areas altered the rationale of governmental activity and introduced a range of new strategic imperatives in the face of industrialization, urbanization, and emergent forms of political agitation. With the expansion of military conscription, for example, the scale of undernourishment and ill health became more readily apparent than it had been in the past. Concerns with public health encompassed not just needs for economic efficiency but also the demographic demands of emerging nationalist ideologies. In one sense the urban population was increasingly regarded as a collective statistical entity, but in another sense the more communal sensory experience of the past was increasingly challenged by new attitudes towards privacy and social distinction. Changing attitudes towards health, hygiene, and cleanliness, for example, involved an emphasis on increasingly individualized forms of identity and a growing cultural emphasis on the redefinition of the domestic arena.[5] The emergence of new social formations also coincided with intensified forms of spatial differentiation so that the vertical segregation of the congested preindustrial city was increasingly superseded by the horizontal segregation of the expanding industrial metropolis.

The place of water within the 19th-century

city reflects an ambiguity between the strategic needs of the modern state and the development of reformist dimensions to urban political discourse. The demonstration of linkages between contaminated water and ill health played a pivotal role in fostering the political demands of the burgeoning public health movement for the physical reconstruction of cities even if the rationale for improving water infrastructure rested on a wider set of factors at best only tangentially related to human health. Many industries in the 19th-century city, such as chemical works, breweries, tanneries, and distilleries, relied on pure and reliable water supplies and demanded action from municipal authorities to tackle the deteriorating situation. In addition to industrial needs for water, the constant threat of fire provided a further spur to action, not least because of the growing political power of the insurance industry.[6]

The rapid growth of 19th-century cities quickly overwhelmed the historic reliance on wells, water vendors, and other sources and led to the introduction of centralized water supply systems in, for example, Paris in 1802, London in 1808, and Berlin in 1856. Yet this shift towards more elaborate systems of water supply introduced new tensions over how the costs of these infrastructure projects would be borne. The transformation of the modern city would have been impossible without the innovative use of financial instruments such as municipal bonds to enable the completion of ambitious engineering projects without imposing substantial additional tax burdens. In the 1830s, for example, New York City issued bonds to enable the completion of the Croton Aqueduct to solve the city's chronic water shortages, and in the 1850s Berlin drew not just on British engineering expertise to develop its water supply but also on the financial resources of the London capital markets.[7] Municipal bond markets weathered the economic turbulence of the 1870s and played a pivotal role in enabling the development of infrastructure networks: by 1905, for example, waterworks constituted the largest component of municipal debt for US cities.[8] These and other financial mechanisms enabled the flow of capital to be channeled into the built environment and also underpinned the growing interconnections between urbanization and international finance.

In addition to new methods of financing public works, the reconstruction of cities also required the establishment of new policy instruments such as the power of eminent domain and other planning mechanisms that enabled a strategic urban vision

to override multifarious private interests. Wealthy residents with their own wells, for example, had frequently sought to organize petitions against the development of municipal water systems that they regarded as expensive and unnecessary. Furthermore, the construction of large-scale hydraulic engineering projects required the acquisition of private lands both for the completion of new infrastructure and also to protect public water systems from contamination with agricultural wastes or other possible sources of pollution. In the case of water, a critical trend from the middle decades of the 19th century onwards was the replacement of inadequate private water companies by public ownership. Private companies routinely exploited their monopoly of individual supply networks by refusing to extend services to outlying districts or by making excessive charges for poor-quality services. In cities such as Los Angeles and New Orleans, for example, the charters of private water companies were revoked under public pressure to allow the development of municipal water services. In the USA some 43% of waterworks were publicly owned in 1890 compared with over 70% by the 1920s as networks expanded to include poorer or more distant neighborhoods.[9] The trend towards the municipalization of water supply involved bringing diverse private operators under the control of the local state to produce more unified, centralized, and democratically accountable forms of service provision.[10] By the 1920s and 1930s, however, the emerging bacteriological city of the late 19th century was metamorphosing into a full-fledged technocratic paradigm for modern governance, and so political changes in the urban arena became a progenitor of wider regional and national goals for public policy. In the US, for example, the New Deal saw a vast expansion in the federal role for water management, ranging from the construction of immense dams and river diversion schemes to the complex reconstruction of flood defenses.[11]

The development of the bacteriological city required the introduction of new forms of technical and managerial expertise in urban government. The replacement of miscellaneous administrative bodies such as parishes and vestries with more centralized approaches to urban management necessitated the expansion of state bureaucracies, such that the development of cities became an interrelated facet of the growing political power of the nation state. Yet the relationship between technical knowledge and municipal reform remained a complex arena where rival technological solutions to the problems of urban sanitation became repeatedly entwined in political conflicts over the autonomy of professional expertise in urban policy: engineers, for instance, frequently expressed their frustration at the fiscal and political barriers to the completion of their work—a sentiment which finds its clearest expression in the ambivalence of colonial urban administrations towards the latest advances in engineering science. In the British colonies, for example, the so-called Manchester doctrine of minimal financial support ensured that comprehensive engineering solutions to problems of ill health and insalubrious urban conditions would never be implemented.[12] In the case of 19th-century Bombay there were decades of discussions between engineers, physicians, and colonial administrators, but little progress towards an integrated sanitation system was ever achieved. By the 1860s the situation was becoming critical as the city's economic boom encouraged vast waves of migration and intense overcrowding. In 1863 the leading British civil engineer Robert Rawlinson called for a modern sewer system to be constructed in Bombay "according to true scientific principles."[13] Yet in a colonial context these advocates for urban improvement operated within a political arena where the nascent forms of citizenship and political reform enjoyed in Europe or North America had only limited significance. The emerging bacteriological city was a technical adjunct to capitalist urbanization, yet its full realization was in conflict with the marginal status of the colonial city, and so moralistic and "neo-miasmic" discourses persisted in preference to any universalist response to the modernization of urban infrastructure. Bombay, like many other colonial cities, experienced a catastrophic decline in urban environmental conditions culminating in an outbreak of bubonic plague in 1896 that was to last more than 15 years and cause immense economic disruption and loss of life.[14]

The emergence of more systematic approaches to the understanding of disease, poverty, and urban labor markets contributed to a rationalization of urban policy so that new analytical methods could be applied to public administration. Changing conceptions of disease epidemiology played a critical role within this transition by introducing a collective conception of human health that began to displace the earlier holistic emphasis on the susceptibilities of individuals or the miasmic focus on physical attributes of cities such as drainage or ventilation.[15] It is in this context that public health advocates such as Rudolf Virchow and

Robert Koch sought to use scientific advances—albeit within a positivist frame—as a means to underpin political demands for social reform that extended far beyond a purely utilitarian or technical agenda. Yet the prevailing view of public health, epitomized by Chadwickian sanitarianism, rested on a restricted conception of urban reform as the modernization of urban infrastructure rather than any wider critique of the process of capitalist urbanization itself.[16] In broad terms we can conceive of the modernization of industrial cities as a shift from the "private city" to the "public city," whereby fragmentary, piecemeal, and highly localized solutions to the problems of water and sanitation were superseded by the promotion of more complex kinds of coordination between political and economic interests. This transition was in fact a double movement, in which public activities such as washing were increasingly restricted to the private sphere whereas privately organized access to potable water or sanitation was gradually incorporated into a centralized, networked, and municipally controlled metropolitan form.

Fractured modernities
The hydrological transformation of the 19th-century city involved the gradual displacement of the "organic city" with its emphasis on the utilization of human wastes for agriculture. Yet elements of this earlier phase persisted into the second half of the 19th century before the epidemiological advances of the 1880s assured the ascendancy of contagionist ideas in public health thinking.[17] In the pre-bacteriological age, for example, it was far easier for figures such as Justus von Liebig and Edwin Chadwick to argue for a continuation in the agricultural uses of human waste and elaborate on complex schemes for the diversion of new sewer outlets to farms in the vicinity of the city.[18] Yet their cyclical conception of a rational urban order, founded on organicist and utilitarian conceptions of nature, conflicted with the underlying dynamics of the capitalist city and the development of a cultural appropriation of nature rooted in leisure rather than the needs of agriculture. The growing popularity of washing, for example, began to threaten the sanitary arrangements of the preindustrial city by flooding cesspits and diluting the nitrogen content of human manure at the same time that the production of artificial fertilizers was becoming more widespread.

The increasing quantities of human waste being discharged into rivers—either directly or through connections to the sewer system—provoked widespread opposition from agricultural, industrial, and fishing interests dependent on clean water as well as from "river fanatics" who insisted on using rivers for drinking water.[20] In the wake of the Hamburg cholera outbreak of 1892, in which nearly 10,000 people died, there was a ferocious standoff between the miasmic theories of Max von Pettenkofer and his allies, which found favor with the ruling elites, and the contagionist arguments of Robert Koch, who emphatically blamed the contamination of water supply for the spread of cholera and called on the German authorities to take decisive action. Pettenkofer denied that drinking water was involved in the spread of disease and insisted instead that the "cholera miasma" originated from localized changes in groundwater levels. In contrast, Koch demanded that the structure of municipal government be altered so that the implementation of public health measures such as the regular monitoring of drinking-water quality became an integral and continuous aspect of governmental activities.[21] The gradual acceptance of contagionist conceptions of disease epidemiology undermined the last vestiges of an organic conception of the modern metropolis and rendered human feces not only a focus of abjection but also a source of danger to public health. In Paris, for example, new legislation in 1894 made the connection of individual dwellings to the main drainage system mandatory as the introduction of tout-à-l'égout replaced the complicated and increasingly unworkable sanitary arrangements of the Haussmann era.[22] Yet throughout much of the global South this last phase in the modernization of water infrastructure remains only partially completed: in many cities, neither comprehensive sewer systems nor wastewater treatment works were ever introduced, and even in Europe and North America the deficiencies of existing water treatment systems have been the focus of new waves of legislation and political contestation since the 1980s.

The spread of these technological networks and new plumbing innovations within the home remained highly uneven in different national and cultural contexts and was largely restricted to middle-class households until the wider diffusion of prosperity during the 20th century: the general introduction of water closets, for example, was limited before the 1880s and bathrooms only became a standard domestic fixture after 1914.[23] When the historian Patrick Joyce refers to the sanitary or hydraulic city as "a dominant social imaginary of the city," he presents a highly generalized interpretation of a medley of different developments: the differential experience of the modern city is obscured by an abstract account of the governmental strategies of political liberalism.[24] Used in a neo-Foucauldian context by Joyce and others, the term "liberalism" denotes the attempt to regulate human behavior through indirect means rather than through more direct forms of state intervention: the growing popularity of new plumbing technologies exemplifies this dynamic by inculcating new washing habits through the co-evolution of society and technological networks.[25] Yet we could argue à la Joyce that the colonial city—with its indirect modes of governmentality—marks the acme of a political strategy to govern through the complex appropriation of existing power structures and social mores in order to combine fiscal austerity with various forms of ideological legitimation.[26] Joyce is also right to highlight not just the technical and governmental parallels between what he terms "colonial and metropolitan governmentality" but also the derogatory hierarchies of human worth that were applied both to the slum dwellers of Europe and the native populations of colonial cities. What he describes as "dislocated liberalism" usefully captures the sense of a governmental regime at the margins of its own internal logic in a colonial context where the political and economic exigencies of rapid urbanization could not be masked by any nationalist appeal to modernization and in which cities would emerge as the loci for nascent independence movements.[27]

Until recently the lagging levels of connection to modern water supply and sanitation systems in the cities of the global South were widely perceived as a temporary phenomenon to be overcome through ambitious efforts at urban planning and reconstruction. In reality, of course, the technocratic ideal that drove the development of the bacteriological city conflicted with the political and economic dynamics behind capitalist urbanization: a tension that was largely masked within the metropolitan core of Europe and North America but which was clearly manifest within colonial cities from the outset. Far from a singular modernity, the development of urban technological networks since the 19th century has generated a diversity of urban forms ranging from the fully connected metropolis of the Fordist era to an array of hybrid entities incorporating a palimpsest of different socio-technological arrangements. The contemporary transition away from the bacteriological city can only be fully appreciated in the context of the innate weaknesses within this centralized technological model. In the last thirty years, the municipal dominance in urban water provision has come under pressure from a number of different quarters: the anomalies within the universalist ideal, where it has been only partially implemented, have been exposed through the so-called "brown agenda" and demands to extend global access to water and sanitation; the integrated model of service provision has been extensively fractured through the splintering and disaggregation of technical networks to produce new inequalities; <u>expert-led approaches to civil engineering and urban planning have been extensively challenged by an emphasis on expanded public participation and a widening array of different interest groups; and the resurgence of private provision, in conjunction with new patterns of capital investment, is generating a different kind of urban landscape to the more ostensibly homogeneous technological landscapes of the past.</u>

Conclusions

<u>The complex interactions between disease, water, and urban infrastructure reveal that while the "bacteriological city" may represent an abstract ideal for the organizational structure of the modern city, it has never fully corresponded with urban realities because of the political and economic tensions that underlie the processes of capitalist urbanization.</u> These anomalies that pervade the technological structure of the modern city become most strikingly represented in the marginal spaces of the city and in those cities that are themselves marginal within the global economy. In the rapidly growing cities of the global South, for example, <u>the dilapidated or never completed infrastructure systems of the bacteriological era have been superseded by a proliferation of alternative networks.</u> By exploring the history of water infrastructure beyond the metropolitan core of Europe and North America, we can uncover fresh insights into the limitations of the bacteriological city as a universal model and also disentangle some of the political tensions underlying the introduction of technological networks in the capitalist city.

The modernization of urban infrastructure required an institutional context that could facilitate the flow of capital into the built environment yet this historic dynamic has been neglected by neo-Foucauldian interpretations of liberal "governmentality." The political dimensions of urban technological networks encompass not just the interface between technology and the body but also the evolving institutional context for the shaping of cities themselves. The bacteriological city emerged out of a synthesis between the scientific and political dimensions of modernity, and so technological characteristics of the networked modern city became characteristic features of a more rationalized urban form. Yet the degree to which these achievements have tended to be associated with individual engineers rather than any more enduring political philosophy underlies the extent to which the sanitarian emphasis of the bacteriological city foreclosed wider political considerations transforming issues such as citizenship rights to basic services into more narrowly technical questions. We can argue that the public realm under the age of the "heroic engineer" remained only tangentially linked to the city as a whole, as evidenced by the extensive fracturing of technocratic planning ideals in the last quarter of the 20th century. Rather than being a teleological conception of urban change, the bacteriological city was one of a number of possible manifestations of urban form in spite of its aura of permanence and universality. The triumph of the 19th-century technocratic vision did not completely preclude its alternatives: the "discontents" associated with the bacteriological city extend to those voices, both now and in the past, who distrust an extended role for the state in urban governance as well as to those critics of the inherent inequities engendered by capitalist urbanization. In reality, the bacteriological city has proved to be a transitional phase: even at its acme, in the middle decades of the 20th century, the techno-managerialist urban paradigm displayed a series of fiscal and ideological weaknesses that would not be fully revealed until the political and economic turbulence of the late 1960s and 1970s. Though most contemporary cities remain dependent on the technological networks built up under the political aegis of the bacteriological era, <u>these increasingly dilapidated urban infrastructures serve as a poignant symbol of the fragility and historical specificity of metropolitan urban form.</u>

Notes ---

1. *Stadtbad Oderberger Straße: Porträt eines historischen Bades* (Berlin: Gesellschaft der behutsamen Stadterneuerung, 2001).

2. See, for example, Stephen Graham and Simon Marvin, *Splintering Urbanism: Networked infrastructures, technological mobilities, and the urban condition* (London and New York: Routledge, 2001); Elisabeth Heidenreich, *Fließräume: Die Vernetzung von Natur, Raum und Gesellschaft seit dem 19. Jahrhundert* (Frankfurt: Campus, 2004); Maria Kaïka and Erik Swyngedouw, "Fetishizing the modern city: The phantasmagoria of urban technological networks," *International Journal of Urban and Regional Research* 24 (2000): 120-38.

3. In an urban context, the advent of neo-liberalism is associated in particular with a diminution in the role of the state in the coordination and provision of collective services, yet this process exhibits wide variations in its scope and timing ranging from new strategies for urban regeneration to vast programs of divestment in public utilities. See, for example, Neil Brenner and Nik Theodore, "Cities and the geographies of 'actually existing neoliberalism,'" *Antipode* 34 (2002): 349-79.

4. Compare, for example, Graham and Marvin, *Splintering Urbanism*; Patrick Joyce, *The Rule of Freedom: Liberalism and the modern city* (London and New York: Verso, 2003); and Martin Melosi, *The Sanitary City: Urban infrastructure in America from colonial times to the present* (Baltimore: Johns Hopkins University Press, 2000).

5. Alain Corbin, *The Foul and the Fragrant: Odor and the French social imagination* (1982; Cambridge, MA: Harvard University Press, 1986); Michel Foucault, "The politics of health in the 18th century," in Paul Rabinow (ed.), *The Foucault Reader*, trans. Christian Hubert (Harmondsworth: Penguin, 1984): 277, originally published in Colin Gordon (ed.), *Power / Knowledge* (New York: Pantheon, 1980). See also Susanne Frank, *Stadtplanung im Geschlechterkampf: Stadt und Geschlecht in der Großstadtentwicklung des 19. und 20. Jahrhunderts* (Opladen: Leske und Budrich, 2003) and E. Lupton and Janice A. Miller, *The Bathroom, the Kitchen, and the Aesthetics of Waste: A process of elimination* (New York: Princeton Architectural Press, 1992).

6. See, for example, Matthew Gandy, *Concrete and Clay: Reworking nature in New York City* (Cambridge, MA: The MIT Press, 2002).

7. See Hilmar Bärthel, *Wasser für Berlin* (Berlin: Verlag für Bauwesen, 1997) and Heinrich Tepasse, *Stadttechnik im Städtebau Berlins 19. Jahrhundert* (Berlin: Gebr. Mann, 2001).

8. David Cutler and Grant Miller, "Water, water everywhere: Municipal finance and water supply in American cities" (working paper 11096, National Bureau of Economic Research, Cambridge, MA, 2005); John Teaford, *The Unheralded Triumph: City government in America, 1870-1900* (Baltimore, MD: Johns Hopkins University Press, 1984).

9. Cutler and Miller, "Water, water everywhere"; William L. Kahrl, *Water and Power: The controversy over Los Angeles's water supply in the Owens Valley* (Berkeley: University of California Press, 1982); Werner Troesken, "Typhoid rates and the public acquisition of private waterworks, 1880-1920," *Journal of Economic History* 59 (1999): 927-48; Werner Troesken and Rick Geddes, "Municipalizing American waterworks, 1897-1915," *Journal of Law, Economics, and Organization* 19 (2003): 373-400; John Walton, *Western Times and Water Wars: State, culture, and rebellion in California* (Berkeley: University of California Press, 1992).

10. See, for example, Martin Daunton, "Public place and private space: The Victorian city and working-class housing," in D. Fraser and A. Sutcliffe (eds.), *The Pursuit of Urban History* (London: Edward Arnold, 1983); Richard J. Evans, *Death in Hamburg: Society and politics in the cholera years 1830-1910* (Oxford: Oxford University Press, 1987); P. Penzo, "L'urbanistica e l'amministrazione socialista a Bologna, 1914-1920," *Storia Urbana* 18, no. 66 (1994): 109-43; John V. Pickstone, "Dearth, Dirt, and fever epidemics: Rewriting the history of British 'public health,' 1780-1850," in Terence Osborne and Paul Slack (eds.), *Epidemics and Ideas: Essays on the historical perception of epidemics* (Cambridge: Cambridge University Press, 1992).

11. Improved flood control was a key part of the New Deal agenda for water management in the wake of extensive loss of life and damage to property in low-lying parts of the Gulf Coast during the 1920s and earlier. On the history of US water resources policy see, for example, David Lewis Feldman, *Water Resources Management: In search of an environmental ethic* (Baltimore: Johns Hopkins University Press, 1991).

12. See, for example, A. Aderibigbe, "Expansion of the Lagos protectorate, 1863-1900" (unpublished PhD dissertation, University of London, 1959).

13. "Increased comfort and cleanliness lead to health and lengthened human life," wrote Rawlinson, "and such improvements ought in their results to be at least an equivalent for the annual money value of the works and costs of management." To Sir Charles Wood, Principal Secretary of State for India from Robert Rawlinson, Report on the proposed scheme of main sewerage and drainage submitted to the Municipal Commissioners of Bombay, dated April 1863, in the Maharashtra State Archives, Mumbai. See also Mariam Dossal, *Imperial Designs and Indian Realities: The planning of Bombay City, 1845-1875* (Bombay: Oxford University Press, 1991) and J. A. Jones, *A Manual of Hygiene, Sanitation, and Sanitary Engineering with Special References to Indian Conditions* (Madras: Government Press, 1896).

14. See David Arnold, *Colonizing the Body: State medicine and epidemic disease in nineteenth-century India* (Berkeley, CA: University of California Press, 1993); Ira Klein, "Urban development and death: Bombay City, 1870-1914," *Modern Asian Studies* 20 (1986): 725-54..

15. See Joyce, *The Rule of Freedom*.

16. See Christopher Hamlin, *Public Health and Social Justice in the Age of Chadwick* (Cambridge: Cambridge University Press, 1998).

17. On the history of water supply see, for example, Jean-Paul Goubert, *The Conquest of Water: The advent of health in the industrial age*, trans. A. Wilson (1986; Oxford: Polity Press, 1989); Andre Guillerme, "Sottosuolo e construzione della città / Underground and construction of the city," *Casabella: International Architectural*

Review (1988) 542/543: 118; and Charles D. Jacobson and Joel A. Tarr, "The development of water works in the United States," *Rassegna: Themes in Architecture* 57 (1994): 37–41.

18. See Dominique Laporte, *History of Shit*, trans. Nadia Benabid and Rodolphe el-Khoury (1978; Cambridge, MA: The MIT Press, 2000); John von Simson, *Kanalisation und Städtehygiene im 19 Jahrhundert* (Düsseldorf: Verein Deutsche Ingenieure, 1983).

19. See Matthew Gandy, "Rethinking urban metabolism: Water, space and the modern city," *City* 8 (2004): 371–87.

20. See Jürgen Büschenfeld, *Flüsse und Kloaken: Umweltfragen im Zeitalter der Industrialisierung,* 1870–1918 (Stuttgart: Klett-Cotta, 1997).

21. Evans, *Death in Hamburg.* Political attempts to refute Koch's research into the epidemiology of cholera also acquired an international dimension. See Mariko Ogawa, "Uneasy bedfellows: Science and politics in the refutation of Koch's bacterial theory of cholera," *Bulletin of the History of Medicine* 74 (2000): 671-707.

22. Baron Haussmann, for example, refused to allow human wastes to enter the newly constructed sewers of Second Empire Paris. See Matthew Gandy, "The Paris sewers and the rationalization of urban space," *Transactions of the Institute of British Geographers* 24: 1 (1999): 23–44; Gérard Jacquemet, "Urbanisme Parisien: La bataille du tout-a-l'égout a la fin du XIXe siècle," *Revue d'histoire moderne et contemporaine* 26 (1979): 505-48.

23. See, for example, D. Glassberg, "The public bath movement in America," *American Studies* 20 (1979): 5-21; J.-P. Goubert, "Wasser und Intimhygiene am Beispiels Frankreichs," in Bernd Busch and Larissa Förster (eds.), Wasser, trans. Stefan Barmann (Köln: Wienand, 2000): 168–76. In private rented sectors of the housing market in some European cities, we find that many households lacked bathrooms as recently as the 1970s or even later in the case of Altbau [old build] apartments in the former German Democratic Republic.

24. Joyce, *The Rule of Freedom,* 245.

25. Thomas Osborne, "Security and Vitality: Drains, liberalism, and power in the nineteenth century," in Andrew Barry, Thomas Osborne, and Nikolas Rose (eds.), *Foucault and Political Reason: Liberalism, neo-liberalism, and the rationalities of government* (London: UCL Press, 1996): 115. See also Christopher Otter, "Making liberalism durable: Vision and civility in the late Victorian city," *Social History* 27 (2002): 1–15. For recent debates over "governmentality" and the application of Foucault's ideas to urban theory, see, for example, M. Dean, *Governmentality: Power and rule in modern society* (London: Sage, 1999); Colin Gordon, "Governmental rationality: An introduction" in G. Burchell, C. Gordon, and P. Miller (eds.), *The Foucault Effect: Studies in governmentality* (Hemel Hempstead: Harvester Wheatsheaf, 1991): 1–52; and Nikolas Rose, *Powers of Freedom: Reframing political thought* (Cambridge: Cambridge University Press, 1999).

26. See also Mahmood Mamdani, *Citizen and Subject: Contemporary Africa and the legacy of late colonialism* (Princeton, NJ: Princeton University Press, 1996).

27. Joyce, *The Rule of Freedom,* 250.

The following text is taken from the chapter "Bringing the Serpent's Tail into the Serpent's Mouth: Edwin Chadwick and the 'Sanitary Idea' in England," from the book *The Sanitary City: Urban Infrastructure in America from Colonial Times to the Present* by Martin V. Melosi, copyright © 2000.

BRINGING THE SERPENT'S TAIL INTO THE SERPENT'S MOUTH: EDWIN CHADWICK AND THE 'SANITARY IDEA' IN ENGLAND

From *The Sanitary City: Urban Infrastructure in America from Colonial Times to the Present*, 2000.

Martin V. Melosi

Mid-19th-century England's "sanitary idea" made popular the notion that the physical environment exercised a profound influence over the well-being of the individual, that health depended upon sanitation. This concept reshaped thinking about the delivery of pure water, the removal of sewage, and the collection and disposal of refuse. As one writer put it, the greatest service of the sanitary idea was "in replacing … fatalism by a new faith in the power of scientific control of the physical environment."[1]

The advent of the sanitary idea offered a clearer rationale and newer strategies for improving sanitary services first in England, and then throughout the world. Historian Ann F. La Berge, however, argued persuasively that late 18th- and early 19th-century France was first to provide a model for public health "theoretically, institutionally, and practically." She claimed that, beginning in Paris, the French were the leaders in public health theory and reform until the 1830s, and that in the 1820s and 1830s there was "considerable cross-fertilization of ideas between public health advocates in Britain and France." By the 1850s, the British claimed preeminence in public health practices, such as sewerage and water supply, and leadership in the field passed to them.[2]

The ascendancy of British public health practices was a triumph, albeit a temporary one, for the theory of disease causation embedded in the sanitary idea. One of the key results was the transformation of Victorian cities from their Dickensian bleakness into more livable environments. Historian Asa Briggs described the achievement in a cautiously romantic way:

> The building of the cities was a characteristic Victorian achievement, impressive in scale but limited in vision, creating new opportunities but also providing massive new problems. Perhaps their outstanding feature was hidden from public view—their hidden networks of pipes and drains and sewers, one of the biggest technical and social achievements of the age, a sanitary "system" more comprehensive than the transport system.[3]

London, however, which benefited noticeably from the sanitary reforms at mid-century, would not address many health problems in the poorer sections much before the 1870s.[4] Nonetheless, an era of disease prevention through environmental sanitation was clearly underway.

The greatest exponent and popularizer of the sanitary idea was barrister-turned-sanitarian Edwin Chadwick. He was born on January 24, 1800, near Manchester, and was the oldest son of James Chadwick, "an outspoken radical, Francophile, and follower of Thomas Paine." James became editor of the Statesman in 1812, soon after moving his family to London following the death of Edwin's mother, and after suffering some economic setbacks. In 1816 James became editor of the Western Times, then remarried in Exeter, and began raising more children. Edwin had fewer contacts with his father at this point, especially after James and his new family moved to New York in the late 1830s.[5]

Edwin had limited educational opportunities as a child and young adult, and grew to resent "the classically educated elite." Largely self-taught, he entered an attorney's office as an apprentice at age 18. After five years he set his sights higher and was admitted to the Middle Temple in 1823 to begin preparation to become a barrister. Chadwick helped support himself by writing brief articles for several metropolitan newspapers, and developed associations with a number of legal and medical students in London. In 1824, he met Dr. Thomas Southwood Smith and later met John Stuart Mill—both Philosophical Radicals—and acquired his first knowledge of Benthamism.[6]

The acknowledged leader of the Philosophical Radicals, Jeremy Bentham, was a jurist, legal reformer, and utilitarian philosopher. He criticized traditional ideas of constitutional law and asserted the theory, as one writer stated, that "right actions are those which are most useful for the promotion of general happiness." He was especially interested in applying his ideas to penal reform.[7]

From Bentham and economist David Ricardo, Chadwick acquired—or at least reinforced—his belief in an activist central government, an idea found in several of his writings. As his reputation grew among the Benthamites, Chadwick was asked to open a debate on the poor laws at the London Debating Society in November 1829. Two years later, he became secretary for Bentham himself, assisting the aging utilitarian in the drafting of his Constitutional Code. Although they differed on several issues, Chadwick continued to share with Bentham a commitment to efficiency and the authoritarian nature of the state.[8]

The condition of the London slums became a major research interest for Chadwick, and while pursuing this work he contracted typhus, from which he fully recovered. In 1832, the year Bentham died, Chadwick was appointed to a commission inquiring into the state of the English Poor Laws, resulting in the publication of the 1834 Poor Law Report. After more than two decades of public service, Chadwick returned to private life in 1854. Ostensibly for reasons of health, he resigned from the General Board of Health, but did so with the fervent blessing of his many antagonists.[9]

Critics regarded Chadwick's social views as repressive, but his stature as an expert in poor law reform grew nonetheless. When three Poor Law commissionerships became available in 1834 for the purpose of implementing the new law, Chadwick assumed he would receive one. But, all too typically, birth and wealth directed appointments of this type, and he was overlooked. His rigid stance on key issues also made him less attractive to a Whig government already operating on wobbly legs. Regarded as indispensable to the work, Chadwick was offered the position as secretary to the new Poor Law Commission. He initially balked at the offer, but reconsidered after being convinced that his power would be greater than his title. Although he played an integral role on the commission, Chadwick never acquired the decision-making influence he expected. When his advice was taken, particularly on workhouses and policing, he often believed that it was carried out badly. Critics, however, continued to identify him with oppression of the poor.[10]

Throughout his career, Chadwick battled with public officials, physicians, and engineers. Some considered him a martinet who did not work well with others. "No one ever accused Chadwick of having a heart," one observer noted.[11] He challenged those criticisms, not able to understand why his commitment to social change inspired such emotional reactions.

Chadwick ultimately turned to a field of government action where he built his most lasting legacy in public health, despite his image of being insensitive to the needs of the poor and his growing disdain for physicians who cared little for preventive medicine. In the wake of the influenza epidemic of 1837-38, the commission was ordered to inquire into the relationship between pauperism and sanitary conditions. Chadwick was given the responsibility to carry out the work. To help him in the task, he enlisted the aid of three doctors known for their devotion to environmental influences on health—James Kay, Neil Arnott, and Thomas Southwood Smith. It was not easy for Chadwick to share this opportunity to conduct a major national sanitary inquiry, especially with men who were rapidly becoming prominent spokespersons for health reform in their own right—and three physicians, at that.[12]

Chadwick, nevertheless, brought great attention to the ravages of poverty and to the dismal health conditions of the industrial cities with the *Report on the Sanitary Condition of the Laboring Population of Great Britain* (1842). The document was widely disseminated, selling more copies than any previous government publication. It was well researched and well argued, painting a vivid picture of urban blight and making emphatic the case for disease prevention.[13] As a good Benthamite, Chadwick also prescribed to the notion of "civic economy," which, in this case, suggested that it was more expensive to create disease than to attempt to prevent it.[14]

The compilation of the report was the handiwork of several members of the commission, although Chadwick liked to claim it as his alone. It was more precisely the culmination of an emerging movement in public health rather than the brainchild of any one reformer.[15] As historian Anthony S. Wohl stated, "Public health ...became a kind of 'fundamental' reform, an underpinning and sine qua non for all other reforms."[16]

What made the report so radical was its denial of disease in fatalistic terms, as God's will, and also its rejection of a more current view that poverty was the main cause of ill health.[17] It turned that argument on its head, stating that ill health was a cause of poverty because disease had environmental roots. The report thus proved to be a forceful indictment of unsanitary living conditions in the industrial slums, as well as a severe criticism of physicians ignorant of the causes of contagion and of the moribund local health boards.[18] As one observer noted, the government was controlled by "a high and haughty class, that gave little close consideration to common human comfort, and took as little care of the public health as it did of the people's education."[19]

The waves of cholera epidemics in England during the early 19th century underscored the powerful language of the report. In the late 1820s, many people accepted chronic dysentery and other endemic diseases as normal, and disregarded warnings by reformers about the health problems mounting in the major cities, especially London. The cholera epidemic of 1831-32 changed all that, and the reformers were taken more seriously. This first of several cholera epidemics to ravage Great Britain took 60,000 lives, many among the poor. Repeated cholera attacks struck the British Isles in 1848-49, 1854, and 1867.

Tainted water was the medium of transmission, and when this cause was identified, the problem could be confronted effectively. While its prevention was learned to be relatively simple, and although it was not as statistically significant as typhus or consumption, cholera terrified people because it hit suddenly and violently and was extremely contagious.[20]

Without an understanding of the bacteriological origin of communicable diseases, English sanitarians of the period drew an immediate correlation between pollution and disease.[21] Bill Luckin rightly characterized Chadwick's view as "proto-environmentalism," because it identified an environmental causation for disease, but without understanding the role of pathogenic organisms as the precise cause or a well-developed notion of pertinent ecological factors contributing to disease.[22] In addition, Chadwick was typical of the Benthamites and middle-class reformers, as Margaret Pelling stated, in having "limited popular sympathies and no egalitarianism, except that which might lie in the assumption that all groups had a natural capacity for happiness ... and a natural life-cycle which ought not to be cut off or disrupted at the point of greatest skill and productivity. Pauperism and disease were alike gratuitous and preventable."[23]

The so-called filth, or miasmatic, theory dominated the thinking of sanitarians until late into the century. Because disease was understood to arise from putrefying organic wastes, bad smells (miasmas), and sewer gases—and could not be transmitted from person to person—the filth theory is described as anticontagionist. Many of the diseases these early sanitarians confronted were intestinal; thus, environmental sanitation was credited with substantial success.[24]

An address given by Dr. Benjamin W. Richardson before the Sanitary Institute of Great Britain on July 5, 1877, captured the spirit of the filth theory. He said in part:

With the progress of sanitary science we must expect to see preventive medicine taking an ascendancy. Cure will cease, prevention will grow. Humanly-made epidemics, like the great plague of London, which was planted and reared in the rush-covered floors of domiciles saturated with the organic refuse for years, or like the modern typhoid, which is fed by streams of drinking water uncleansed from human excreta, such self-made epidemics will be prevented by simple mechanical skill. Diseases imposed by indulgence in harmful pleasures and appetites, or by physical overwork and shock, will be removed by effect of moral influences and knowledge of cause; and gradually, I believe, those persistent evils, which like the lightning-stroke, come without human ordinance or fault, will be placed also under some protecting care, and if not removed, reduced to a short calendar.[25]

Having provided a rudimentary environmental context for identifying the cause of disease and ill health in his report, Chadwick began to shape an administrative structure for addressing the problems and a technical response to implement new methods of disease prevention. By emphasizing environmental as opposed to personal aspects of hygiene, he envisioned doctors and other medical personnel extending their roles beyond specific treatment of sick individuals to a broader range of social action, especially inoculation programs and environmental sanitation. As a Benthamite, he refused to accept contemporary laissez-faire thinking with respect to industry or the life of the citizenry. Thus it followed that the rights of the few were outweighed by the needs of the many. Local control of sanitary services, therefore, had to be supervised by a strong central authority. Among other things, this meant the hiring of paid inspectors—a precedent-setting notion predating modern civil service.[26]

For Chadwick, the appropriate technological response for dealing with unhealthy conditions was to be found in improved public works, including waterworks, sewers, paved streets, and ventilated buildings.[27] He proposed a hydraulic (or arterial-venous) system that would bring potable water into homes equipped with waterclosets, and then would carry effluent out to public sewer lines, ultimately to be deposited as "liquid manures" onto neighboring agricultural fields.[28]

In conceiving this plan, Chadwick was strongly influenced by John Roe, a railway and canal engineer, who had been surveyor for the Holborn and Finsbury Sewers Commission. Roe introduced Chadwick to what one writer called "all of the evils of the sewers of the day." He took Chadwick down into the sewers themselves to show him the vermin and the crumbling brickwork of the old drains. One of Roe's solutions was to use a constant flow of water in concert with sewers that were small and egg-shaped, rather than round or flat brick, to increase the velocity of the flow and thus to increase the sewer's carrying capacity. This concept fit well into Chadwick's evolving arterial system, although it broke with much of the conventional thinking of the day.[29] With the addition of the fertilization phase, Chadwick noted, "we complete the circle, and realize the Egyptian type of eternity by bringing as it were the serpent's tail into the serpent's mouth."[30]

Many of Chadwick's contemporaries, including those with predictable vested interests, claimed that his scheme was technically impractical, too costly, and difficult to administer. In the end, his comprehensive hydraulic system was not adopted. It fell to the increasing leverage of local authorities, who were unwilling to allow the state to dictate the type or extent of their public works programs, and to incrementalism because of budget limits or other constraints and priorities.

Chadwick's report and his subsequent attempts to implement the hydraulic system nonetheless marked a turning point in the establishment of modern sanitary services. For the first time, four essential criteria necessary for citywide service were united: a clear environmental context (health depended on sanitation), an administrative structure (the need for centralized public control), a substantial technical response (investment in new infrastructure), and recognition of the need for breadth of service delivery (via public health concerns). Not until the passage of the 1875 Public Health Law would these criteria converge, but Chadwick helped set them in motion.

The development of the new services occurred against a backdrop of important public health legislation in the mid- to late-19th century. The Public Health Act of 1848 was viewed by many as the culmination of Chadwick's sanitary work, although his influence on the movement itself had waned by the late 1840s. The act, nonetheless, marked the first time that the British government took responsibility for protecting the health of its citizenry. Along with the Sanitary Act of 1866, the 1848 law was the departure point for developing modern legal machinery for dealing with sanitation—no longer to be subject only to common-law notions of nuisance.[31] The legislation introduced the legal concept of the "statutory nuisance" to cover, in theory at least, all of the environmental ills that Chadwick and his cohorts had identified in the 1842 report.[32]

In another sense, the 1848 act failed to meet Chadwick's own expectations about administering sanitation reform. Anticentralizers won concessions, since the act failed to establish a national framework of local authorities. Local boards of health were to be established only if more than 10% of the ratepayers petitioned for one or if the death rate exceeded 23 per 1,000 persons. Also, boards were permitted (but not required) to appoint medical officers and were permitted (but not obliged) to undertake paving, sewerage, and water-supply programs.[33] A relatively weak Central Board of Health was created, but its term initially was limited to five years.[34]

In some respects Chadwick's own actions, or inactions, contributed to the success of the anticentralizers. The report set forth no clear idea on how central authority would be utilized to implement the sanitary plan, or how it was superior to local action past and present. Chadwick biographer Anthony

Brundage suggested that Chadwick's reluctance to deal with [central coordinating authority] directly perhaps derived from his fear of appearing once again as the power-hungry centralizer, a charge that had been hurled at him repeatedly." His first priority was to gain support for his analysis of public health conditions, then "there would follow as a matter of course an administrative machine that only Chadwick could properly direct." If this was his plan, it failed to materialize and the anticentralizers won the day.[35]

Several pieces of legislation followed the 1848 act, many of which addressed the specific issues related to sanitary services, especially sewerage. However, the true consummation of the work of the sanitary movement was the Public Health Act of 1875, which came two decades after Chadwick's withdrawal from public participation in the movement. Almost all earlier legislation was consolidated (twenty-two acts of Parliament) and extended in what must be regarded as a remarkably comprehensive sanitary code. With the enactment of the 1875 act, further sanitary legislation ceased for many years. The new law was the broadest articulation of public health thinking prior to the emergence of bacteriology in the late 19th century, and provided major impetus for the principles inherent in environmental sanitation.[36]

For more than a decade after the publication of the 1842 report, Chadwick's public life consisted of attempting to establish a new sanitary system, grappling with legislation, and generally trying to sustain his career in the field. Historian Christopher Hamlin persuasively argued that the debate over Chadwick's arterial plan ended "not ... with the triumph of Chadwick's system, but with an affirmation of the flexible, client-driven practice that characterized British engineering." In other words, it was the practice of British engineers to deal with problems defined by their clients—not to advocate their own technical visions— and to operate in an environment where there was no single technical solution. Chadwick's system assumed a single solution to the problem of urban sanitation and, as Hamlin concluded, "So well integrated were the components of Chadwick's 'arterial-venous' system that there was no clear place to begin a design.[37]

The debate over an integrated sanitary system, nonetheless, had a bearing on the development of English sanitary services and the diffusion of English public works concepts throughout the world. While an abundant, pure water supply was central to the Chadwickian system, and was a subject of continuing concern in England, primary attention turned to sewage disposal and sewage utilization. Refuse disposal, more narrowly focused on public street cleansing and private responsibility for disposal, was never an integral part of Chadwick's sanitation plans. In some respects, garbage and other solid wastes were treated more as nuisances than health threats at this time, compared with disease-bearing effluent in sewers or watercourses.[38]

The principles of acquiring and distributing pure sources of water had been essentially worked out earlier in the century. While few communities satisfied their water needs by 1850, the approaches and mechanisms for doing so seemed to be well understood. If the water-supply issue found a place in the debate over sewerage, it had to do with the changing circumstances for sewage disposal aggravated by increased water usage. New central supplies were changing the habits of many urbanites in the 19th century as rising demand (washing, bathing, waterclosets) led to greater pumpage. The old system of sewage disposal simply was incapable of handling the amounts of effluent leaving the homes and businesses that had piped-in water.[39]

The Health of Towns Commission became the lightning rod for debate over sewage between 1842 to 1845, marking the beginning of the gradual adoption of the water-carriage system.[40] The problem of sewage flow had interested Chadwick primarily because existing public sewers and cesspools allowed surface water and what it carried simply to seep into the ground rather than depositing the material in some outfall, but also because Chadwick's arterial system theoretically offered a way to capture the sewage for use on agricultural lands, providing possible revenue for a variety of urban improvements.[41]

Chadwick required the help of engineers to implement his system. Civil engineering had only recently emerged as a recognizable branch of engineering proper in Great Britain. The chartering of the Institution of Civil Engineers in London in 1828 marked public recognition of that fact. What distinguished "civil" from other branches of engineering was the focus on infrastructure such as roads, bridges, and tunnels—and soon water supply and wastewater systems. The commercial and industrial expansion of Great Britain in the 18th century stimulated the development of new projects outside of government. Until the 1750s, the state—particularly the military— was the major patron of engineering throughout Europe.[42] Prior to the 19th century, however, British engineers engaged on "civilian" (civil) projects had begun to exchange ideas and develop a professional identity.[43]

Although the civil-engineering profession was relatively new in the mid-19h century, Chadwick's and Roe's notions concerning the arterial system challenged current engineering expertise. There were seven sewer commissions in Greater London (regarded by Chadwick, with some justification, as corrupt and inefficient). They were appointed bodies, which often protected the interest of architects, builders, and surveyors who served on the commissions. They were also "quasi-judicial bodies" which administered rather than built or controlled the sewers in their districts. Despite the fact that the link between the commissions and civil engineers was tenuous, Chadwick criticized most engineers for not embracing his hydraulic system, as he had criticized physicians for not practicing preventive medicine.

The Metropolitan Sanitary Commission, established in 1847 to reform the sewer administration of London, became Chadwick's vehicle for attacking the more traditional engineers. Because of his earlier reformist history, and through the encouragement of some of his recent followers, Chadwick's assault not only focused on the technical inferiority of his opponents' work, but on their failure to meet the moral obligations of sanitary improvement. He even abandoned some of his old allies, including Roe, whom he replaced for a time with John Phillips. The new technical expert was a bricklayer, self-taught in hydraulics, who had risen to surveyor of the Westminster Commission. Egg-shaped sewers were out, glazed earthenware pipes were in. Increasing the velocity of flow became Chadwick's chief obsession.

The 1849 Metropolitan Sewers Commission, also dominated by Chadwick, was established to build a system of sewers for Greater London. Roe was back in the fold by then, responsible for sewer-flow experiments. At the same time, the arterial system was under testing. But moving from design to construction proved difficult. The data collected did not support the approach Chadwick wanted to employ. His engineers found it necessary to deviate from his principles to achieve a practical solution to problems they encountered in construction. Chadwick's approach also was undertaken outside of London. The work was sanctioned by the General Board of Health, the body charged with building sanitary works and, according to Hamlin, "Chadwick's last stronghold." By 1852, when enough sewers were completed to evaluate the approach, the results were disappointing. Blockages and breaks in the lines were common.

Chadwick then returned to his proposal for strong administrative control, arguing that the way to keep the sewers clear was to educate the public on how to use the lines and to regularly inspect household connections. Most engineers and responsible administrative bodies did not look upon these ideas with much enthusiasm, and thus refused to endorse them. The gap between intention and execution doomed Chadwick's sewage program—at least in terms of winning over engineers to its unique solution to the problem of waste removal. Hamlin argued that Chadwick tried to change the purpose of sewers "from removing surface and soil moisture to spiriting away wastes." While both functions were required, Chadwickians were "reluctant to acknowledge a need for separate sanitary and storm sewers."[44]

Criticizing Chadwick for advocating a complementary system of disposal over a separate and unique one, however, ignores the fact that the sewer design debate was essential for determining how to cope with both liquid wastes and the increasing volume of water. Chadwick had made a strong case against viewing sewers only for drainage purposes, thus making sewerage a major issue in the battle for good sanitation. While his technical scheme, his abrasive assault on engineers, and his authoritarian approach to management worked against his free-flowing arterial system, he irreparably altered the perspective on the value of sewers.

Nonetheless, an incremental approach to building sewer lines, as opposed to developing an integrated water and sewer system, dominated public works practices for the remainder of the 19th and into the 20th-century. Chadwick's failure to complete a successful pilot program eroded his credibility. His plan's greatest shortcoming was its ambitiousness. Investments to be made on a technical system of this scale are difficult to imagine in an era when local authorities were just beginning to take command of their public works, and when the national government was just authorizing new sanitary rules.

While most English towns eschewed the particulars of Chadwick's integrated approach, sewer development per se flourished as the era of water-carriage systems began by mid-century.[45] Water-carriage systems achieved central importance especially after 1847 when Parliament gave local authorities power to discharge sewage directly into rivers or the sea. Prior legislation seemed to be more concerned, if concerned at all, with minimizing soil pollution near dwellings and abating nuisances in street drains.[46]

The spread of water-carriage systems demanded attention be given not only to drainage technology, but to outfalls, to the location of discharge points, and eventually to new and different forms of water pollution. Cities such as Birmingham and Manchester began to grapple with these issues relatively early.[47]

Following an epidemic of cholera, Parliament passed the Nuisance Removal Act of 1855, which established the Metropolitan Board of Works to develop an adequate sewerage system for London. The law made emphatic the requirement that sewers be constructed and sewer disposal be developed in some satisfactory manner. Joseph William Bazalgette became chief engineer to the Metropolitan Board of Works in that year and began the sewage project for London in 1859. The main drainage was virtually completed by 1865.[48]

Bazalgette had been a consulting engineer at Westminster, mainly involved with railways. At age 28 he gave up his work because of poor health. The following year, he returned as a staff member for London's Metropolitan Sewer Commission and soon was appointed engineer-in-chief.[49] Opposed to the Chadwickian system, Bazalgette instead proposed a series of main intercepting sewers for London, which ran east-west to catch discharges before they entered the Thames.[50] The discharge would then be redirected into outfalls far down river from the city. Flow by gravity was possible during low tide, but pumps would need to be employed during high tide to force the sewage into the river.

Location of the outfall was a point of controversy. While receiving support from the Metropolitan Board, Bazalgette's ambitious proposal made in 1856 was initially rejected. Two years later, the government reversed itself largely because of the so-called Great Stink of 1858. Hot weather and the use of thousands of waterclosets created an ungodly stench lasting two years, caused by putrefying sewage caught in the tidal reach of the river. Crews on boats suffered from headaches and nausea, and sessions in Parliament were made bearable only by hanging sheets soaked in chloride of lime from each open window.[51] Dr. William Budd observed:

> For the first time in the history of man, the sewage of nearly three millions of people had been brought to seethe and ferment under a burning sun, in one vast open cloaca lying in their midst. The result we all know. Stench so foul, we may well believe, had never before ascended to pollute this lower air. Never before, at least, had a stink risen to the height of an historic event. [52]

Where engineering drawings and verbal persuasion had failed, an assault on the nostrils gave Bazalgette his victory. During the following two decades, approximately 83 miles of sewers were laid, draining 100 square miles of the city.[53]

The increased use of waterclosets and the implementation of the water-carriage system redirected pollution problems away from households and into the rivers and streams. The concern over tainted water supplies of a generation earlier was revisited in Great Britain by the adoption of technologies meant to vastly improve sanitary conditions in the cities. Chadwick's hydraulic system provided a theoretical framework for understanding the need to design a complete sewage-disposal mechanism built into a city's sanitation infrastructure. But the more common piecemeal development, such as Bazalgette installed, left "end-of-the-pipe" issues as an afterthought. The battles over sewerage, however, did produce salutary results in reducing the death rates. A study of twelve large towns in Great Britain before and after the adoption of sewerage systems indicated a drop from 26 per 1,000 deaths to 17 per 1,000 deaths.[54]

Optimism over the use of sewage as agricultural fertilizer never materialized to the extent that Chadwick envisioned, even though his idea was supported by the Local Government Board and many official studies. By 1880, about 100 towns had tried irrigation on sewage farms. The crops responding to repeated applications were limited, and land along the periphery of towns once only relegated to sewage irrigation often became more valuable for other purposes.[55] Several English farmers of the time turned to new concentrated fertilizers, such as artificial superphosphate or South American guano, rather than to the constant-flowing diluted sewage fertilizer that, as Hamlin stated, "kept coming whether it was needed or not." Other available solutions were expensive and not necessarily effective, such as the precipitation or flocculation process by which chemicals were added to liquid sewage to recover fertilizing material, or the collection of wastes in a dry form.[56]

Not until the 1880s and 1890s did Victorians begin to understand sewage treatment as a biological process. William Joseph Didbin, a chemist employed by London's Metropolitan Board of Works, was one of the first people to suspect that bacteria might be related to sewage purification.[57]

More often than not, sewage pollution became

as much a political and jurisdictional issue as a technical problem. Pollution of streams and rivers was particularly critical in Great Britain because the land area of the country was so limited, the population along watercourses was dense, and sources of water were shared by several municipalities along the banks.[58] Discharging sewage into watercourses was cheap and practical for many towns that already invested the ratepayers' money in a water-carriage system. Once waste was flushed down a watercloset or otherwise left a house through pipes, the interest of the citizenry in final disposal dropped off sharply—especially if their waste floated downstream. Landowners along the rivers and downstream towns, of course, did not take kindly to such benign neglect.

Concern over river pollution predated the commission's report by twenty-five or thirty years, but the most intense debates began in the 1850s. Bill Luckin has drawn an interesting distinction between the debates over river pollution in London and those in northern industrial areas of England. The dominant economic and social structure in London at mid-century was decidedly nonindustrial; in particular, the city lacked a unified Liberal manufacturing class. As a result, Londoners looked upon river pollution from the vantage point of commerce and consumption rather than production. The growing suburban ring also insulated the city from rural economic interests and values. Thus, in London the most offensive source of pollution was human waste, not manufacturing waste. Attention turned primarily to the problem of waterclosets and household effluent, and the issue pitted a variety of local vested interests against each other.

In the industrial North and the west Midlands, the struggle between the new industrial bourgeoisie and those in government wanting to limit the power of the manufacturing class—or to gain some control over manufactures—focused attention on industrial pollution instead of human wastes. In essence, class interests as well as jurisdictional disputes complicated the process of confronting river pollution in 19th-century Great Britain.[59]

Because the issue of river pollution had national significance, inquiries, investigations, and new legislation were promoted to address the growing problem. A Royal Sewage Commission was appointed in 1857 to ascertain how to safeguard rivers and how to determine the best methods of disposing sewage. It stated that "the increasing pollution of the rivers and streams of the country is an evil of national importance,

which urgently demands the application of remedial measures." It recognized that "this evil has largely increased with the growing cleanliness and internal improvements of towns as regards water supply and drainage; that its increase will continue to be in direct proportion to such improvements." The final report in 1865 recommended land treatment of sewage, but it was unenthusiastic about the profitability of sewage farming. The report asserted that towns causing pollution should cease to do so, and suggested that where cesspools were a health hazard, they should be replaced by a more modern sewerage system. Such a conclusion did not effectively address the polluting capabilities of water-carriage systems or provide an effective mechanism for change.[60]

The Rivers Pollution Prevention Act was passed eleven years later and became the basic water pollution law for seventy-five years. The law contained several safeguards and reservations to protect industrial interests, however, and the lines of authority to implement it were distributed among several authorities.[61]

The English experience with the development of water-carriage systems had mixed results. Homes and commercial establishments with piped-in water now had a means to discharge its effluent efficiently—clearly an advantage over the older cesspools and cesspits. But the cost of this new inner-city efficiency was displaced pollution problems along almost every major watercourse. Increased pollution load threatened the purity of the water supply, aggravated relations between upstream and downstream communities, and undermined a variety of competing uses for those watercourses.

Accompanying a cartoon of the filthy state of the Thames published in *Punch* in 1855 was a poem on "King Thames" which read in part:
King Thames was a rare old fellow,
He lay in his bed of slime,
And his face was disgustingly yellow,
Except where 'twas black with slime.
Hurrah! Hurrah! for the slush and slime.[62]

A year earlier, the widely recorded Broad Street Pump episode took place. Dr. John Snow, a London physician, had been studying the causes of cholera. In an 1849 pamphlet, "On the Mode of Communication of Cholera," he hypothesized that the disease was caused by an organic poison that could be discharged in human feces. If the infected feces entered the public water supply, he believed, an epidemic was sure to follow.

While investigating a severe outbreak of cholera near Broad Street, he learned that a workshop in the same area (which had its own well) reported no cases of the disease among its employees. This led Snow to seek out the polluted well. Simply by breaking the pump handle he ended the scourge. Snow had made the link between polluted water and epidemic disease, and helped to curb the spread of cholera and inspire others to study waterborne transmission of disease.[63]

The not -so-good news of events like the Broad Street episode was the continued threat to health posed by tainted water supplies. In this case, government authorities learned from the cholera epidemics in the late 1840s and early 1850s. In 1852, Parliament passed the Metropolis Water Act, which required that all water drawn from the Thames (and other rivers supplying the city with water) must be filtered by January 1856. In 1855, responding directly to Broad Street, municipal authorities in London required all water companies to supply filtered water. Within ten years, various British and European cities installed filters or filter galleries.[64] By the standards of the day, filtering water was the best way to insure a "safe" water supply. Not until the 1880s and the advent of bacteriology would it become apparent that filtration, while valuable in combating many pollutants, could not in and of itself stave off waterborne disease.[65]

In grappling with efforts to secure a pure water supply and to provide effective sewage disposal in the 19th century, the English strongly influenced the development of sanitary systems on the Continent and especially in North America. Technological improvements, such as waterclosets, pumps, sewage piping, and slow sand filters set the standard for the day. Legislative and court action established benchmarks or, in some cases, demonstrated the limits of governmental and judicial action in an era when laissez-faire notions were competing with demand for greater utilities regulation and other service functions.

The development of new sanitary services in England in the mid-19th century left an important legacy, one that profoundly influenced the United States. First, the correlation between the sanitary idea and the need for proper sanitation infrastructure and services became gospel. A pure water supply was not just a convenience, but a necessity for good health. Whatever its variety of uses—for fire protection, for industrial production, or for personal consumption—acquiring sufficient quantities of water had to be matched with acquiring a high quality of water (at least in contemporary terms). Proper sewerage also had powerful links to health. Removal of wastes from the home would insure protection from waterborne diseases and putrefying of wastes that were understood to cause disease.

The second legacy was less definitive, but still vastly significant: the manner in which new sanitary services were delivered. Chadwick's hydraulic system, in retrospect, was clearly the most environmentally sophisticated notion of its time. The linking of sanitary functions into a singular, closed system was elegant, but proved impractical.

Many engineers were not comfortable with developing a system that fell outside their normal relationship with clients, where clients dictated the project and the engineers fulfilled the specific requests. Many engineers also were not convinced that the Chadwickian system could work if implemented, even if they had the authority and the funds to carry it out. Financial resources for such a massive construction project would have been extremely difficult to accumulate, let alone allocate even if costs were amortized over several years.

Most significantly, the kind of centralized authority that Chadwick envisioned to implement his system did not exist in a society infused with strong decentralizing tendencies, and where private companies played such a significant role in the delivery of services, most especially water supply. Because of the complex issues involved and the many vested interests, control and management of sanitary services in England were the product of a shared authority among local government, private companies, and Parliament— varying in approach, depending on the location.

Unfortunately, the failure to integrate—or at least coordinate—sanitary services may have delayed improvements in public health for some years. J. A. Hassan made that very point:

> The direct environmental benefits of increased water deliveries for sanitary purposes were probably limited before the whole range of water services, including sewage treatment and river conservancy, were modernized. Indeed, a poorly balanced program of sanitary reform might exacerbate public health problems in the short-term. 'The spread of piped water among the comfortable dirtied the environment of the poor.'[66]

At the very least, however, Chadwick's vision of bringing the serpent's tail into the serpent's mouth initiated a dialogue in which modern sanitary services could be addressed and evaluated in a meaningful way. He helped to establish an environmental context for public health reform through the sanitary idea. He raised the important question of developing a workable administrative structure to implement change. And he linked the demand for a pure and abundant water supply with the necessity to devise a complementary wastewater evacuation system.

While the English wrestled with converting the sanitary idea into a plan of action at home, it also helped to set in motion a sanitation revolution in other parts of the world, including the United States.

Notes ---

1. Charles-Edward Amory Winslow, *The Conquest of Epidemic Disease* (1943; New York: Hafner, 1967), 243.

2. Ann F. La Berge, *Mission and Method: The Early 19th-Century French Public Health Movement* (New York: Cambridge UP, 1992), xiii, 3, 283-84.

3. Asa Briggs, *Victorian Cities* (1963; Berkeley: University of California Press, 1993), 16-17.

4. Ibid., 19.

5. Anthony Brundage, *England's "Prussian Minister": Edwin Chadwick and the Politics of Government Growth, 1832-1854* (University Park: Pennsylvania State UP, 1988), 4·

6. Ibid., 4-5.

7. D. D. Raphael, "Jeremy Bentham," *The Collegiate Encyclopedia*, vol. 2 (New York: Grolier, 1971), 518-19.

8. Brundage, *England's "Prussian Minister,"* 7-11.

9. Ibid., 150-57. In 1889, Chadwick received a knighthood, which Brundage argued, "was a belated recognition of his many important contributions. It was not, however, a vindication of his concept of the role and structure of government, for his autocratic, highly centralized approach had been repudiated decades before." See p. 171.

10. Ibid., 35-77.

11. C. Fraser Brockington, *The Health of the Community,* 3d ed. (London: J. and A. Churchill, 1965), 29, 32-33; Margaret Pelling, *Cholera, Fever, and English Medicine, 1825-1865* (London: Oxford UP, 1978), 7; Winslow, *Conquest of Epidemic Disease,* 242-43; W. M. Frazer, *A History of English Public Health, 1834-1939* (London: Bailliere, Tindall and Cox, 1950), 13, 15.

12. Brundage, *England's "Prussian Minister,"* 79-81. See also other biographies of Chadwick, especially S. E. Finer, *The Life and Times of Sir Edwin Chadwick* (London: Methuen, 1952), and R.A. Lewis, *Edwin Chadwick and the Public Health Movement, 1832-1854* (London: Longmans, 1952).

13. Brundage, *England's "Prussian Minister,"* 83-84.

14. See Briggs, *Victorian Cities,* 21; John H. Ellis, *Yellow Fever and Public Health in the New South* (Lexington: UP of Kentucky, 1992), 3, 5-6.

15. Edwin Chadwick, *Report on the Sanitary Condition of the Laboring Population of Great Britain,* ed. with an introduction by M. W. Flinn (Edinburgh: UP, 1965), 1. Studies by some of Chadwick's colleagues predate the 1842 report, and also helped to focus attention on public health problems. For example, Neil Arnott and James Kay issued *Prevalence of Certain Physical Causes of Fever in the Metropolis;* Thomas Southwood Smith contributed *Some of the Physical Causes of Sickness and Mortality (Which Are Capable of Removal by Sanitary Regulations)* and *Prevalence of Fever in Twenty Metropolitan Unions and Parishes;* and the report of the Select Committee of the House of Commons on Burials in Towns. See C. W. Hutt and H. Hyslop Thompson (eds.), *Principles and Practices of Preventive Medicine,* vol. 1 (London: Methuen, 1935), 7.

16. Anthony S. Wohl, *Endangered Lives: Public Health in Victorian Britain* (Cambridge, Mass.: J. M. Dent, 1983), 7.

17. The relationship between personal well-being and the environment was lost on physicians prior to this time, but it had not dominated thinking about disease causation until after the 1842 report. See Roy Porter, "Cleaning Up the Great Wen: Public Health in 18th-Century London," in W. F. Bynum and Roy Porter (eds.), *Living and Dying in London* (London: Wellcome Institute for the History of Medicine, 1991), 69.

18. Chadwick's criticism of physicians in high administrative positions contrasted with his respect for Poor Law medical officers. See Pelling, *Cholera, Fever, and English Medicine,* 12-13.

19. D. B. Eaton, "Sanitary Legislation in England and New York,' paper read before the Public Health Association of New York, 1872, 6. In the early 19th century, public health was regarded as a local concern, and local boards of health tended to be ad hoc bodies created in times of epidemic. While reformers often derided borough governments as corrupt or inactive, the lack of funds made sustaining a board of health difficult. As M. W. Flinn stated, "The question of public health, in other words, raises important points of political principle. The fact that by the 20th century this difficult issue was decided in favor of state or local authority coercion of the individual should not allow us to overlook the long and frequently passionate struggles on this question during the 19th century which had to be resolved before local authorities could be forced and adequately equipped to tackle their public health problems successfully. There was, moreover, a reluctance to spend ratepayers' money on services which would not bring some obvious and immediate benefit to the ratepayers themselves; and ratepaying was normally confined to the occupants of better-quality housing." See M. W. Flinn, *Public Health Reform in Britain* (London: Macmillan, 1968), 14.

20. Daniel E. Lipschutz, "The

Water Question in London, 1827-1831," *Bulletin of the History of Medicine* 2 (Sept./Oct.1968): 510, 523-25; M. W. Flinn, introduction to *Report on the Sanitary Condition of the Laboring Population of Great Britain* (hereafter Flinn, introduction) 9-10; Brian Read, *Healthy Cities: A Study of Urban Hygiene* (Glasgow: Blackie, 1970), 9.

21. William Hobson (ed.), *The Theory and Practice of Public Health* (New York: Oxford UP, 1979), 4.

22. Bill Luckin, *Pollution and Control: A Social History of the Thames in the 19th Century* (Bristol: Adam Hilger, 1986), 4.

23. Pelling, *Cholera, Fever, and English Medicine*, 10.

24. Exactly how "filth" influenced disease transmission was the basis for speculation in these years. As Christopher Hamlin argued, Chadwick and Southwood Smith were concerned mainly with "toxic or asphyxiating concentrations of the inorganic products of decay." German chemist Justus von Liebig pioneered the notion of putrefaction itself as "the quintessential pathological process." Since it was impossible to tell when putrefaction was in its "pathological mode," sanitarians were obliged to remove decomposing matter and prevent its decay. Liebig's "zymotic analogy' came to dominate British sanitary science beginning in the mid-1850s, and proved to be an important link to the development of the germ theory of disease. See Christopher Hamlin, "Providence and Putrefaction: Victorian Sanitarians and the Natural Theology of Health and Disease," *Victorian Studies* 28 (Spring 1985): 381-86.

25. Benjamin W. Richardson, "The Future of Sanitary Science: Political, Medical, Social," *Nature* (July 5, 1877): 189-90.

26. Brockington, *Health of the Community*, 30; Frazer, *History of English Public Health*, 14- 15; Hobson, *Theory and Practice*, 4; *Felling, Cholera, Fever, and English Medicine*, 12; Flinn, introduction, 58-67.

27. Christopher Hamlin, "Edwin Chadwick and the Engineers, 1842-1854: *Systems and Antisystems in the Pipe-and-Brick Sewers War*," Technology and Culture 33 (Oct. 1992): 680.

28. Chadwick believed that the system could be funded by 30-year loans. See Finer, Life and Times, 226-29.

29. Read, *Healthy Cities*, 1012; W.H.G. Armytage, *A Social History of Engineering* (London: Faber and Faber, 1976), 140-41; Lewis, Edwin Chadwick, 33,52-53,58-59, 105. According to Brundage, aside from the technical promise of the new system, "it recommended itself to Chadwick for quite another reason—it would tend to divert the sanitary movement away from the medical concerns and solutions to one in which the emerging science of civil engineering would play a key role." This may be true, but whether it was intentional on Chadwick's part is uncertain. See Brundage, *England's "Prussian Minister,"* 81-82.

30. Quoted in Finer, *Life and Times*, 222.

31. See chapter 1 [of Melosi, The Sanitary City] for a discussion of nuisance law.

32. This legal refinement alone was not sufficient to protect the public from environmental hazards.

33. On local control of sanitary services, see Christopher Hamlin, "Muddling in Bumbledom: On the Enormity of Large Sanitary Improvements in Four British Towns, 1855-1885," *Victorian Studies* 32 (Autumn 1988): 59-60. Through an examination of six large municipal sanitary

improvement projects undertaken in British towns between 1860 and 1885, Hamlin attempts to reverse the view that local governments in England were "neglectful, incompetent, and obstructive" in efforts to bring about sanitary reform in the 19th century. See especially pp. 78-83.

34. Flinn, *Public Health Reform*, 31-32; Flinn, introduction, 1; Frazer, *History of English Public Health*, 108, no, 135; Albert Palmberg, *A Treatise on Public Health and Its Applications in Different European Countries* (London, 1895), 7-8; Brockington, Health of the Community, 35-38; Arthur J. Martin, *The Work of the Sanitary Engineer* (London: MacDonald and Evans, 1935), 5; Eaton, "Sanitary Legislation in England and New York," 15-18; Brundage, *England's "Prussian Minister,"* 113-33.

35. Brundage, *England's "Prussian Minister,"* 85. Brundage devoted a whole chapter to a curious phase of Chadwick's life in the mid-1840s. With a sense that the Tory government in power would not likely move ahead with sanitary reform—and thus would give him little chance for advancement—Chadwick became involved in the Towns Improvement Company scheme in 1844 and 1845. The company was meant to deliver services to towns in lieu of public actions. Chadwick's participation in private enterprise appeared to fly in the face of his commitment to Benthamism. Brundage argued that Chadwick's inability to find a place as an important public official in the Peel government essentially led him to pursue pragmatism over principle—at least until the company failed to attract major investors and an opportunity in government reemerged in the form of commissioner on the General Board of Health. See pp.101-12.

36. George Rosen, *A History of Public Health* (New York: MD Publications, 1958), 232; George

Newman, *The Building of a Nation's Health* (London: Macmillan, 1939), 24; Armytage, *Social History of Engineering*, 142, 244; Brockington, *Health of the Community*, 39-40; Martin, *Work of the Sanitary Engineer*, 6-8.

37. Hamlin, "Edwin Chadwick," 682, 695.

38. According to Brian Read, refuse disposal by public authorities "had scarcely begun by the 1850s" and did not become general until the Public Health Act of 1875 made it compulsory. In the wealthier neighborhoods, private contractors—called "nightmen"—often collected night soil. It was typical for many people to store refuse in an ash pit, a structure made of brick or slate. Some communities—especially areas inhabited by the middle and upper classes—had regular street cleansing, and some mechanical sweeping machines were used in Chadwick's day, but it was more typical for streets to be cared for erratically and often to be used as dumps for all kinds of wastes. See Read, *Healthy Cities*, 57-60. See also Flinn, *Public Health Reform*, 11. John F. J. Sykes, *Public Health Problems* (London, 1892), 291; Albert Palmberg, *Treatise on Public Health and Its Applications in Different European Countries*, 116-17, 209; John V. Pickstone, "Dearth, Dirt, and Fever Epidemics: Rewriting the History of British 'Public Health, 1780-1850," in Terrence Ranger and Paul Slack (eds.), *Epidemics and Ideas: Essays on the Historical Perception of Pestilence* (Cambridge: Cambridge UP, 1992), 137.

39. See Flinn, *Public Health Reform*, 17-18.

40. Charles J. Merdinger, "Civil Engineering Through the Ages," *Transactions of the ASCECT* (1953): 22, 98.

41. Hamlin, "Edwin Chadwick," 683-84.

42. The French engineering tradition was older than the English, being a well-established profession in France by 1800. The civil engineering tradition, however, was better and earlier developed in Great Britain than on the Continent, and engineering was open to all classes. See Terry S. Reynolds, (ed.), *The Engineer in America* (Chicago: University of Chicago Press, 1991), 7-9.

43. Ibid., 8-9; Merdinger, "Civil Engineering Through the Ages," 3-4, 18-19; Martin, *Work of the Sanitary Engineer,* 24-26. Early in the history of Rome, the "aediles"—a combination of sanitarian and architect engineer—were responsible for supervising and maintaining the aqueducts and sewers of the city. Since they were employed by the state, they cannot be considered "civil" engineers in the same sense as the 19th-century British engineers under discussion. See Freedman, *Sanitarian's Handbook.*

44. Hamlin, "Edwin Chadwick," 681-706.

45. In 1842, Hamburg became the first city in Germany to introduce a well-designed sewerage system. The first modern main sewer was built in Paris along the Rue de Rivoli in 1851, and in 1856 the new sewer system planned by Belgrand was adopted. There were no modern sewers in Rome until 1871. See William Paul Gerhard, *Sanitation and Sanitary Engineering* (New York, 1909), 100.

46. George W. Fuller and James R. McClintock, *Solving Sewage Problems* (New York: McGraw- Hill, 1926), 3, 22-23; Hamlin, "Providence and Putrefaction." 393; Gerhard, *Sanitation and Sanitary Engineering*, 100.

47. Flinn, *Public Health Reform,* 40-41, 44.

48. Richard Shelton Kirby and Philip Gustave Laurson, *The Early Years of Modern Civil Engineering* (New Haven: Yale University Press, 1932), 231; H. B. Hommon, "Brief History of Sewage and Waste Disposal," *Pacific Municipalities* 42 (May 1928): 161; Leonard Metcalf and Harrison P. Eddy, *American Sewerage Practice,* vol. 1 (New York, 1914), 5, 10.

49. Read, *Healthy Cities,* 15, 20.

50. Previously, London's sewers ran north-south above the river and south-north below the city, discharging directly into the Thames.

51. Armytage, *Social History of Engineering,* 141; L.T.C. Rolt, *Victorian Engineering* (London: Allen Lane, 1970), 143; Flinn, *Public Health Reform,* 37-40; Fred B. Welch, "History of Sanitation," paper read at the First General Meeting of the Wisconsin Section of the National Association of Sanitarians, Inc., Milwaukee, Dec. 1944, 43, 45.

52. Quoted in Harold Farnsworth Gray, "Sewerage in Ancient and Medieval Times," *Sewage Works Journal* 12 (Sept.1940): 945.

53. Armytage, *Social History of Engineering,* 141.

54. S. H. Adams, *Modern Sewage Disposal and Hygienics* (London: E. and F. N. Spon, 1930), 52.

55. Nicholas Goddard, "Nineteenth-Century Recycling: The Victorians and the Agricultural Utilization of Sewage," *History Today* 31 (June 1981): 36.

56. Christopher Hamlin, "William Didbin and the Idea of Biological Sewage Treatment," *Technology and Culture* 29 (April1988): 191-92; Hamlin, "Providence and Putrefaction," 393-94.

57. Hamlin, "William Didbin," 189-218; Read, *Healthy Cities,* 26-33.

58. J. J. Cosgrove, *History of Sanitation* (Pittsburgh, 1909), 113.

59. Luckin, *Pollution and Control,* 49,141-43.

60. Metcalf and Eddy, *American Sewerage Practice,* 1:1-2; Kirby and Laurson, *Early Years,* 235; T.H.P. Veal, *The Disposal of Sewage* (London: Chapman and Hall, 1956), 2-4, 16-17; Frazer, *History of English Public Health,* 225.

61. Elizabeth Porter, *Water Management in England and Wales* (Cambridge: Cambridge UP, 1978), 26; F.T.K. Pentelow, *River Purification: A Legal and Scientific View of the Last 100 Years* (London: Edward Arnold, 1953), 9; Fuller and McClintock, *Solving Sewage Problems,* 29-30; William Oswald Skeat, ed., *Manual of British Water Engineering Practices,* vol. 1 (Cambridge: Heffer, 1969), 9; "The London Water Supply," *Engineering Magazine* 2 (Jan. 1870): 82-83; Clement Higgins, *A Treatise on the Law Relating to the Pollution and Obstruction of Watercourses* (London, 1877), 1-2; Julius W. Adams, *Sewers and Drains for Populous Districts* (New York, 1880), 39; W. Santo Crimp, *Sewage Disposal Works* (London, 1894), 7-30.

62. Quoted in H. W. Dickinson, *Water Supply of Greater London* (London: Newcomen Society at the Courier Press, 1954), 106.

63. Stuart Galishoff, "Triumph and Failure: The American Response to the Urban Water Supply Problem, 1860-1923," in Martin V. Melosi (ed.), *Pollution and Reform in American Cities, 1870-1930* (Austin: University of Texas Press, 1980), 38.

64. The water supply of Berlin was filtered by 1856. Filtration of water in Paris began as early as 1826. See M. N. Baker, "Sketch of the History of Water Treatment," *AWWA* 26 (July 1934): 905.

65. See Luckin, *Pollution and Control,* 35-37, 41, 45, 48. For a careful study of the development of water analysis, see Christopher Hamlin, *A Science of Impurity: Water Analysis in 19th- Century Britain* (Berkeley: University of California Press, 1990). According to Hamlin, prior to the 1890s experts claimed to be analyzing water, but without any sound scientific basis for determining contaminants. In most cases, analyses were not done by disinterested public health experts, but by those engaged in competition with each other and representing different interests. See pp. 3-9.

66. J. A. Hassan, "The Growth and Impact of the British Water Industry in the 19th Century," *Economic History Review* 38 (Nov. 1985): 543.

ACT TWO:
THE
CONTAMINATED
DEFEND

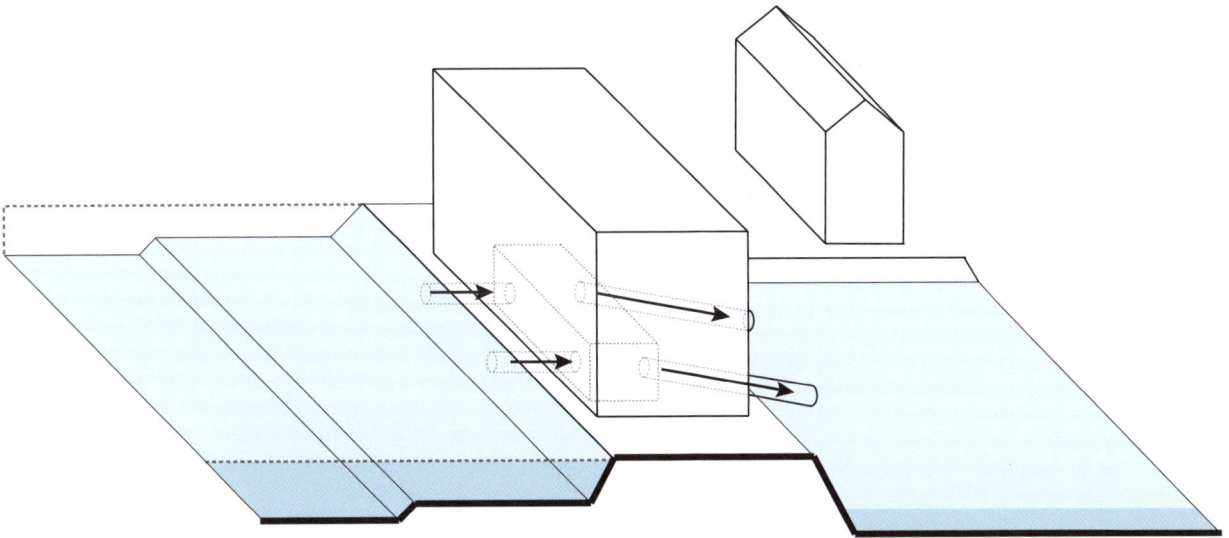

To construct a mechanism to keep contaminated water away from people, buildings, and cities.

Index of tools | shipping barricade, encircling booms, controlled burn, chemical dispersant, man-made sand berms

DEEPWATER HORIZON OIL SPILL

British Petroleum (Bp), Noaa, Etc. April 2010

Benjamin Gregory

On the evening of 20 April 2010, pressure caused by an escaped methane bubble in the drilling column of the Deepwater Horizon semi-submersible offshore drilling unit resulted in an explosion which engulfed the rig and ruptured the well-head at the sea's floor. Over two months, an estimated 4.9 million barrels of oil billowed from the earth's crust and into the Gulf of Mexico. It was the largest accidental marine oil spill in history, the complete damage of which has yet to be accounted for.

In the following days, BP officials and government authorities scrambled to find ways to stop the oil from leaking, and to clean the oil that had escaped into the water of the Gulf. The effort was on a scale that had never before been undertaken. Some 7,000 ships were deployed to contain the escaped oil. Hundreds of miles of booms were laid out, encircling the oil that had made it to the surface, and then skimming it into tankers. Other patches of oil were disposed of with a controlled burn, setting the water of the Gulf alight. Large plumes were doused with chemical dispersant, separating the oil into less harmful droplets, which were easier to absorb naturally. All this was done in an effort to keep the oil from inundating the sensitive beaches and estuaries along the Gulf's coastline. In total, 980,000 barrels were recovered, and 1,030,000 barrels were destroyed. The remaining 2.8 million barrels were either dispersed naturally or unaccounted for altogether. In comparison, the Exxon Valdez oil spill released about 750,000 barrels of oil. To prevent the oil from reaching land, a series of man-made sand berms

was proposed along the barrier islands off the shore of Louisiana, stretching 135 miles. Ultimately deemed too invasive and impractical, this large-scale engineering solution was abandoned, and only accounted for a small degree of the total impact the oil would have.

The impacts of the spill have only slowly been realized, and its full extent may never be known. The primary detriment has been towards marine life—the plants, fish, birds, and mammals which depend on the Gulf's waters, and live and feed along its seafloor and coastal estuaries. Unlike many other recorded oil spills, the oil that escaped from the Macondo well contained as much as 40% methane. Dissolved into the sea water, this amount of methane would create large zones of oxygen depletion and could interact with the water to release harmful chemicals, such as benzene, into the atmosphere. Large plumes of oil suffocated wildlife along the shore, in the water, and at the seabed. Massive amounts of chemical dispersant were used to break down the large plumes and make them less harmful. But these smaller droplets of oil made it easier for organisms to ingest them. Studies showed that zooplankton in the region contained oil compounds, which could be passed up the food chain affecting all the wildlife of the region, with unknown consequences, including human ingestion. As the oil combined with particulates in the water, it began precipitating to the sea bed, covering parts with tarry substances and perhaps altering the availability of food which many commercial fish species depend on.

Sources:
Federal Interagency Solutions Group, Oil Budget Calculator Science and Engineering Team; "Oil Budget Calculator: Technical Documentation" November 2010.
New York Times; "Map and Estimates of the Oil Spill in the Gulf of Mexico". August 1, 2010.

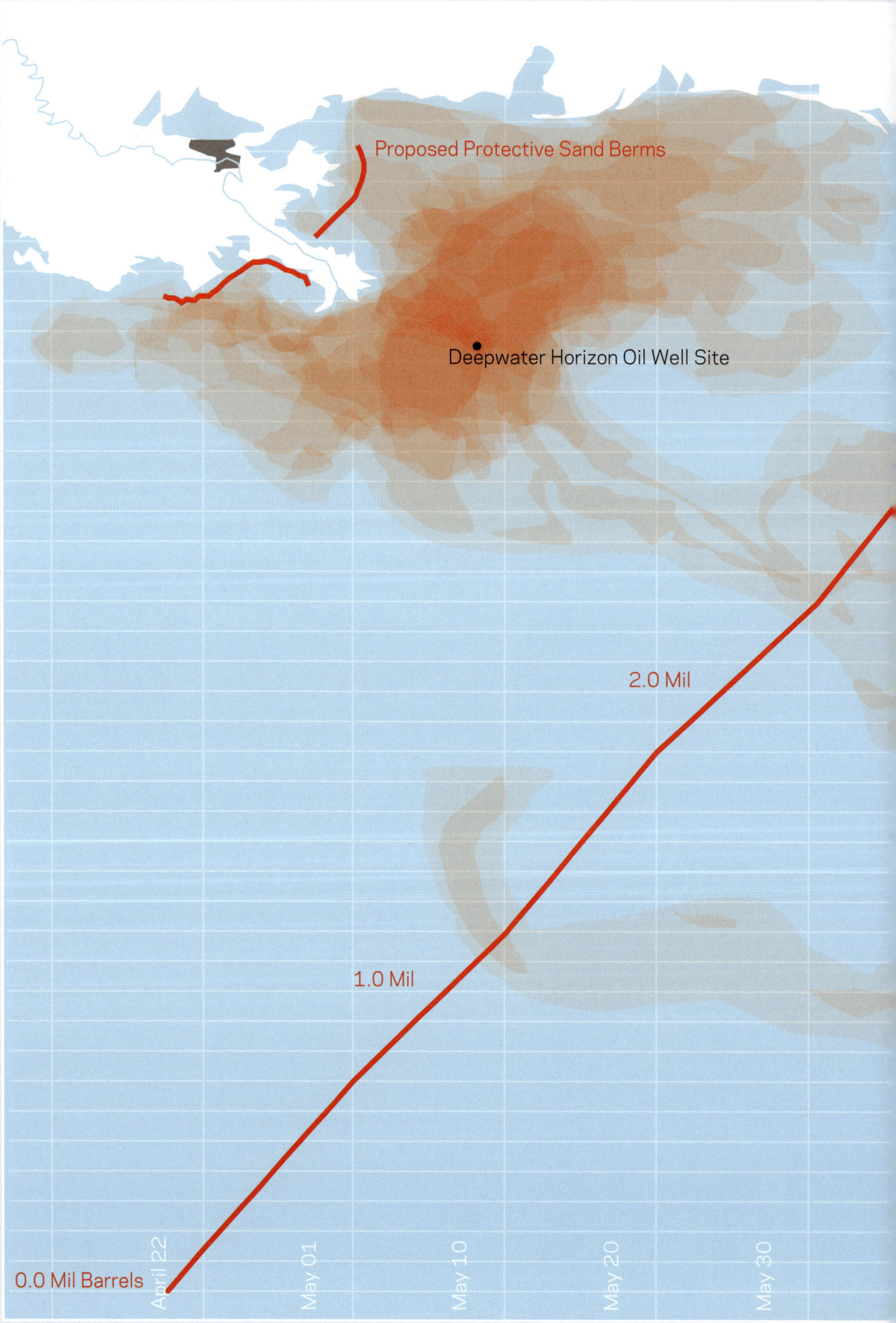

Proposed Protective Sand Berms

Deepwater Horizon Oil Well Site

2.0 Mil

1.0 Mil

0.0 Mil Barrels

April 22

May 01

May 10

May 20

May 30

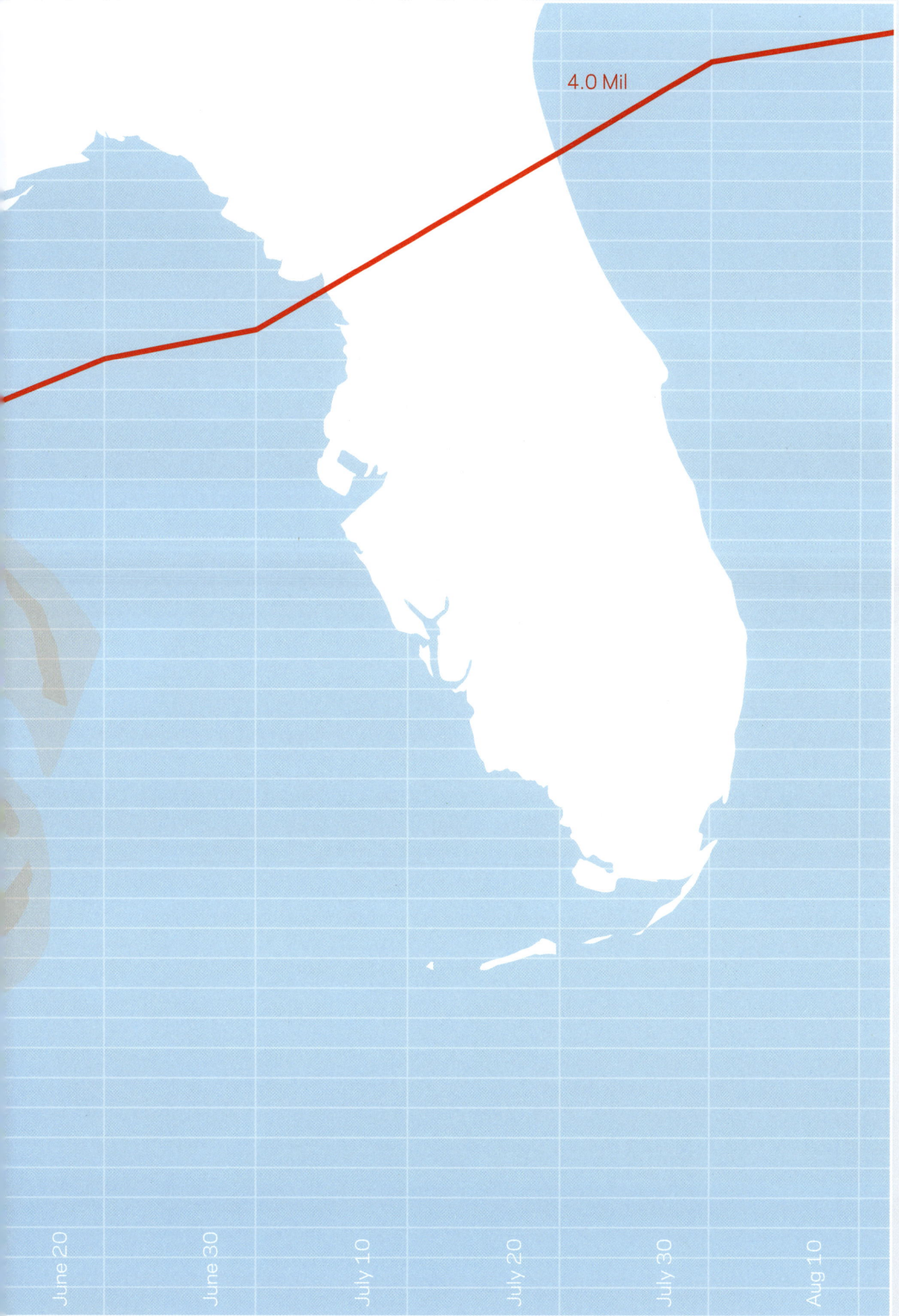

4.0 Mil

June 20

June 30

July 10

July 20

July 30

Aug 10

Like an inverted drop of water, the sliver expresses the workings of the plant below. Its shape creates a curvilinear interior space open to a large window view of the surrounding landscape while its exterior reflects the horizon in the landscape.
Below: Copyright Steven Turner
Above: Copyright Elizabeth Felicella

WHITNEY WATER PURIFICATION FACILITY + PARK

Connecticut, United States. Constructed 2005.

Steven Holl Architects and Michael Van Valkenburgh and Associates

This water purification plant and park uses water and its purification process as the guiding metaphors for its design. Its program consists of water treatment facilities located beneath a public park and a 360-foot-long stainless-steel sliver that encloses the client's public and operational programs. Like an inverted drop of water, the sliver expresses the workings of the plant below. Its shape creates a curvilinear interior space open to a large window view of the surrounding landscape, while its exterior reflects the horizon in the landscape.

The public park comprises six sectors that are analogous to the six stages of the water treatment in the plant. The change in scale from the molecular scale of the purification process below ground to the landscape above is celebrated in an interpretation of microscopic morphologies as landscape sectors. The park's "micro to macro" reinterpretation results in unexpected and challenging material-spatial aspects.

For example, in a field formed by the green roof, which corresponds to ozonation bubbling, there are "bubble" skylight lenses that bring natural light to the treatment plant below. In the landscape area corresponding to filtration, vine wall elements on trellises define a public entrance court. Following the natural laws of gravity, water flows across the site and within the purification plant. As the water courses through its turns and transformations toward its final clean state, it creates

microprogram potentials within the vast space of the new park. Aligned along the base of the sliver are water pumps that distribute clean water to the region.

Given the urgent need to manage and conserve water resources, this project is an example of today's best sustainable design measures and water shed management practices. Indeed, it even includes the enlargement of an existing wetland into a vibrant microenvironment that increases biodiversity.

Low Environmental Impact Technology and Sustainable Design Measures

Gravity Flow Operation
Setting the plant in the ground places the treatment process below lake level; this enables the purification plant to be driven by the lake's gravity pressure, eliminating the need for running energy-consuming pumps.

Renewable Energy
A groundwater heat pump system of 88 wells provides a renewable energy source for heating and cooling the building, and avoids the environmental impact associated with the use of fossil fuel energy, saving 850,000 kilowatt hours annually. The below-grade plant's large thermal mass generates stable temperatures and minimizes the need for

In a field formed by the green
roof, which corresponds to
ozonation bubbling, there
are "bubble" skylight lenses
that bring natural light to
the treatment plant below.
Courtesy Steven Holl.

air-conditioning. A close collaborative effort with the local energy company to minimize energy usage resulted in the plant receiving significant monetary subsidies or Energy Credits. The building systems use no HCFCs, CFCs, or halons.

Erosion and Sedimentation Control

The design team consulted the Connecticut Department of Environmental Protection, US Army Corp of Engineers, and Inland Wetland Committee to develop an extensive erosion-control/ plant dewatering strategy to prevent erosion.

Reduced Site Disturbance/Landscape

The design minimizes site disturbance by preserving existing wetland conditions and natural vegetation. The landscape design supports biodiversity and preserves natural habitats. The existing site wetland has been documented as a recess point for certain species of migrating birds; this important feature has been preserved and enhanced. Trees and bushes provide shading throughout. The majority of the plant species are native grass and low shrubs, which greatly reduce maintenance and irrigation costs.

Stormwater Management

The stormwater drainage system is managed through landscaping as opposed to piping. The surface pond to the east of the project is designed as a catchment area for detaining stormwater. While paved walkways and a plaza are part of the design, they have been minimized in lieu of a net decrease in the rate and quantity of stormwater runoff from the existing developed site.

The public park comprises six sectors that are analogous to the six stages of the water treatment in the plant. Courtesy Steven Holl Architects

ACT TWO: THE CONTAMINATED RETREAT

To move people, buildings, and cities away from contaminated water levels.

--- ---

Index of tools | local initiatives, build-destroy cycle, Re-choreographing debris, demolition rubble, phytoremediation plants.

SEEPING BOUNDARIES: INFORMAL INFRASTRUCTURES OF DIRT, DEMOLITION, AND SEWAGE IN THE WEST BANK

West Bank, Palestine. Speculative.

Suzanne Harris-Brandts

"Seeping Boundaries" explores the social, environmental, and political agency of sewage within the context of both the Middle East water shortages and the specific protracted conditions of Israeli military occupation in the Palestinian West Bank. The overall quantities of such liquid waste in this region are staggering— some 91 million cubic meters are produced every year by both Palestinians and Israeli settlers. On the Palestinian side, 90-95% of this West Bank sewage will not be treated at all due to a lack of functioning infrastructure. The responsibility for providing water infrastructure in the West Bank rests with the occupying Israeli state. However, in the absence of such state-provided projects, local initiatives will need to be introduced to mitigate the severe environmental and health concerns currently facing local Palestinians.

The proposed design therefore suggests a tactical method for intervention, paradoxically co-opting one of the occupation's most destructive forces to address this sewage problem—the Israeli demolition of houses. Throughout the West Bank a network of demolished Palestinian residential sites sit abandoned and are overlooked as a prospective resource for infrastructural development. Because some areas cannot be fully reconstructed due to the perpetual threat of their re-demolition by the Israeli army, the proposed design bypasses this relentless build-destroy cycle by preconceiving the agency that demolition rubble can provide in its current form. As such, reorganized piles of concrete debris are sorted on-site to create a new wastewater filtration system. Phytoremediation plants are added to further cleanse the streams of raw sewage, which would otherwise remain untreated and exposed. Such streams are ultimately rendered suitable for plant irrigation, offsetting the local water deficiencies for agriculture and the high cost associated with buying a trucked water supply.

Politically, the design challenges the severe construction restrictions that lead to the demolition of such privately owned Palestinian property in the West Bank, while bringing to the fore the urgent need for the approval of new infrastructure projects in the area. Activating demolition sites after their post-destruction abandonment, the design calls for the strategic rebuilding of such properties in a way that blurs the distinctions between "waste" and "resource" and between "land utilization" and "land abandonment," challenging the very definition of what constitutes construction and therefore reasserting the territorial claims of the indigenous Palestinian population to these otherwise off-limit sites.

"Rechoreographing" debris, the design reenvisions the potential hidden inside sewage and demolition rubble, seizing an opportunity within the West Bank's current volatile climate to provide water infrastructure by using one form of waste to remediate another.

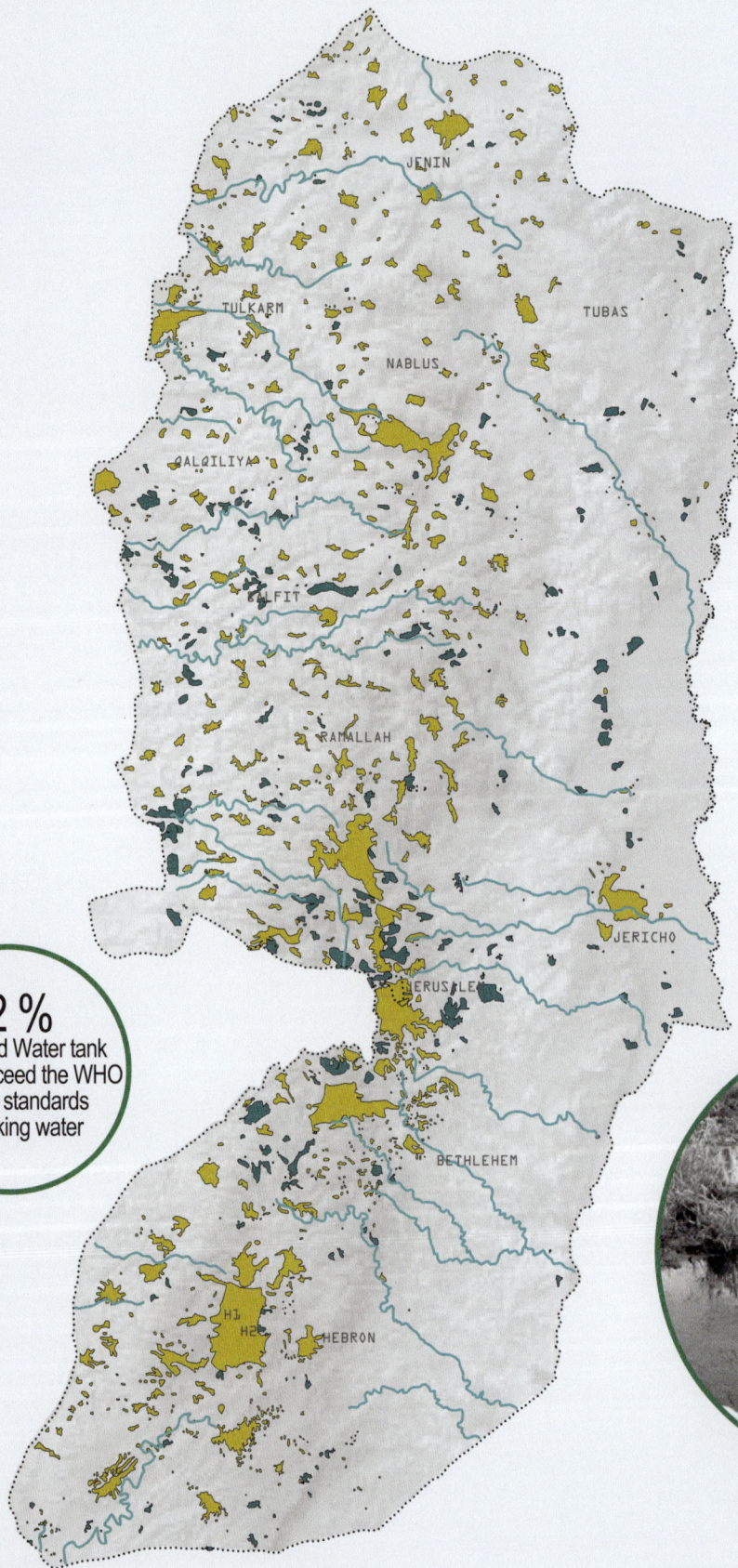

90-95 %
of Palestinian
wastewater is
untreated

JENIN

TULKARM

TUBAS

NABLUS

QALQILIYA

SALFIT

22 %
of Well and Water tank
samples exceed the WHO
bacteria standards
for drinking water

RAMALLAH

JERICHO

JERUSALEM

BETHLEHEM

H1
H2   HEBRON

● Urban Palestinian area   ● Urban Israeli settlement area   ▬ West Bank streams

Comparison of the sewage-
producing urban Palestinian and
Israeli settlement areas of the West
Bank to local natural streams.

BOUNDARY STRETCHED OUTWARD BY
UNAUTHORIZED PALESTINIAN CONSTRUCTION

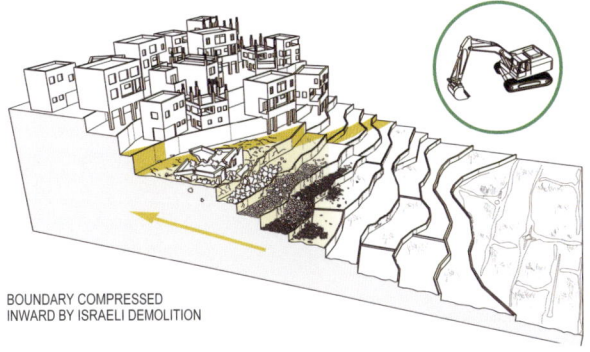

BOUNDARY COMPRESSED
INWARD BY ISRAELI DEMOLITION

BOUNDARY STRETCHED OUTWARD BY
PALESTINIAN DEMOLITION RE-ACTIVATION

Above:
Boundary Manipulations in the Build-Destroy Cycle:
The owners of Palestinian houses demolished at the threshold of Area C in the West Bank can reassert their territorial claims by introducing a new, ambiguous form of infrastructural "construction" vis-a-vis the re-sorted demolition waste.

Below
New ecologies emerge around the site of a house demolition being used for sewage dumping. Plants and debris serve as remediating agents, filtering the wastewater and turning it into a source for agricultural irrigation.

Palestinian House Demolition

Palestinian Village Sewage Stream

Sewage Filtering Phytoremediation Plants

Sewage Filtering Demolition Debris

Coarse Demolition Debris

House Demolition for
construction done without
Israeli approval

Medium-Coarse
Demolition Debris

Fine Demolition Debris

Israeli-Palestinian
jurisdiction boundary line
(Oslo areas A/B/C)

Urban Palestinian Area

Area B (Palestinian contruction approval)

Area C (Israeli contruction approval)

VILLAGE SEWAGE SOURCE

1

2

3

SEWAGE TREATMENT

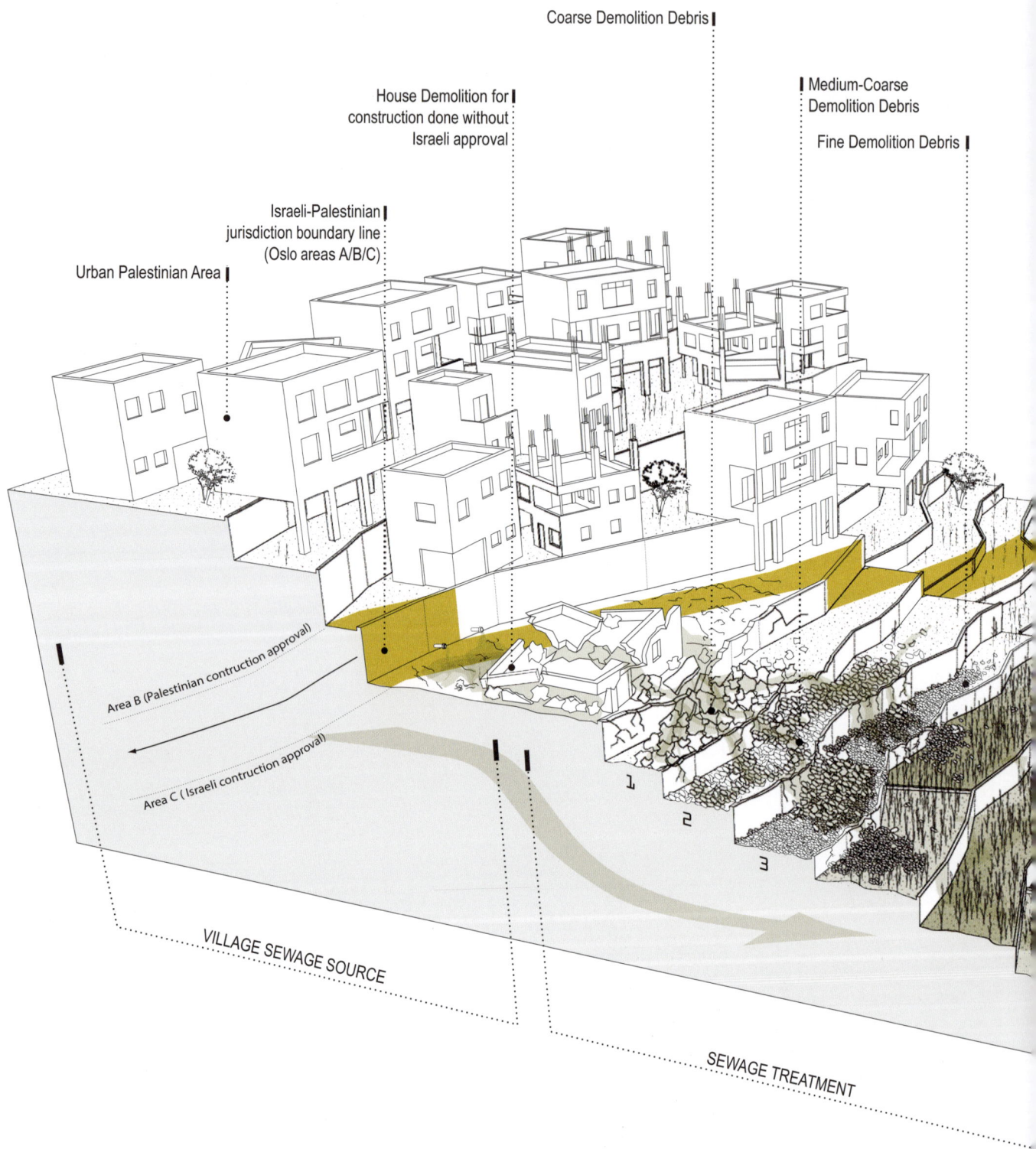

Wastewater Treatment Through
On-Site Demolition Reorganization:
The abandoned site of a demolished
Palestinian house (and its adjacent
affected terraces) is reorganized to
produce a new location of collective
wastewater infrastructure.

Filtering Vegetation

Olive Tree Terraces

Agricultural Fields

1 DEMOLITION BOULDERS

SIZE: LARGE / COARSE

LOCATION: TERRACE ONE

USE: FILTRATION OF LARGE DEBRIS

2 DEMOLITION ROCKS

SIZE: MEDIUM / COARSE

LOCATION: TERRACE TWO

USE: FILTRATION OF PARTICULATE MATTER

3 DEMOLITION DIRT AND GRAVEL

SIZE: SMALL / FINE

LOCATION: TERRACE THREE

USE: FILTRATION OF FINE PARTICULATE MATTER

AGRICULTURAL IRRIGATION

Demolition debris components.

Existing Hydrological Patterns The above aerial photography depicts New Jersey's highly urbanized river and stream systems along the New Jersey Turnpike. In response, the Army Corp. of Engineers has begun to identify at risk river corridors and establish a 300' building setback along these rivers--thus a establishing a strict line between built and hydrologic systems. Photo taken at Linden, New Jersey: Rahway River and Bayway Refinery

DENDRITIC ZONING: ESTABLISHING A NEW URBAN GRADIENT IN THE GARDEN STATE

New Jersey. Speculative, 2012.

Kimberly Garza

New Jersey's rapid urban development has shifted land-use patterns from forestlands and agriculture to suburbs, diminishing the state's vast river network and compromising the quality of its primary residential water source. This project demonstrates the redistribution of land-use patterns with respect to hydrological systems, and utilizes the New Jersey Turnpike Right-of-Way (ROW) as an opportunity to reconceive and deploy binary relationships between built and hydrological systems as an urban and hydrological gradient.

Background and Context

In the 20th century, access to mobility aided in the structuring of contemporary urbanism. Since the creation of the US National Highway System in 1956, highways have become the primary artery in the landscape, linking communities, towns, cities, states, and regions. This vast network has informed settlement patterns, generated and sustained economies, and defined regions. More significantly, in the latter half of the 20th century, highways have aided in the structuring of low-density urbanism. As a result of the 2009 financial crisis, highway rehabilitation has been targeted to stimulate local and regional economies and repair the 60-year-old infrastructure. Current highway planning proposes horizontal and vertical expansion to accommodate increased traffic. For example, the New Jersey Turnpike, a vital corridor for the Northeast region, is undergoing lane-widening projects, impacting adjacent urban and ecological communities.

Challenge

New Jersey, the most densely populated state in the US, is the site of low-density urbanization that resulted, in part, from the establishment of the New Jersey Turnpike in the middle of the 20th century. The state's rapid urban development has shifted land-use patterns from forest lands and agriculture areas to suburbs and urbanized hydrological systems, laying pressure on state agencies to preserve ecologically sensitive sites in the north and south of the state (the Highlands and Pinelands), and targeting central New Jersey as the primary corridor for "smart growth" development. In 2007, urban development surpassed New Jersey's largest land-use type, forestlands. However, central New Jersey is the site of the state's most prime agriculture land, giving the state its nickname, the "Garden State," and is composed of a vast network of rivers and streams—the primary water source for residences. As the area continues to urbanize, with urban growth rates over four times greater than the growth rate of the population between 2002 and 2007, New Jersey's agriculture areas are diminishing and the state's water quality is at risk. In response, New Jersey's Smart Growth plans for the state position urban development towards existing towns and cities and aim to preserve fragmented farm and forestlands; yet while doing so, they inefficiently support the integration of urban and ecological systems. Furthermore, the Army Corps of Engineers has begun to identify threatened stream corridors and to enforce a 300-foot buffer along rivers and

INTERCHANGE 9

INTERCHANGE 8A

INTERCHANGE 8

INTERCHANGE 7A

INTERCHANGE 7

INTERCHANGE 6

Activating The Right Of Way
The central portion of the NJ Turnpike is currently undergoing lane-widening projects to combat increase traffic demands--current expansion and construction highlighted in red. This project utilizes the NJ Turnpike ROW (and its current expansion due to lane widening projects) as an area of opportunity to deploy the redistribution of land use patterns. The design intervention focuses at newly proposed Interchange 8, experiencing significant infrastructural transformation.

streams to combat urban development—establishing a strict line between built hydrological systems.

Solution / Design

Dendritic Zoning demonstrates the redistribution of land-use patterns and transportation networks with respect to hydrological systems situated within a suburban environment, and reconceives existing binary relationships between the built environment and hydrological systems (common in suburban environments) as a hydrological and urban gradient. Rather than a fixed 300-foot buffer along streams and river networks, this approach rezones land uses according to hydrological adaptability. Concurrently, this project utilizes the New Jersey Turnpike ROW (and its current expansion due to lane-widening projects in central New Jersey) as an area of opportunity to redistribute land-use patterns. Specifically, the design intervention focuses on newly proposed

Interchange 8 as an ecological node, with site-specific agriculture (e.g., water-based agriculture: cranberry, blueberry, and rice fields), redistributed golf and recreation fields along expanded stream corridors, and multimodal infrastructure supporting Interchange car, bike, and pedestrian trails.

In a larger context, this project aims to promote the reevaluation of current land-use planning models with respect to regional ecologies and provides ways in which to evaluate regional landscapes and develop local landscape strategies. Extending beyond planning strategies, this project also repositions transportation networks as the platform for a multimodal network that reacts to local landscapes. No longer linear, transportation networks become ecological frameworks—generating new urban ecologies, morphologies, and multiple nodes of mobilities that support contemporary urban life.

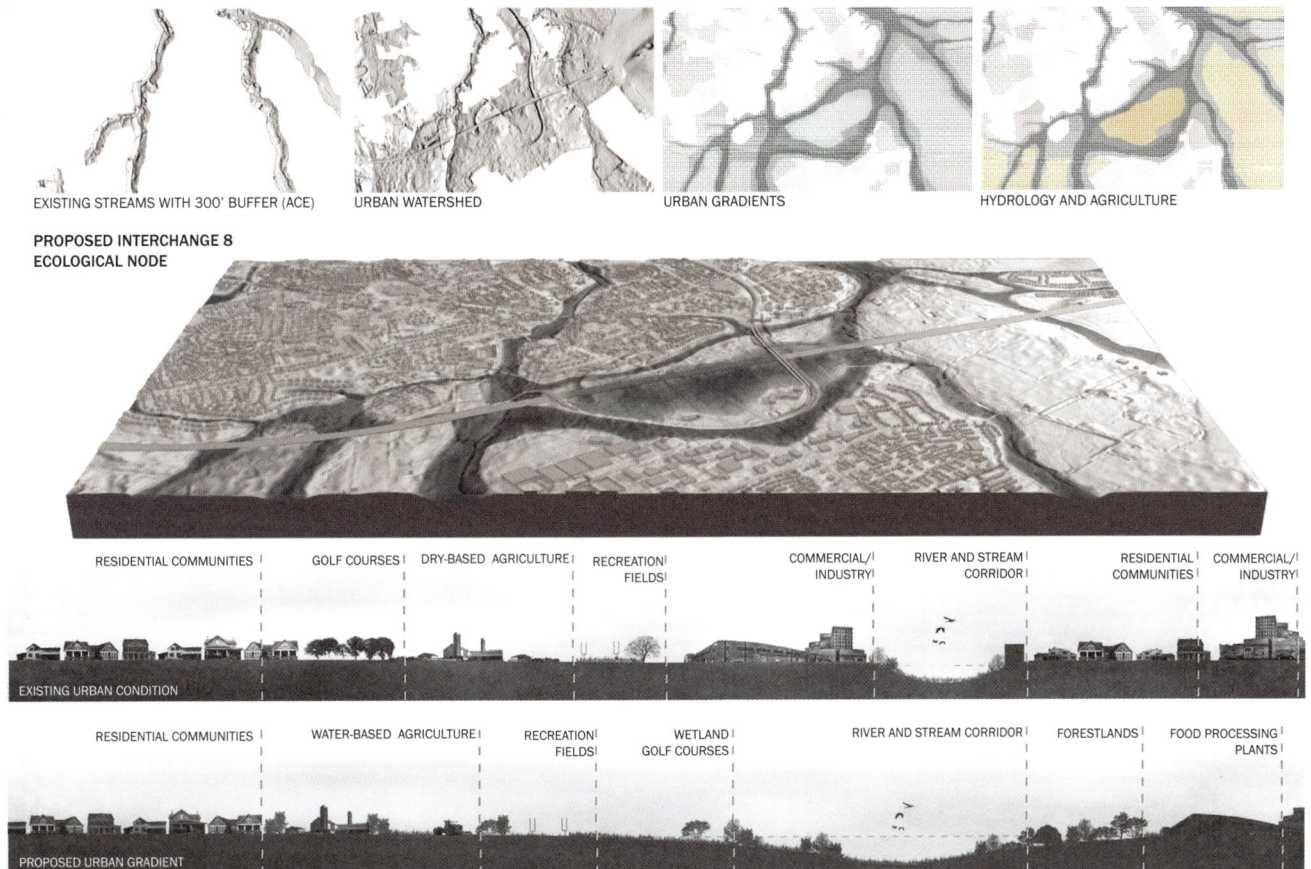

EXISTING STREAMS WITH 300' BUFFER (ACE) · URBAN WATERSHED · URBAN GRADIENTS · HYDROLOGY AND AGRICULTURE

PROPOSED INTERCHANGE 8 ECOLOGICAL NODE

Urban areas are re-organized along a reconceived urban gradient that positions land uses according to an expanded stream corridor based on flood adaptability. Urban areas follow existing dendritic stream patterns

ACT TWO:
THE
CONTAMINATED
ADAPT

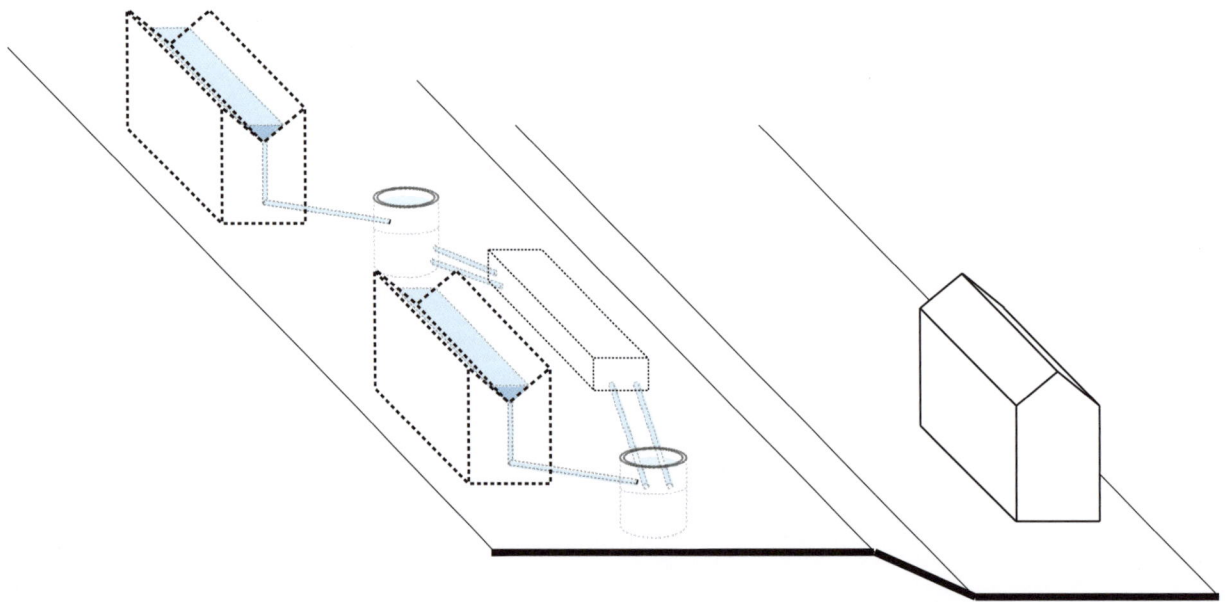

To allow contaminated water to enter the spaces of cities and communities, prompting the built environment to cleanse the water within proximity of point of use.

159

--- ---

Index of tools | stormwater park, green sponges, native biodiversity, filtration ponds, perimeter filtration

A GREEN SPONGE FOR A WATER-RESILIENT CITY: QUNLI STORMWATER PARK

Qunli New Town, Harbin, Heilongjiang Province, China. 2012.

Turenscape

Contemporary cities are not water-resilient, and inundations of surface water pose a substantial problem for them. Landscape architecture can play a key role in addressing this issue. Stormwater parks, connected and integrated into an ecological infrastructure across various scales, can act as green sponges, cleansing and storing urban stormwater.

Due to China's ever expanding urbanization and, arguably, to climate change leading to unpredictable precipitation, urban flooding caused by stormwater has become a global issue. In China, where most cities have monsoon climates, 70% to 80% of the annual precipitation falls in the summer, and in some extreme cases, 20% of the annual rainfall can happen in a single day. Beijing, for example, has an average annual precipitation of only about 500 millimeters (20 inches), but it received 50 to 120 millimeters (2 to 5 inches) of rainfall in just one day in 2011. Serious urban floods have been hitting the major cities in China even in times of normal rainfall, mainly because of an increase in impermeable paved surfaces.

Conventionally, people turn to engineering to solve urban flood problems, installing larger pipes, more powerful pumps, or stronger dikes. This single-minded approach is questionable for a number of reasons:

1. Economics: To construct an underground pipe system with sufficient capacity to drain away the extreme torrential rains is wastefully costly. It will also impose a large management and maintenance burden on future generations.

2. Water Shortages: There is a shortage of fresh water in China. In the metropolitan areas, the drop of the underground water table is a serious issue. Out of more than 660 Chinese cities, 400 are experiencing water shortages. In northern China, for example, the underground water table drops up to 2 meters (6 feet) each year. Beijing has seen its water table drop by 1.5 meters (4 feet), on average, every year for the past three decades, due to overuse of underground water with almost no aquifer recharge. All the stormwater has been drained away through pipes or channeled into rivers.

3. Ecological Services: Engineered stormwater drainage leads to the disappearance of surface-water features including water-based habitats, especially urban wetlands. In addition, much more irrigation is needed for parks and green space in the city when all this rainwater is drained away, and this worsens the water shortage problem.

Using the landscape as a sponge is a good alternative solution for urban stormwater management. An example of this approach is demonstrated in Turenscape's stormwater park in Harbin, which integrates large-scale urban stormwater management with the protection of native habitats, aquifer recharge, recreational use, and aesthetic experience, in all these ways fostering urban development.

Beginning in 2006, a 2,733-hectare (6,753 acres) new urban district, Qunli New Town, was planned for the eastern outskirts of Harbin in northern China. Thirty-two million square meters (344 million square

Using the landscape as a sponge is a good alternative solution for urban storm water management. An example of this approach is demonstrated in Turenscape's storm water park in Harbin.

feet) of building floor area will be constructed in the next 13 to 15 years. More than one-third of a million people are expected to live there. While about 16.4% of the developable land was zoned as permeable green space, the majority of the former flat plain will be covered with impermeable concrete. The annual rainfall there is 567 millimeters (22 inches), with the months of June, July, and August accounting for 60% to 70% of annual precipitation. Floods and waterlogging have occurred frequently in the past.

In mid-2009, Turenscape was commissioned to design a park of 34.2 hectares (84.5 acres) right in the middle of this new town. The site is surrounded on four sides by roads and dense development. This wetland had thereby been severed from its water sources and was under threat. Going beyond the original task of preserving the wetland, Turenscape reconnected water networks and transformed the area into an urban stormwater park that will provide multiple ecosystem services. It collects, cleanses, and stores stormwater and infiltrates it into the aquifer. The stormwater park has not only become a popular urban amenity but has also been upgraded to a protected National Urban Wetland park because of its improvement to ecological and biological conditions. Several design strategies and elements have been employed:

1. The central part of the existing wetland is left alone to allow the natural habitats to continue to evolve.

2. Earth is excavated and used to build up an outer ring—a necklace of ponds and mounds. This ring acts as a stormwater-filtrating and-cleansing buffer zone for the core wetland, and a transition between nature and city. Stormwater from the newly built urban area is collected into a pipe around the perimeter of the wetland and then released evenly into the wetland after having been filtered through the ponds. Native wetland grasses and meadows are grown next to ponds of various depths, and natural processes are initiated. Groves of native silver birch trees (Betula pendula) grow on mounds of various heights and create a dense woodland. A network of paths links the ring of ponds and mounds, allowing visitors the experience of walking through a forest. Platforms and seats are placed near the ponds to enable people to have close contact with nature.

3. A skywalk links the scattered mounds, allowing visitors to have an above-the-wetland experience. Platforms, five pavilions (bamboo, wood, brick, stone, and metal), and two viewing towers (one made of steel and located at the east corner, the other one made of

wood and resembling a tree at the northwest corner) are set on the mounds and connected by the skywalk, providing views into the distance and enabling the observation of nature in the center of the park.

These design strategies have dramatically transformed the site. It is now a park performing many functions: collecting stormwater, and cleansing, storing, and recharging underground aquifers. The park can retain and filtrate up to 500,000 cubic meters of stormwater, and has successfully solved the stormwater inundation problem for an area of 300 square kilometers (10 times the area of the park) in the past two years (2012 and 2013). This data also means that if a city can allocate 10% of its total area as a green sponge area for stormwater management, it can virtually solve the stormwater problem that is commonly seen in contemporary cities. Water quality has also been dramatically improved, from raw stormwater to the water in the central wetland after its being filtrated through the bioswales system at the periphery.

At the same time, the wetland habitat has been restored and native biodiversity has been preserved. Many native species of flora and fauna were observed taking places in the park right in the middle of the city. Through the transformation of this dying wetland, stormwater that frequently causes flooding has now become a positive environmental amenity in the city.

The working wetland park also creates a unique public space for recreation and aesthetic experience in an area of over 30 square kilometers for the residents in the new community. The land value surrounding the park increased dramatically (a 100% increase in the two years after the park was built).

The project was realized through the funding of the local government in charge of the development of Qunli New Town. Because of the application of a minimal design strategy (only the periphery of the park was constructed through simple cut-and-fill), no earth was imported and no dirt was exported, the cost of the project was minimized, and the total investment was 40 million RMB (about $6.7 million in US dollars), which is only about one-third of the budget of a normal urban park built in this region. The vegetation used in this park need low maintenance and are self-reproductive, which dramatically reduce the cost of the park.

Because of these dramatic improvements to the site, the stormwater park now has been listed as a National Urban Wetland Park. This project demonstrates an ecosystem services approach to urban park design, and is a showcase for sustainable

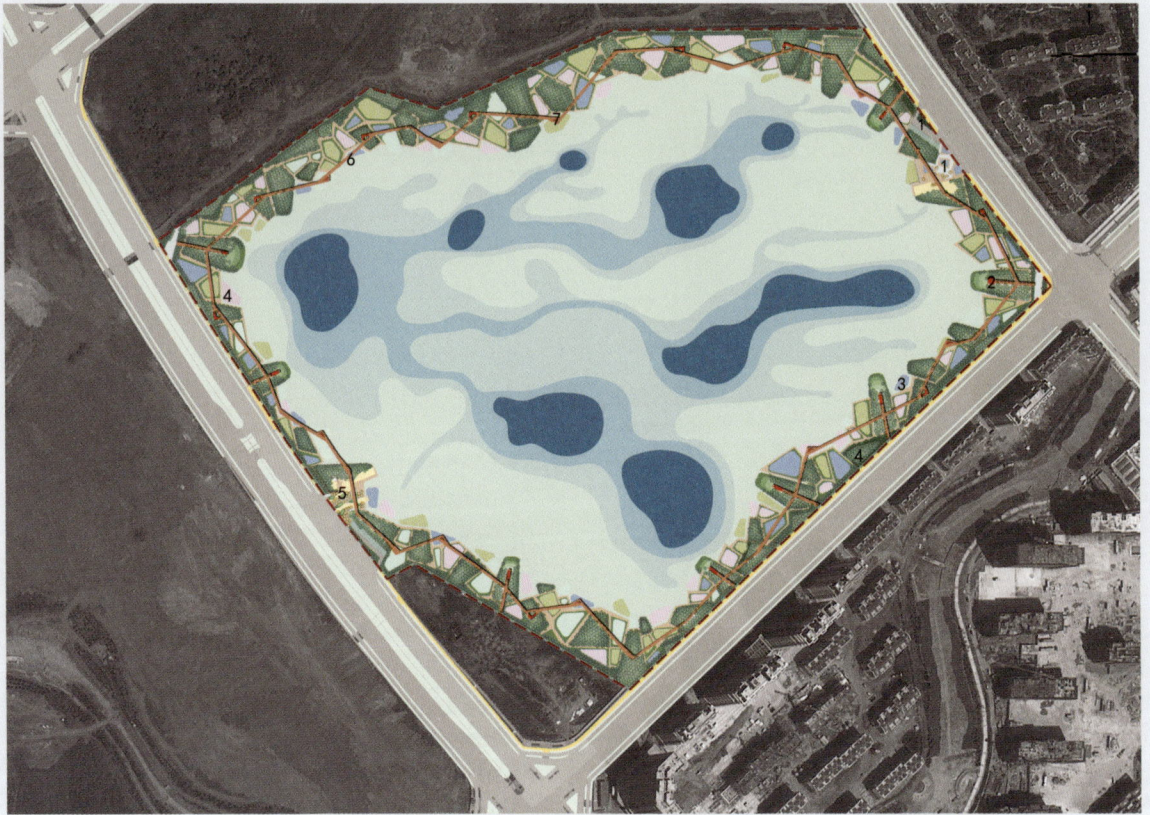

The central part of the existing wetland is left alone to allow the natural habitats to continue to evolve.

urban stormwater management. The sustainable strategies tested in this park are replicable, and the park has now become a model of sustainable urbanism, visited by urban decision makers and professionals from all parts of China and beyond, teaching them how the worrisome urban stormwater issue can be solved through an alternative green approach. The project has now been widely published and referred to as a sustainable practice around the world and has won several important international awards.

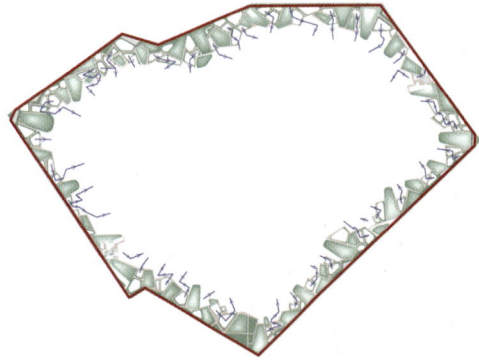

Right:
Earth is excavated and used to build up an outer ring—a necklace of ponds and mounds.

Below:
Elevation and Cross-section of Pavilion

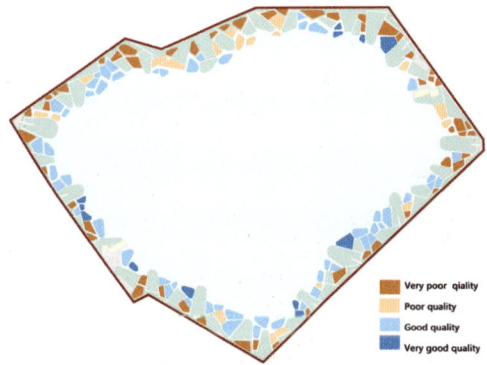

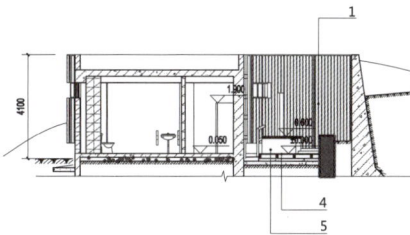

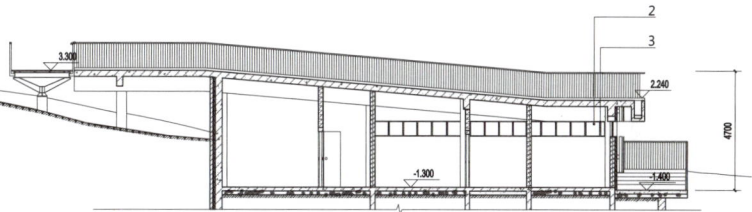

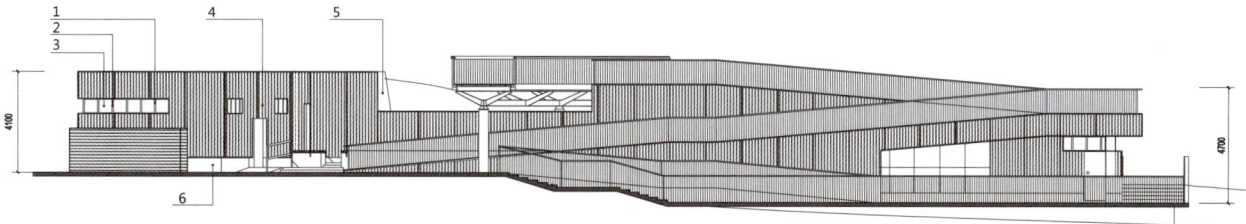

Very poor qiality
Poor quality
Good quality
Very good quality

A skywalk links the scattered mounds, allowing visitors to have an above-the-wetland experience.

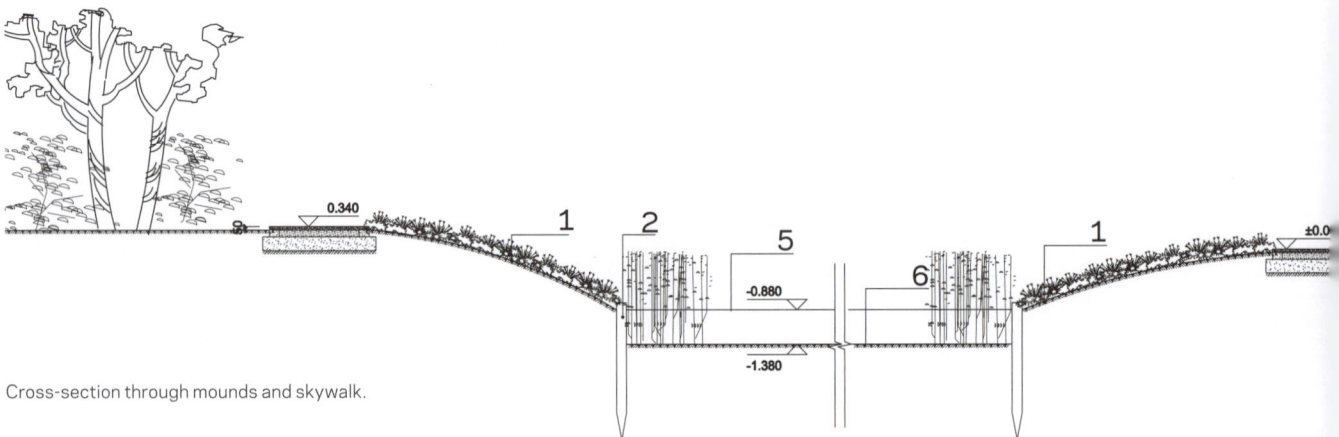

Cross-section through mounds and skywalk.

Sky Walk,Pavillions and Towers

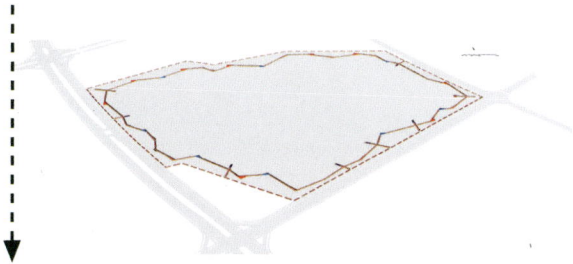

Grounel Level Path Network and Platform

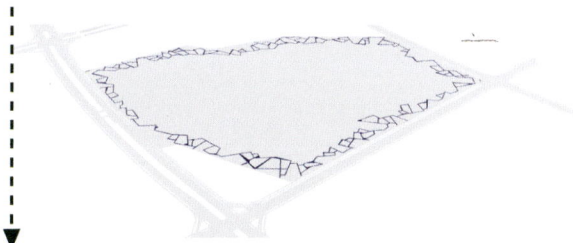

Fill Ring

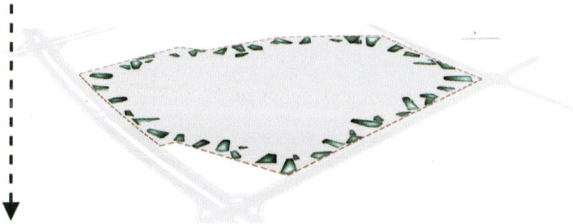

Cut Ring

Exiting Wetland

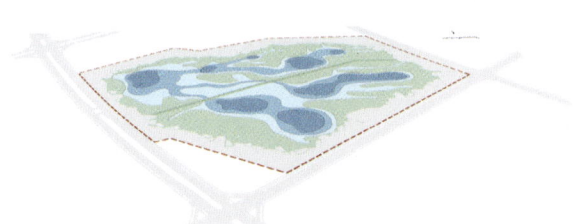

General Plan

Stormwater from the newly built urban area
is collected into a pipe around the perimeter
of the wetland and then released evenly
into the wetland after having being filtered
through the ponds.

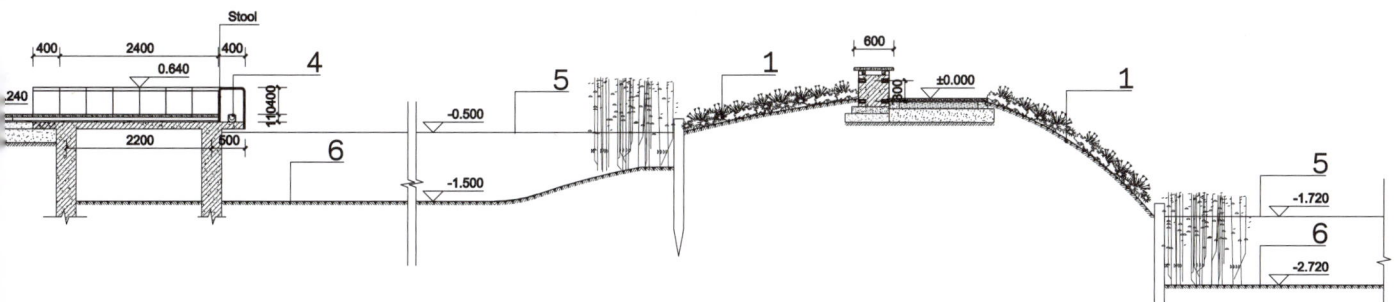

Wildlife and natural processes have
been restored to te site.

SHANGHAI HOUTAN PARK: LANDSCAPE AS A LIVING SYSTEM

Pudong, Shanghai, China. Constructed 2010.

Turenscape

Built on a former industrial brownfield, Houtan Park is a regenerative living landscape on Shanghai's Huangpu riverfront, a narrow, linear 14-hectacre band. Its constructed wetland, ecological flood control, reclaimed industrial materials, and urban agriculture are integral components of a sustainable design to treat polluted river water and recover the degraded waterfront in an aesthetically pleasing way.

The site has multiple challenges including the degraded environment, polluted river water, and flood control that requires over 4 meters of floodwall protecting against a 1,000-year flood event. The objective was to transform the brownfield into a public park while at the same time demonstrating sustainable urbanism and green technologies.

A linear, constructed wetland was designed between the 20- and 1,000-year flood event levees to treat contaminated water from the Huangpu River. Cascades and terraces covered with selected species of wetland plants are used to oxygenate the nutrient-rich water, removing sediments while creating pleasant water features. Each day 2,400 cubic meters of water can be treated to improve from Lower Grade V to Grade III. Crops and native plants were selected to create a productive and low-maintenance landscape. The design preserves, reuses,

and recycles the existing industrial structures and materials to celebrate the site's industrial past.

The park has proven to be a great success in many ways, including providing native habitats for biodiversity and offering recreational uses, as well as demonstrating ecological solutions to cleanse water and manage floods.

Huangpu River

The site

A linear constructed wetland was designed between the 20 and 1000-year flood event levees to treat contaminated water from the Huangpu River.

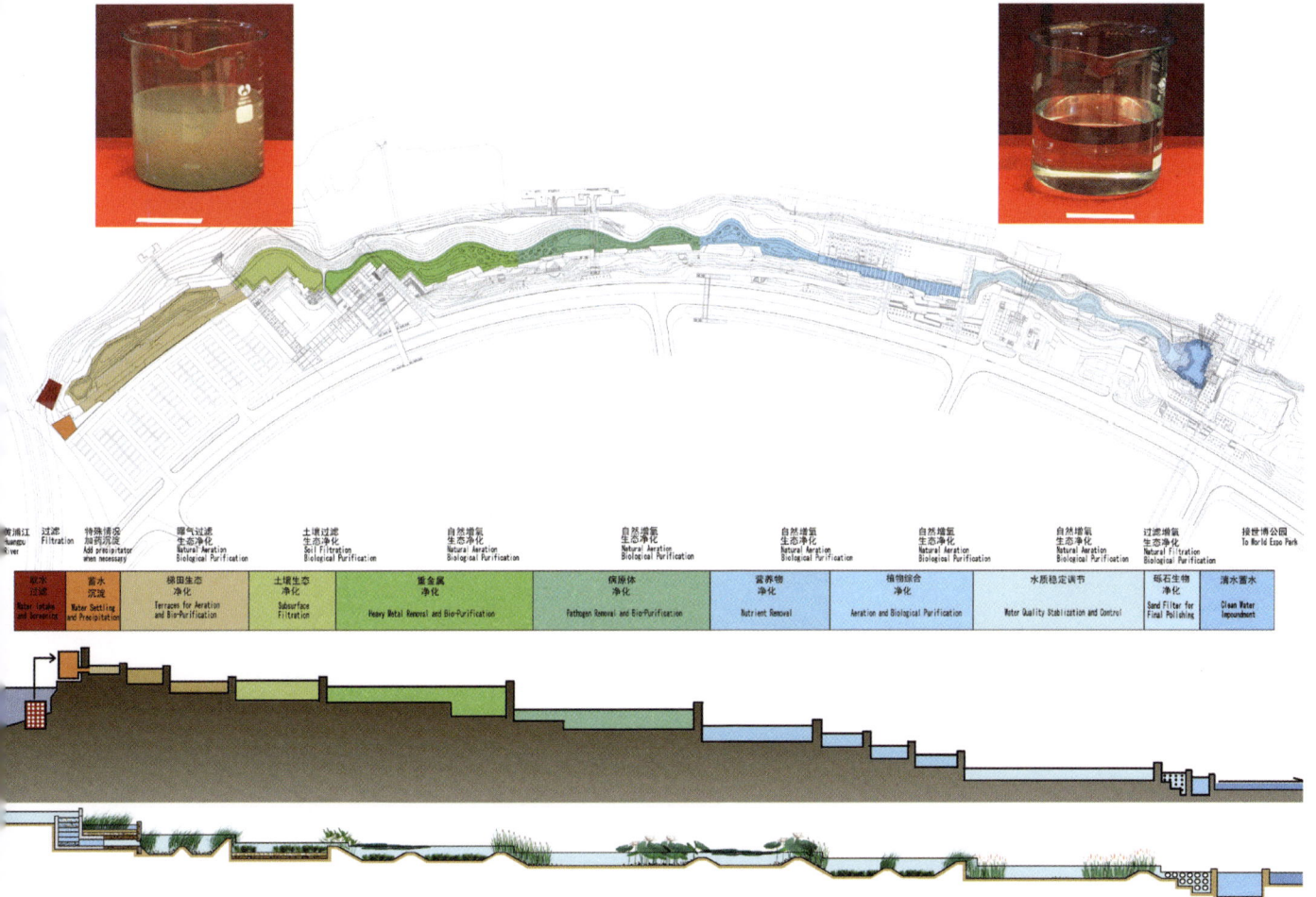

黄浦江 Huangpu River	过滤 Filtration	特殊情况 加药沉淀 Add precipitator when necessary	曝气过滤 生态净化 Natural Aeration Soil Filtration Biological Purification		土壤过滤 生态净化 Soil Filtration Biological Purification	自然增氧 生态净化 Natural Aeration Biological Purification	自然增氧 生态净化 Natural Aeration Biological Purification	自然增氧 生态净化 Natural Aeration Biological Purification	自然增氧 生态净化 Natural Aeration Biological Purification	自然增氧 生态净化 Natural Aeration Biological Purification	过滤增氧 生态净化 Natural Filtration Biological Purification	接世博公园 To World Expo Park
取水 过滤 Water Intake and Screening	蓄水 沉淀 Water Settling and Precipitation	梯田生态 净化 Terraces for Aeration and Bio-Purification	土壤生态 净化 Subsurface Filtration		重金属 净化 Heavy Metal Removal and Bio-Purification	病原体 净化 Pathogen Removal and Bio-Purification	营养物 净化 Nutrient Removal	植物综合 净化 Aeration and Biological Purification	水质稳定调节 Water Quality Stabilization and Control		砾石生物 净化 Sand Filter for Final Polishing	清水蓄水 Clean Water Impoundment

Cascades and terraces covered
with selected species of wetland
plants are used to oxygenate
the nutrient-rich water removing
sediments while creating pleasant
water features.

Above:
The man-made terraced wetland
under construction (view from
southwest corner toward the north).

Below:
Construction photograph of banded
wetland.

Opposite:
The terraced wetland is inspired by
the rice paddies found in traditional
Chinese agriculture, where water
and fertilizers are retained to
nurture the crops slowly. In our
case, the plants absorb elements
from the nutrient-rich river water.

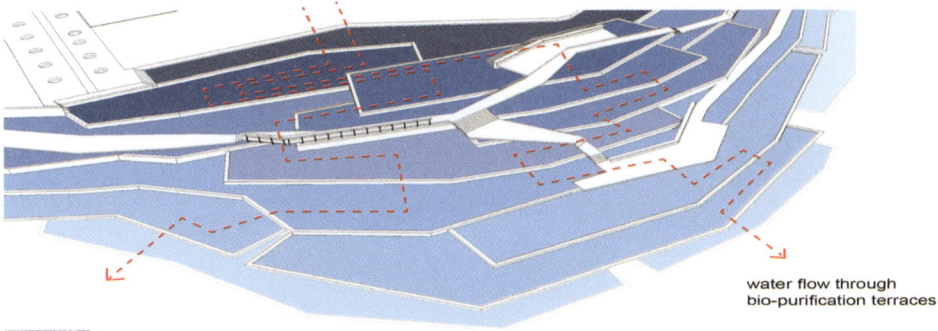

water flow through
bio-purification terraces

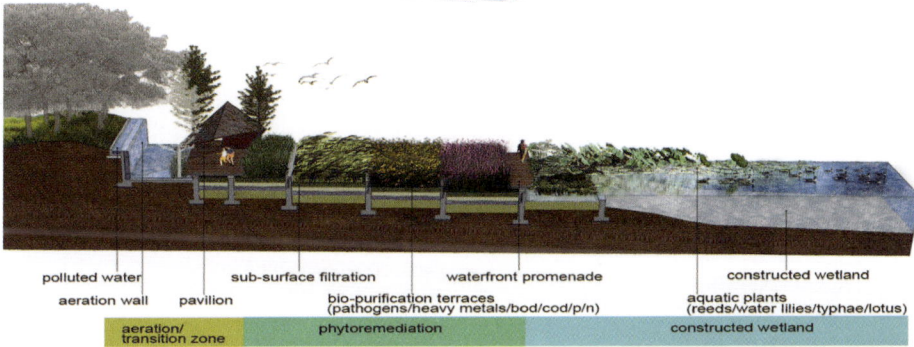

Spring

Summer

polluted water   sub-surface filtration   waterfront promenade   constructed wetland

aeration wall   pavilion   bio-purification terraces   aquatic plants
                                   (pathogens/heavy metals/bod/cod/p/n)   (reeds/water lilies/typhae/lotus)

| aeration/ transition zone | phytoremediation | constructed wetland |

Fall

Site plan showing the water treatment sequence

Source

Flow

Sink

Winter

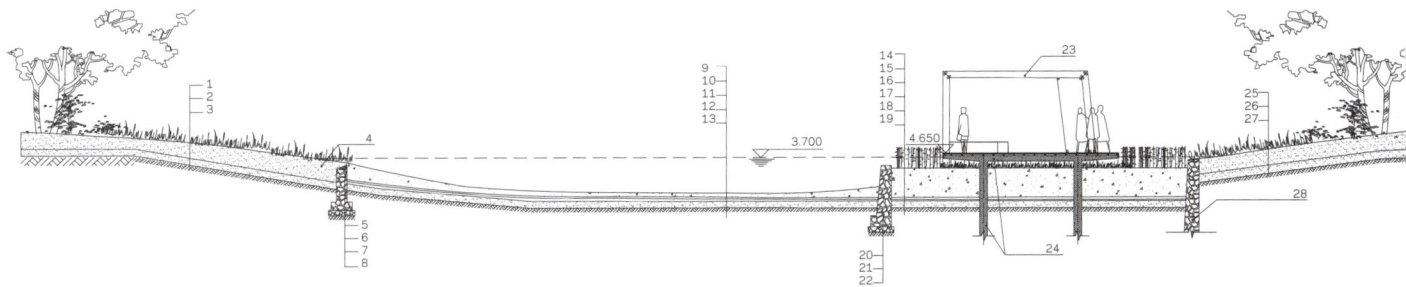

3.700

4.650

01. back fill planting soil layer (minimum thickness 300mm)
02. clay compaction, 250mm (slope ratio ≦ 1:2)
03. prime soil compaction
04. full with mound
05. reinforced concrete top
06. rubble, 400mm
07. graded gravel, 200mm

08. prime soil compaction
09. sand, 150mm (zeolite stone 3-5/m², each piece 80mm-120mm wide)
10. gravel, 50mm
11. sand, 40mm
12. clay compaction, 250mm
13. prime soil compaction
14. wetland plantation

15. sand
16. gravel, 50mm
17. sand, 40mm
18. clay compaction, 250mm
19. prime soil compaction
20. rubble, 400mm
21. graded gravel, 200mm

22. prime soil compaction
23. origami ribbon
24. reinforced concrete pillar and plate
25. back fill planting soil layer (minimum thickness 300mm)
26. clay compaction, 250mm (slope ratio ≦ 1:2)
27. prime soil compaction
28. waterproof mortar bonded rubble retaining wall, 400mm

Inside the Mine

6 Pumpmen and Pipemen 29 Laborers
34 Machine Runners and Scrapers
2 Door Boys 206 Machine Miners
200 Pickminers 7 Mine Foremen
43 Timbermen and Rockmen
20 Motormen 6 Brattccemen
8 Car Handlers 6 Trackmen
2 Drivers 2 Trip riders
7 Asst Mine Foremen
5 Electricians 3 Fire
Bosses 3 Shot Firers

Outside the mine

13 Office Employees
6 Engineers and Foremen
29 Tipplemen
11 Trackmen
3 Machinists
49 Coke
Employees
12 Blacksmiths
4 Carpenters
3 Foremen

A
coal mine
is warm, dark
and often wet
It is devoid of common
outdoor sounds, the
chirping of birds, the rustling
of wind through leaves
It is a strange kind of silence,
the nineteenth-century miner
listened for the cracking of the roof,
a sound that rolled like thunder for
days before a fall
He listened to the significant scuttling
of mine rats
He watched the flame of his lamp for a blue cap,
indicating the presence of gas

Priscilla Lane
Where the Sun Never Shines

VINTONDALE
PA
1923

MINE No. 3

MINE No. 6

MINE No. 3

North Branch

Blacklick Creek

South Branch

Eliza Furnace

Bracken Run

Blacklick Creek

Vinton Colliery Company #6 mine, 1923-24. Coke Breeze is at the upper end of the coke ovens.
Powder magazine is on the lower end of the ovens.

Courtesy of Diane Dusza.

LONG DARK COMMUTE:
Miles of underground mine tunnels
demanded that miners crawl up to
an hour to reach the 'working face'
of shallow coal seams. Outside the
mine portal, 1200 foot-long beehive
coke ovens stretched across a
depleted floodplain.

--- ---

Index of tools | passive treatment system, ecological washing machine, Litmus Garden, settling basins, clarification marsh, emergent wetland, limestone.

VINTONDALE RECLAMATION PARK

Vintondale, Pennsylvania. Constructed 2004.

D.I.R.T. Studio

It was unnerving when the hydrogeologist on the project team cautioned against using the term "water" instead of "unstable aqueous solution" in reference to the benign-looking liquid flowing into a creek in southwestern Pennsylvania. The truly sinister fluid was acid mine drainage (AMD). It's a polluted concoction with a pH as low as 2.5 and laden with heavy metals. It suffocates entire watersheds of coal country, coating streambeds with a toxic orange crust and rendering them lifeless. As an interdisciplinary team of artists, designers, scientists, and historians, D.I.R.T. Studio's charge was to design a system to turn this aquatic killer into a substance we could deem healthy water.

With the genuine participation of local communities, regional watershed groups, and state and federal environmental agencies, the Vintondale Reclamation Park took shape on a fallow 40-acre floodplain where for over a century the Vinton Colliery produced coke to stoke steel furnaces in Pittsburgh. Hundreds of immigrant men and boys crawled and toiled in miles of underground tunnels. After their livelihood was shut down by Mine No. 6 being exhausted, the hollowed-out chambers filled with rain and groundwater. Instead of dark, dusty ore, AMD gushed from the mine portal, forming a ferric oxide scar across the floodplain denuded by muffling deposits of boney, or mine tailings, with more of the murky refuse heaped high above the polluted Blacklick Creek. Residents of Vintondale took the eerily glowing streams and daunting boney piles as a given, sheepishly proud of their industrial heritage, yet painfully confronted with a deadly legacy. D.I.R.T.'s team joined forces with them to, in essence, finish the work of their ancestors. They helped build a passive treatment system to tackle AMD, an ecological washing machine to cleanse the acrid golden poison into a vital, green organism. The wash and rinse cycles were conceived of as the Litmus Garden, the progression made legible with a series of geometric basins flowing alongside the Ghost Town Rail Trail. Full-scale, visible signs of recovery offered the disenfranchised neighbors this biological model of AMD treatment as a catalyst for them to reinvest in their hometowns.

The challenge of the Vintondale project was to make more than just a giant science project in the residents' post-industrial backyard. The passive AMD treatment system had to be a vital part of the town's everyday life; it had to double as a park—a landscape where generations could tell stories about work in the mine in an unapologetic way. A place with ordinary things like a ball field and picnic tables to soften the social toll the mine closure took. A living laboratory where folks from around the region could see that they didn't have to live with streams without fish anymore.

Not that the science wasn't fun. Giving form to the alchemy of acid-to-alkaline was a fierce rally between scientist and designer, engineer and artist.

LONG DARK COMMUTE: Miles of underground mine tunnels demanded that miners crawl up to an hour to reach the 'working face' of shallow coal seams. Outside the mine portal, 1200 foot-long beehive coke ovens stretched across a depleted floodplain.

The enumerated formula took artistic license to render industrial-strength forms and to make the sequence transparent as it fell into place across the floodplain. First, AMD is intercepted before it spills into the creek and the unstable aqueous solution is delivered to the acid basin at pH 2.9, with metals, mainly iron, oxidizing and precipitating into bright bronze sediment. Then, in an alternating pattern of SAPS (sequential alkaline-producing system) and settling basins, limestone, plants, and bacteria work their magic. Framing each pH pond of the Litmus Garden, rows of native trees and shrubs display their fall foliage, color-coordinated with the treatment's progress, from a red maple's October glory to the blue-green of a shedding sycamore. Once the solution cascades over Spillway No. 6, the clarification marsh and emergent wetland provide the final rinse. Spread across the wetland, the 1,400-foot-long footprint of the historic beehive coke ovens surfaces as an earthen plinth of wildlife habitat. Finally, the water—yes, water—is released into the Blacklick Creek.

This project conquered AMD in only one creek, renewed one landscape of a single town. But D.I.R.T. documented the entire process of coalition-building, the science and design of the treatment system, and

funding and implementation, all with the intent of creating a prototype for the region as well as for other abandoned coal-mine land and post-industrial towns further afield. Ultimately the Vintondale project is about having no fear to embrace the good, the bad, and the ugly of the industries that built this country but also trashed its liquid life-source. The guts to hug the Rust Belt comes from trusting that the closer technologies come to natural models, the easier it is to imagine and actually make straightforward humble places to, after a hard day's work, just sit and fish.

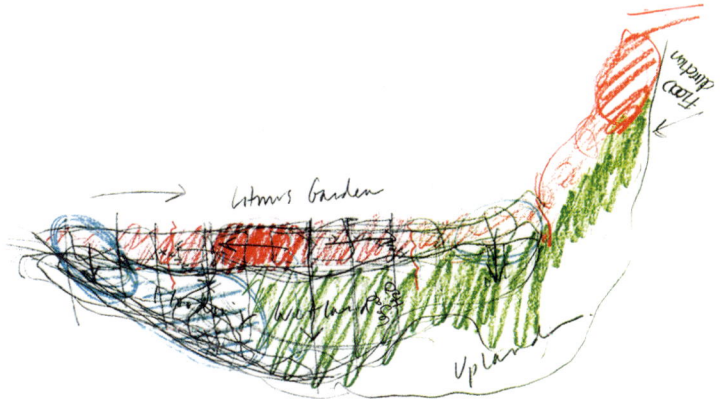

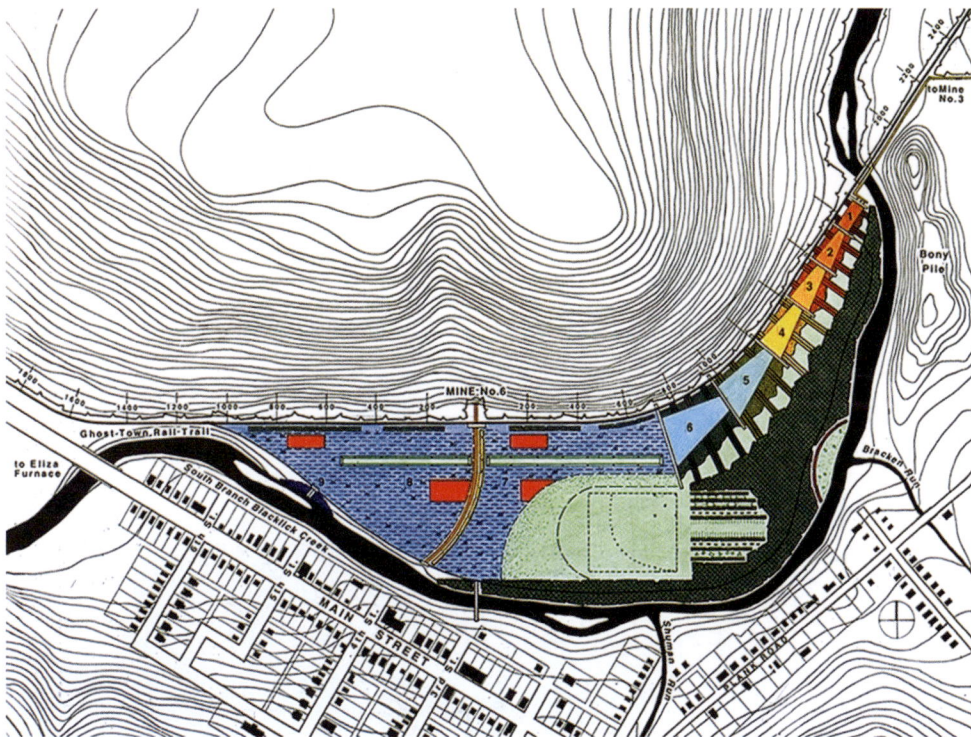

ECOLOGICAL WASHING MACHINE: Working closely with a hydrologic engineer, the design team envisioned making the acid mine drainage treatment process legible through geometric ponds followed by a final rinse through a clarification marsh.

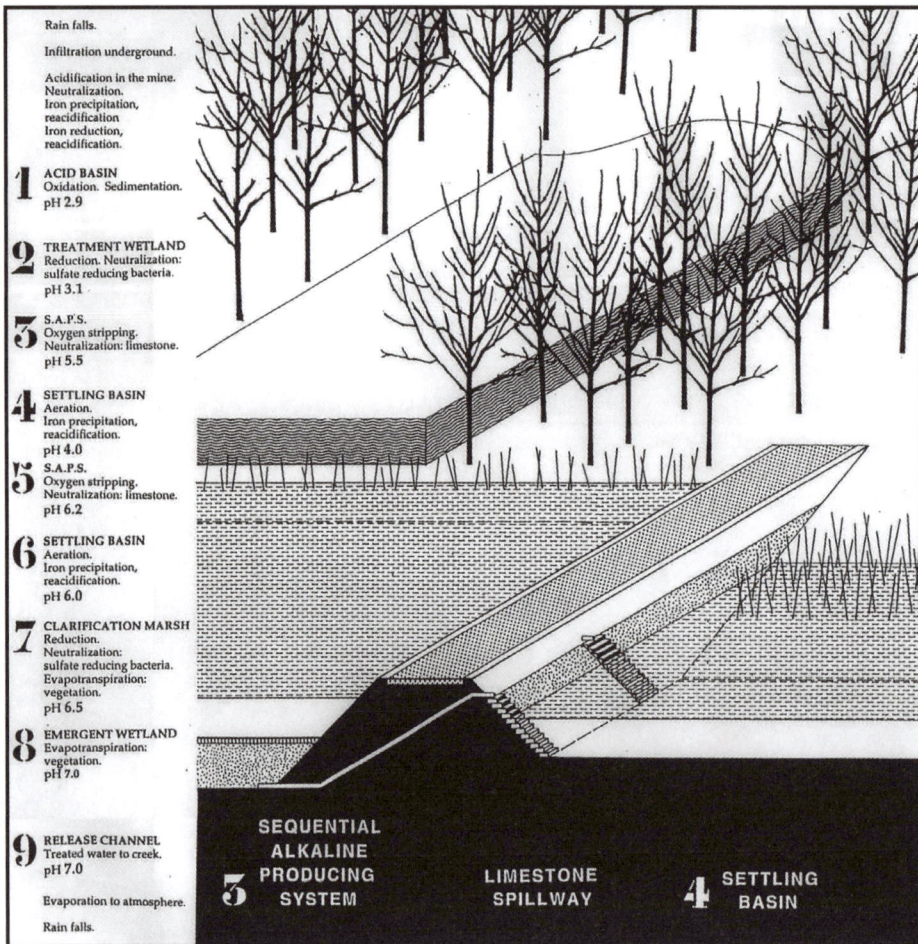

Rain falls.

Infiltration underground.

Acidification in the mine.
Neutralization.
Iron precipitation,
reacidification
Iron reduction,
reacidification

1 ACID BASIN
Oxidation. Sedimentation.
pH 2.9

2 TREATMENT WETLAND
Reduction. Neutralization:
sulfate reducing bacteria.
pH 3.1

3 S.A.P.S.
Oxygen stripping.
Neutralization: limestone.
pH 5.5

4 SETTLING BASIN
Aeration.
Iron precipitation,
reacidification.
pH 4.0

5 S.A.P.S.
Oxygen stripping.
Neutralization: limestone.
pH 6.2

6 SETTLING BASIN
Aeration.
Iron precipitation,
reacidification.
pH 6.0

7 CLARIFICATION MARSH
Reduction.
Neutralization:
sulfate reducing bacteria.
Evapotranspiration:
vegetation.
pH 6.5

8 EMERGENT WETLAND
Evapotranspiration:
vegetation.
pH 7.0

9 RELEASE CHANNEL
Treated water to creek.
pH 7.0

Evaporation to atmosphere.

Rain falls.

3 SEQUENTIAL
ALKALINE
PRODUCING
SYSTEM

LIMESTONE
SPILLWAY

4 SETTLING
BASIN

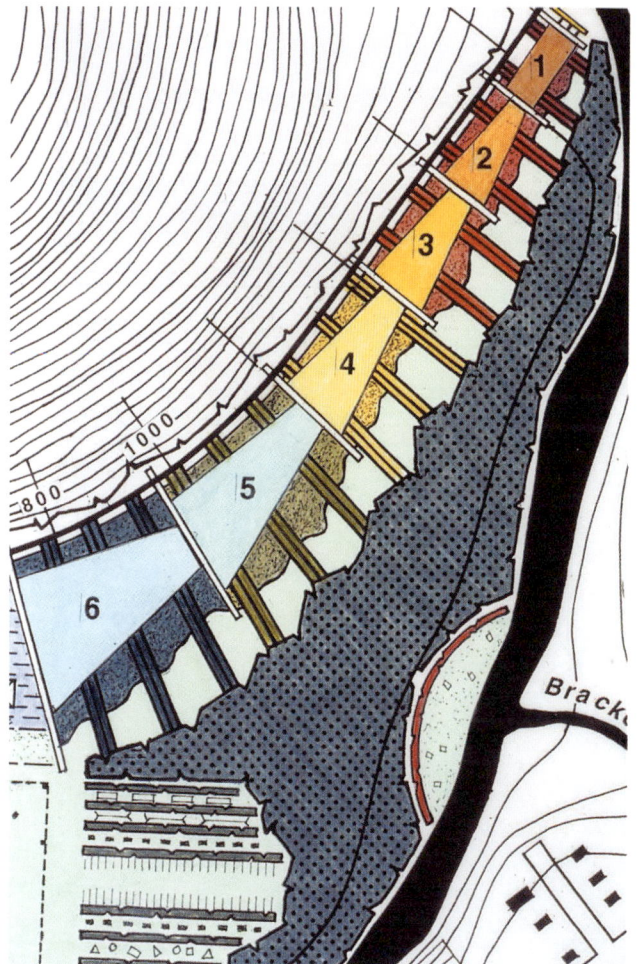

BLACK MOUNTAIN (above left): From atop what locals call "bony piles," an orange Blacklick Creek glows from acid mine drainage laden with heavy metals.

HEAVY METAL (above right): When acid mine drainage exits the coal tunnels, metals - primarily iron - deposit a deadly substance nicknamed "yellow boy."

TREATMENT TRAIN (left): As heavy metals precipitate, acid mine drainage cascades over limestone spillways into increasingly alkaline basins that are flanked by native trees and shrub masses.

LIQUID OF ANOTHER COLOR (right): Acid mine drainage, technically an "unstable aqueous solution," cannot be properly call "water" until it is metal-free and a healthy pH in basin #6.

Landscaped green roof provides for gathering/special event space while acting as an observation deck with amazing views towards Manhattan

The facade employs structural mullions which also act as solar screens

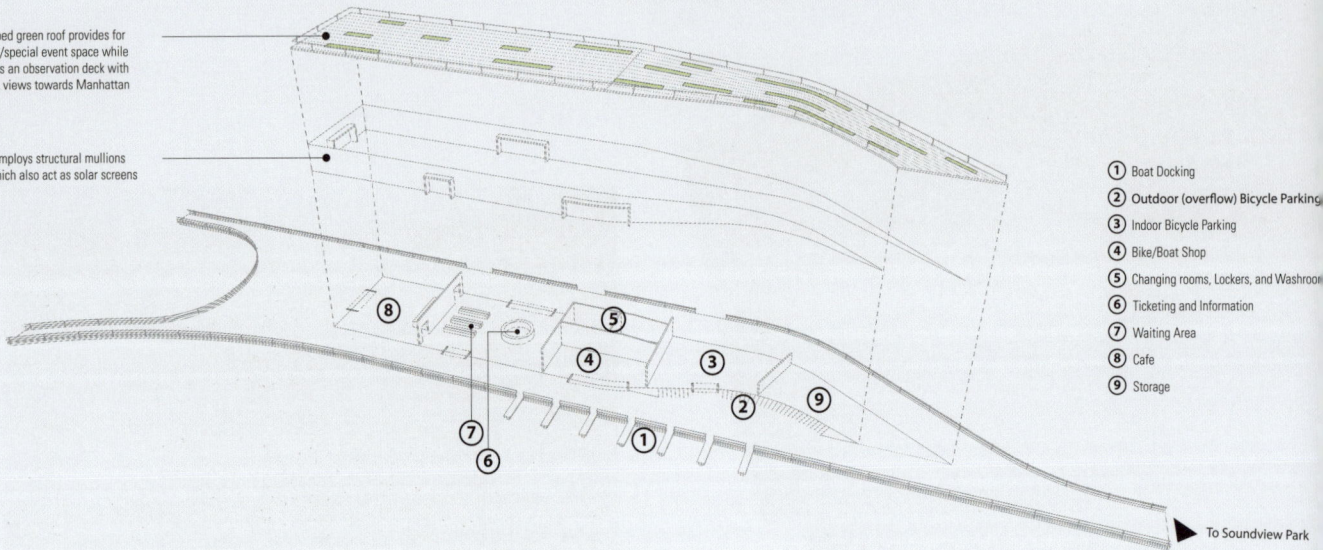

1. Boat Docking
2. Outdoor (overflow) Bicycle Parking
3. Indoor Bicycle Parking
4. Bike/Boat Shop
5. Changing rooms, Lockers, and Washroom
6. Ticketing and Information
7. Waiting Area
8. Cafe
9. Storage

To Soundview Park

Above:
A section perspective of one of the BTT arms.

Below:
Bird's-eye view of the Bronx Blue Terminal

Index of tools | flexible landscape, liquid network, co-habitation, mobile cellular network, symbiotic activities, floating pods, artificial islands, marine habitat stimulator.

PARALLEL NETWORKS

New York City. Speculative, 2012.

Op.N (Ali Fard and Ghazal Jafari)

New York City's relationship with water has been essential to its historical development and growth. Traditionally, water for New York has been a source of connectivity and large-scale economic and industrial development, which have slowly transformed the once diverse water's edge into industrial hardscapes. The partial migration of industrial operations from the waterfront to the outskirts of New York's populated region has been motivating the conversion of the water's edge to a new extended urban space. This spatio-economic transformation, coupled with an awareness of environmental complexities, is offering a new potential for reimagining a resilient, yet productive, landscape capable of accommodating a diverse range of urban programming.

The five boroughs of New York are separated by an expansive body of water, consisting of the Upper Bay, the rivers, and canals. This "blue network" is home to overlapping ecologies that are often understood and analyzed in isolation from each other. Ecologies of shipping and service industries, marine transportation, wildlife habitat, urban recreation, and cultural activities, are constitutive elements of New York's blue network and act as a liquid connective tissue. The operations and activities of these networked ecologies shape and transform the spatial characteristic of New York's wet landscape, or what is referred to here as the sixth borough.

By analyzing the overlapping ecologies of the sixth borough, this project attempts to highlight a series of opportunities for synergetic operation of these networks. Parallel Networks proposes a process of spatial production, which not only allows for the co-habitation of these networks, but also leverages extra economic, ecological, and cultural values. This strategy introduces a new understanding of these overlapping networks: the operations and spatial transformations of these networks are highly interconnected and inherently inseparable.

Unlike typical urban infrastructural development, this project is envisioned as a flexible landscape that is grown, shaped, and defined over time as a result of the operation of an expandable and mobile cellular network. Hence, rather than fixing the infrastructure to the ground and in time, the project explores the means by which a flexible system can stimulate and intervene in the existing natural processes of the wet landscape, to gradually grow a flexible infrastructure as a platform for symbiotic activities of the current networks.

Ecologies of the Blue Network
Today, the ports of New Jersey and New York together make up the third-largest port complex in the United States, feeding a service industry and freight transport structure that stretches from New York to Canada. This port complex will continue to grow in size and

Blue Network

Urban Activities
- Education | Research
- Production
- Recreation
- Logistics

Transportation
- Energy
- Maintenance

Waterscapes
- Flood Protective
- Remediating
- Wild life habitat

Tunnel Exhaust   CO2 collector

CO^2

Carbon Dioxide exhaust from numerous vehicular tunnels in the area will be collected and transported to algae cultivation areas in Upper New York Bay, using special pods.

Brine shrimp → Wild life habitat

Produced Brine shrimp in nursery pods is a dominant food for shore birds. It can increase the attraction of birds to floating wetlands of Bronx Blue Terminal.

Algae will be delivered to newly created biopower plants in Cleantech Industry City in Queens. The algae will be harvested and its oil extracted to be used as biofuel.

Algae Production → Bio Fuel

Gowanus Canal
Production Ponds
Commercial/ Recreational unit
Conduits
Outdoor Leisure spaces

Gowanus Canal Production Ponds: Enclosed water pools with various water depths, are separated from major water bodies by levees (conduits). This landscape makes it possible to control pollution, salinity, and other environmental conditions, necessary for various aquatic farming. Conduits circulate water through the ponds and contain monitoring, desalination, and remediation facilities.

CO^2 → Algae Production
Collected Carbon Dioxide   Algae Production

Collected CO2 from tunnels will be fed into movable floating bioreactors to cultivate algae which will be used to produce biofuel.

$ Food
Egg   Spat   Mature oyster   Coral reef

Nursery pods can provide the requisite water condition for accelerated production of commercial oysters. The shell is then thrown on the reef island area to nourish the islands

Algae + Brine shrimp

Brine shrimp → Fish

Nursery pods containing algae, provide the proper water salinity for brine shrimp production.

Produced Brine shrimp in nursery pods can be used as nourishment in fish farms of Gowanus canal.

Legend
- Electric Ferry
- Algae Bioreactor pods
- CO2 collecting pods
- Brine shrimp and oyster nursery pods
- Vehicular tunnel exhaust/CO2 collection node
- Algae cultivation cluster
- Oyster and brine shrimp nursery
- New Transit Hubs

0 1 2 5 10 KM

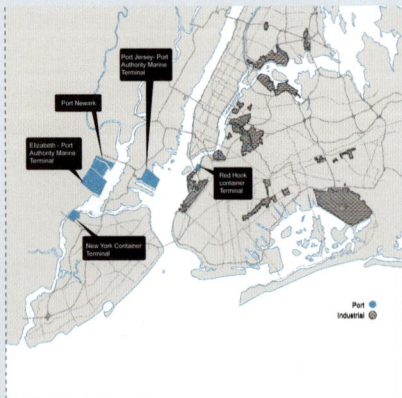

Industrial and Shipping Activity
- Port Jersey- Port Authority Marine Terminal
- Port Newark
- Elizabeth - Port Authority Marine Terminal
- Red Hook container Terminal
- New York Container Terminal
- Port
- Industrial

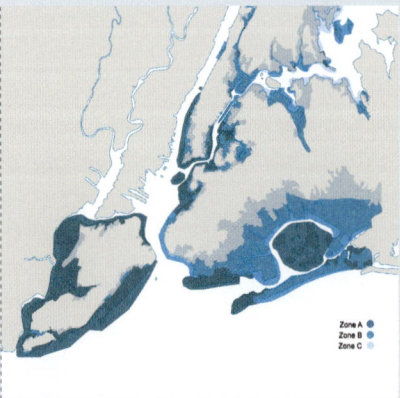

Storm Vulnerability
- Zone A
- Zone B
- Zone C

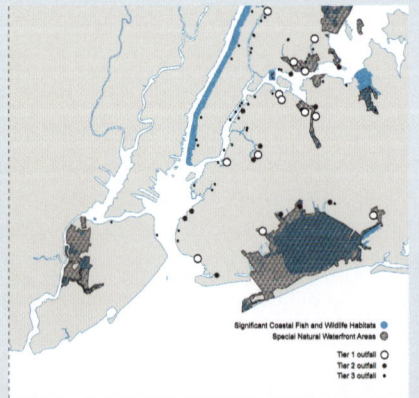

Sewer Overflow Outfalls & Natural Waterfront Areas
- Significant Coastal Fish and Wildlife Habitats
- Special Natural Waterfront Areas
- Tier 1 outfall
- Tier 2 outfall
- Tier 3 outfall

Above:
Plan showing the proposed networks operating within the sixth borough

Below:
Contextual maps of NYC waters

operational scale. Following the enlargement of the Panama Canal, which will make passage of larger cargo ships possible, New York ports, like many other ports in the region, are planning to enhance their physical infrastructure and technical equipment to accommodate the new panamax vessels. These shifts will also include the deepening of major shipping routes in Kill Van Kull and Arthur Kill Channels. New York's growing shipping industry along with its related businesses and training facilities are in need of space close to water and industrial areas. These include mooring docks, structures for servicing and maintenance, parking areas for barges and tugboats, educational centers, and other facilities. In addition, the distribution of goods and materials to the regional warehouses and retail centers is currently performed mainly by trucks, which aside from the obvious environmental pollution, has proven to be a less economic and less efficient mode of transportation when compared to water-based shipping modes. Once a major producer of oysters, New York's water bodies will soon be saturated with marine traffic. Furthermore, this water body is surrounded by hard-edged landfills and has lost its flexibility in dealing with water fluxes and flood events. As a result, the sixth borough is not protected against the continuously rising sea levels and is prone to seasonal and storm-related flooding. Possible inundation, coupled with increasing water pollution, threatens the remainder of the wildlife habitat in the region.

On the other hand, New York's population is set to reach 8.7 million by 2020, and 9.1 million by 2030 (Vision 2020). Similar to many other North American cities, New York has experienced a gradual shift in its waterfront character. The industries that have long occupied the water's edge are migrating, and the leftover industrial brown fields are being designed to host the city's growth. New York's waters can once again be publicly activated to act not only as a connector but also as a destination for a variety of activities, such as energy production, recreation, and education among others. Through this transformation, water will be imagined as a new public space, able to deal with flux and change, and easily adaptable to future uses.

Parallel networks

New York City, like many other metropolitan areas, was founded in a context that is environmentally and socially in constant flux. Therefore, the blue network needs to take into account not only what is necessary

for the region's economic livelihood, but also the environmental factors and socio-cultural transformation of the region. This is possible through coupling or bundling of transportation infrastructure with water-based urban programming for production, recreation, living, protection, research, and rehabitation. As this is an expansive infrastructural project, a systematic thinking remedy—instead of an all-encompassing super-structural remedy—can effectively enhance the performative qualities of the network. In this proposal this systematic thinking is manifested through a cellular infrastructure. This cellular system allows for incremental implementation by various public and private entities, while allowing for easy maintenance, since each cell can be removed and repaired without jeopardizing the overall health of the network. The cells of the blue network are translated to floating pods of various scales and functions, able to accumulate and disband, and to be moved from one location to another. This system promotes a need-based growth over time, allowing for opportunities to experiment, while keeping the integrity of the network intact. Furthermore, pods are designed to forge symbiotic relationships with the context they are located in.

Two test sites are introduced for a more detailed examination of this design strategy. NY Gaia, located in Upper New York Bay, is a productive landmark, a center for clean energy production, and an operational node for marine transportation activities around New York. The Gaia will feed other clean-tech industries with energy through wind power and bio-fuel harvested from large-scale algae cultivation. In addition, much-needed maintenance and training facilities for marine transportation will be located adjacent to reef islands, which are artificial islands that grow over time. The performance of pods in this area can be summarized in the following order: Collection pods collect CO2 from the tunnels of New York City. The collected CO2 is brought to the reef islands to nourish the algae pods. Algae pods are relocated to potential clean-tech industries along the shore for energy production. The algae is also used as a fertilizer in nursery pods that hold and accelerate brine shrimp and oyster production. Brine shrimp pods are relocated to Gowanus Canal to nourish the potential aquatic farming. Oyster shells or spat is dumped in Upper New York Bay to feed and build natural islands that are both a flood defense system and a marine habitat stimulator.

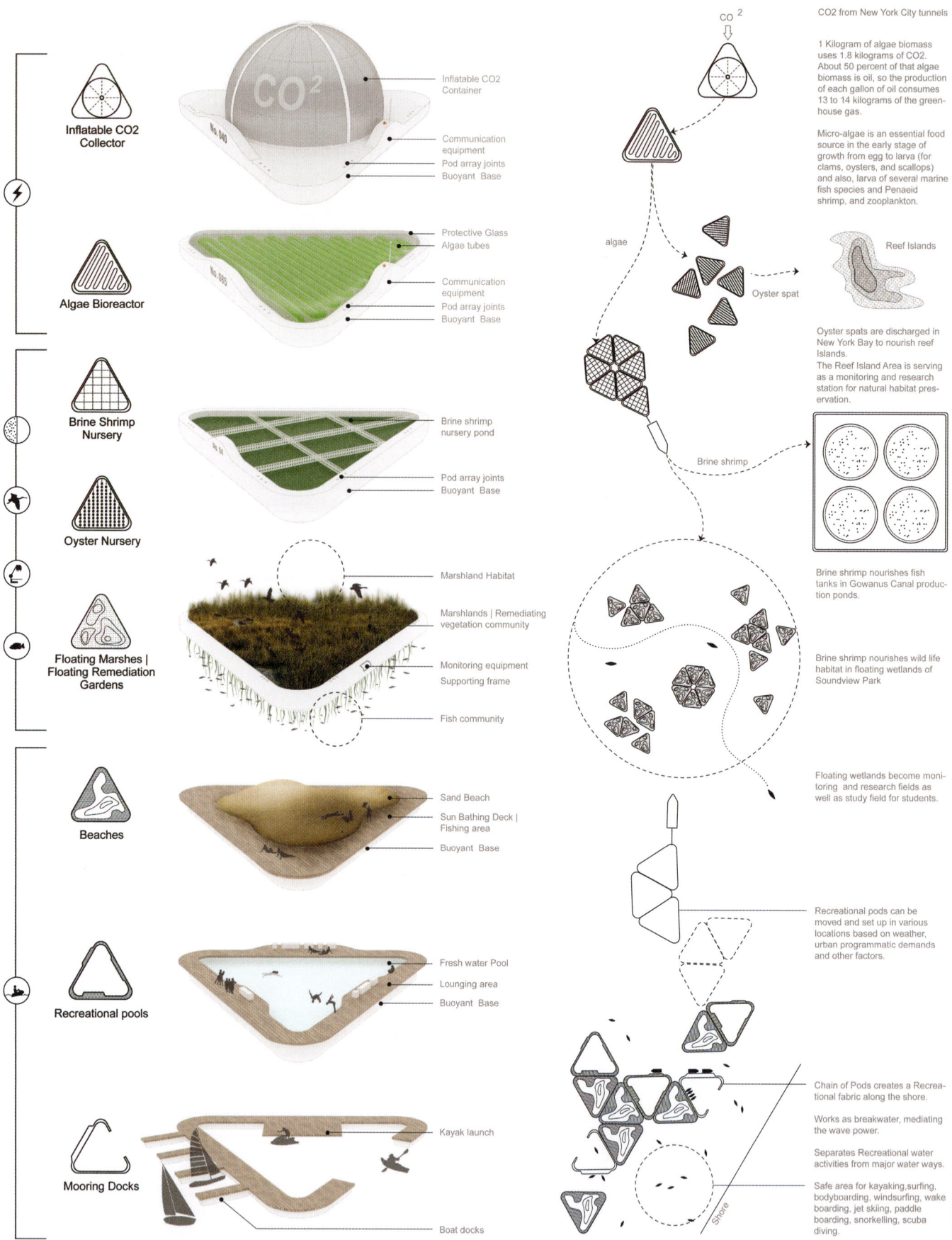

Inflatable CO2 Collector

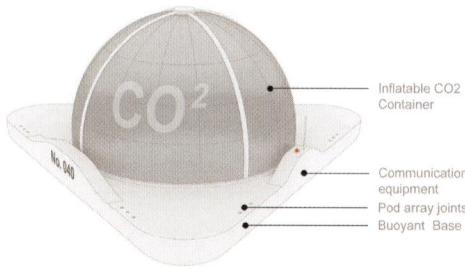

- Inflatable CO2 Container
- Communication equipment
- Pod array joints
- Buoyant Base

No. 040

CO$_2$

Algae Bioreactor

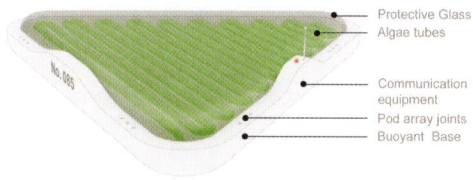

- Protective Glass
- Algae tubes
- Communication equipment
- Pod array joints
- Buoyant Base

No. 086

Brine Shrimp Nursery

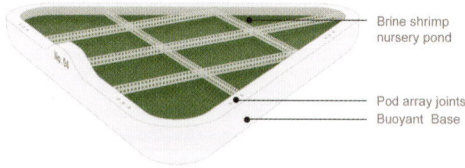

- Brine shrimp nursery pond
- Pod array joints
- Buoyant Base

No. 04

Oyster Nursery

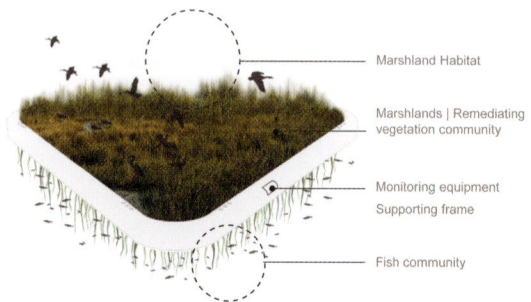

Floating Marshes | Floating Remediation Gardens

- Marshland Habitat
- Marshlands | Remediating vegetation community
- Monitoring equipment
- Supporting frame
- Fish community

Beaches

- Sand Beach
- Sun Bathing Deck | Fishing area
- Buoyant Base

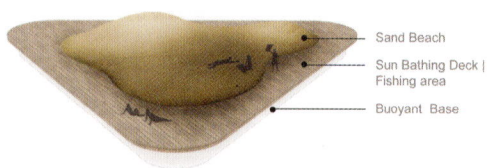

Recreational pools

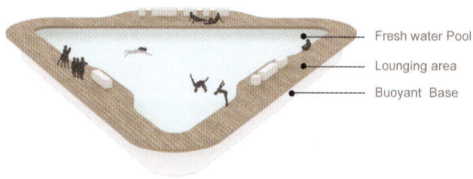

- Fresh water Pool
- Lounging area
- Buoyant Base

Mooring Docks

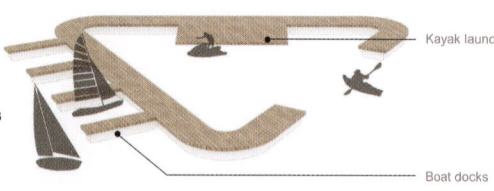

- Kayak launch
- Boat docks

CO$_2$ from New York City tunnels

1 Kilogram of algae biomass uses 1.8 kilograms of CO2. About 50 percent of that algae biomass is oil, so the production of each gallon of oil consumes 13 to 14 kilograms of the greenhouse gas.

Micro-algae is an essential food source in the early stage of growth from egg to larva (for clams, oysters, and scallops) and also, larva of several marine fish species and Penaeid shrimp, and zooplankton.

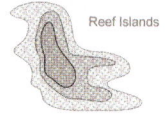

algae

Reef Islands

Oyster spat

Oyster spats are discharged in New York Bay to nourish reef Islands.
The Reef Island Area is serving as a monitoring and research station for natural habitat preservation.

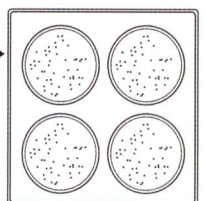

Brine shrimp

Brine shrimp nourishes fish tanks in Gowanus Canal production ponds.

Brine shrimp nourishes wild life habitat in floating wetlands of Soundview Park

Floating wetlands become monitoring and research fields as well as study field for students.

Recreational pods can be moved and set up in various locations based on weather, urban programmatic demands and other factors.

Chain of Pods creates a Recreational fabric along the shore.

Works as breakwater, mediating the wave power.

Separates Recreational water activities from major water ways.

Safe area for kayaking, surfing, bodyboarding, windsurfing, wake boarding, jet skiing, paddle boarding, snorkelling, scuba diving.

Shore

The typological breakdown of the proposed cellular infrastructure, various pod types and the operational logic of their organization

183

Islands form around wind turbines, which are located in the middle of the bay to free the urban context of their noise pollution.

Bronx Blue Terminal (BBT), located at the mouth of the Bronx River, will provide the Bronx with a ferry terminal. In addition to being a transit hub, BBT will act as a recreation, research, and education node within the blue network, building on the educational programs already underway in the area. Habitat preservation and regeneration is another part of the mandate for BBT. The performance of pods in this area can be summarized in the following order: Floating wetland pods recreate a natural habitat that has been replaced by landfills along the Soundview Park shore in the past. Planted with remediating plants, this floating park can also remediate the water of metals and other toxins. This complex creates a potent field of research for the study of wetland ecosystems, hydrology, vegetation, and native wildlife species. The ferry terminal extends out of Soundview Park, creating two paths of recreation, with activities including fishing, observation decks, kayaking, sailboat mooring areas, bike parking, beaches, and seating. Floating recreation

pods, accessible by kayaks and boats, complement these recreation paths as "off-shore" recreational nodes. The symbiotic operations of these parallel networks, mediated through a cellular infrastructural system, act as a catalyst for the development of a blue network between the five New York boroughs. This network is as much a performative waterscape as it is a connective urban tissue.

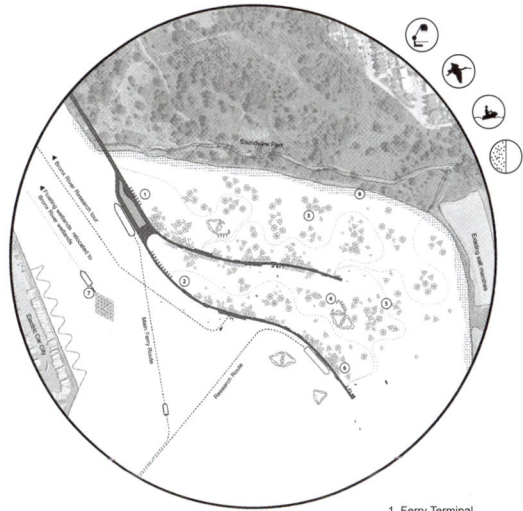

1. Ferry Terminal
2. Recreation | Observation Path
3. Floating wetlands
4. Recreation Pods
5. Kayaking area
6. Recycled barges: re-purposed as floating research units, traveling from one research base to another.
7. Bio-fuel unloading deck
8. Reconstructed wetland strip along the shore

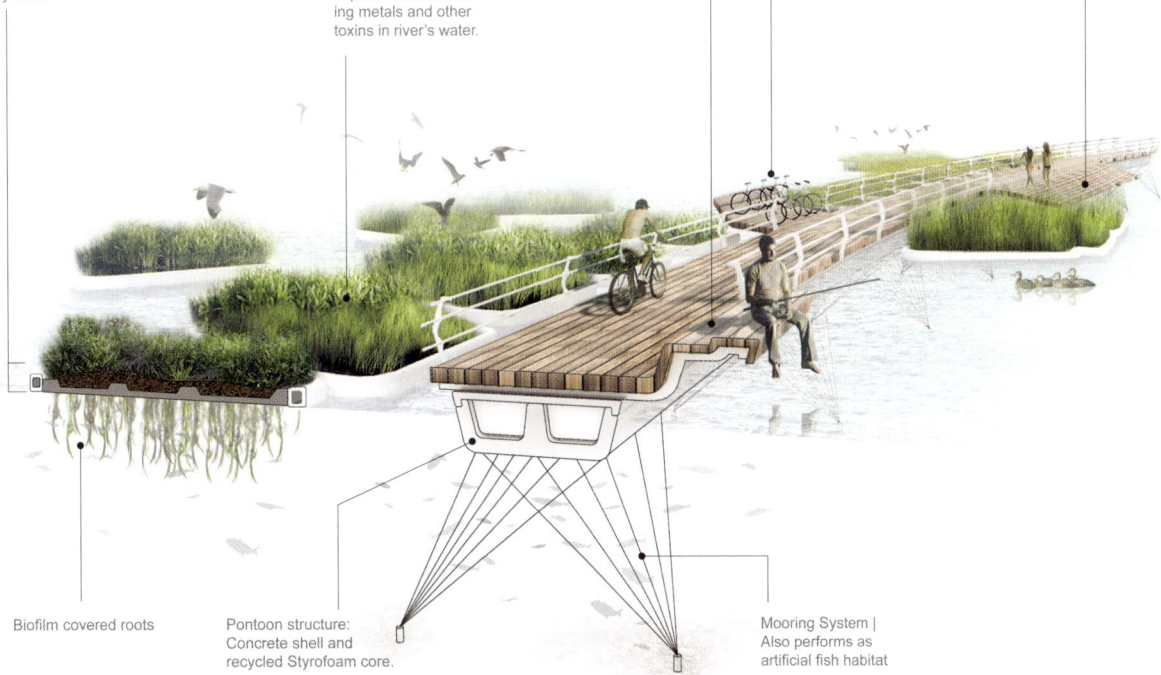

Planting Media +
Floating mat: intertwined fine polymer strands Made of recycled plastic and Recycled Styrofoam

Floating Wetland
These pods can also be vegetated with remediating plants, capable of remediating metals and other toxins in river's water.

Bench   Bike Parking

Dipped Platform to make access to water

Biofilm covered roots

Pontoon structure:
Concrete shell and recycled Styrofoam core.

Mooring System |
Also performs as artificial fish habitat

Above:
Plan of the proposed Bronx Blue Terminal, first case study site

Below:
View from Soundview Park towards BBT and an exploded diagram of BBT's spatial organization

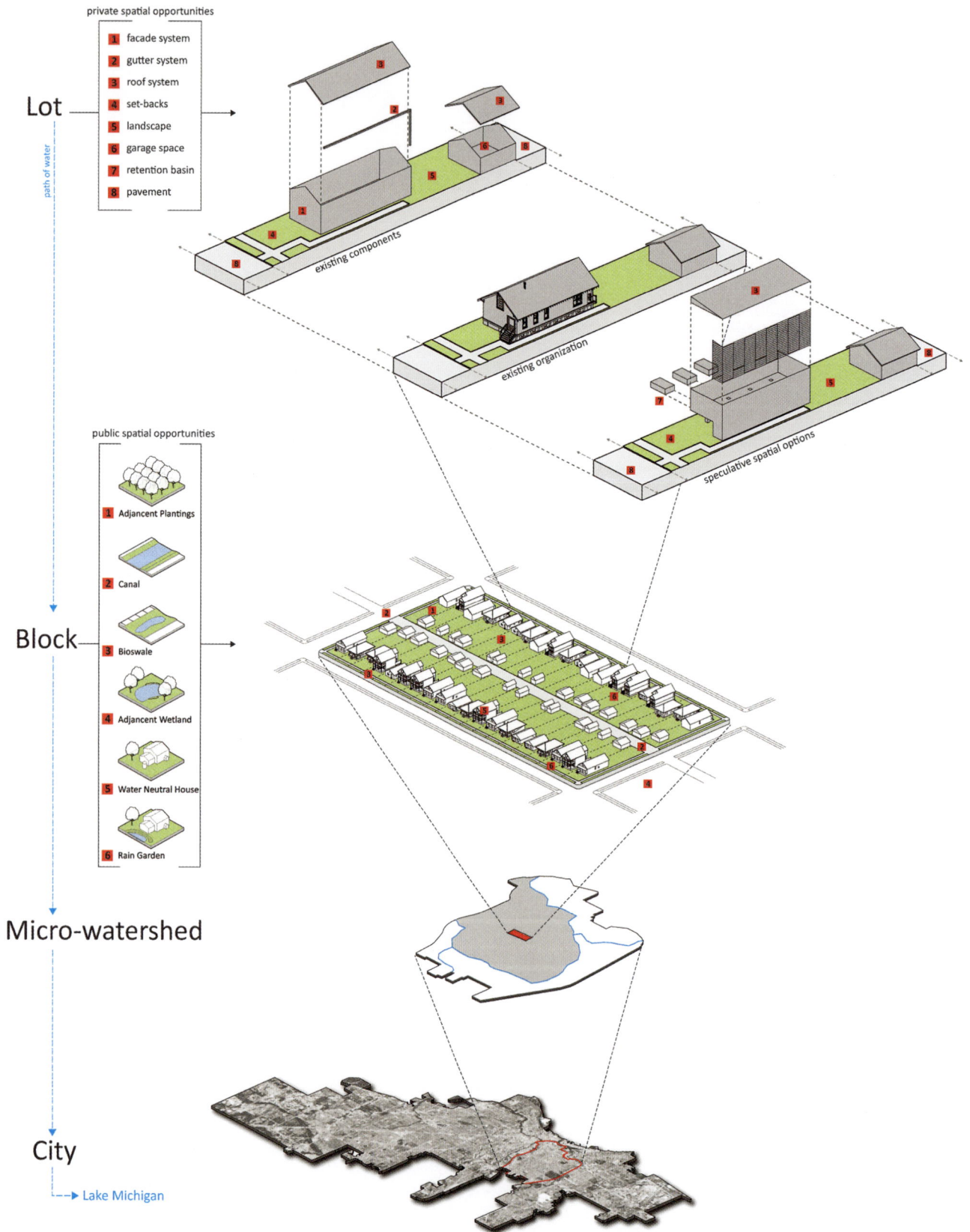

private spatial opportunities

1	facade system
2	gutter system
3	roof system
4	set-backs
5	landscape
6	garage space
7	retention basin
8	pavement

Lot

path of water

existing components

existing organization

speculative spatial options

public spatial opportunities

1	Adjacent Plantings
2	Canal
3	Bioswale
4	Adjacent Wetland
5	Water Neutral House
6	Rain Garden

Block

Micro-watershed

City

Lake Michigan

Conceptual relationship between strategies, scales, and goals defining spatial architectural interventions.

WATER CORE HOME

Milwaukee / Great Lakes. Speculative, 2012.

Dan Williamson, David Karle, and Sarah Thomas Karle

The Great Lakes represent the largest collection of surface fresh water in the world. Pollution in the lakes has been a primary by-product of over a century of urbanization, and only since the passing of the Clean Water Act of 1972 have environmentally significant investigations occurred. Currently, the United States Environmental Protection Agency has identified 84 areas of concern in the Great Lakes Watershed due to non-point source combine sewer overflows (CSO). Combine sewer overflow into surface water violently damages ecosystems, impairing critical environmental, social, and economical uses of the water by communities. CSO contamination is not isolated to the Great Lakes Watershed; it impacts water quality in over 700 communities across the United States. American cities are grappling with declining centralized water systems confined by an inefficient linear process of fast conveyance and end-of-the-pipe treatment. This outdated model of urban infrastructure calls for a new paradigm of decentralized urban water management systems with cyclical strategies for water storage and conveyance.

The city of Milwaukee serves as a prototypical site for investigating decentralized urban water management strategies. EPA identified areas of concern in the city that contribute over a billion gallons of CSO yearly into Lake Michigan. A decentralized water management system was proposed for one of the city's micro-watersheds, with the goal of capturing and treating

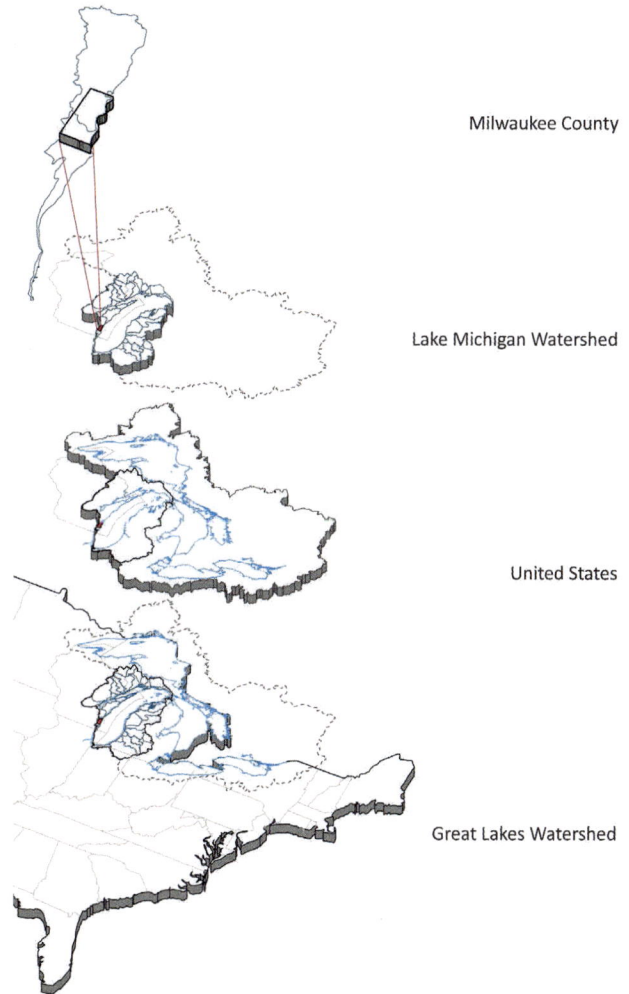

the first inch of water from any wet weather event. Expanding beyond the current EPA-recommended Green Infrastructures strategies, this proposal investigates the design of the single-family home as part of a decentralized micro-watershed management system. The Water Core Home leveraged internal and external spatial strategies that aimed to reduce the quantity and improve the quality of the water before it reached the combined sewer system. This proposal challenges existing urban residential blocks, lots, and homes relying on combined sewer systems to be spatially reconfigured and to architecturally address water performance.

Milwaukee County

Lake Michigan Watershed

United States

Great Lakes Watershed

Multiple agencies of various scales are aimed at reducing the number of combined sewer overflows impacting the Great Lakes Watershed.

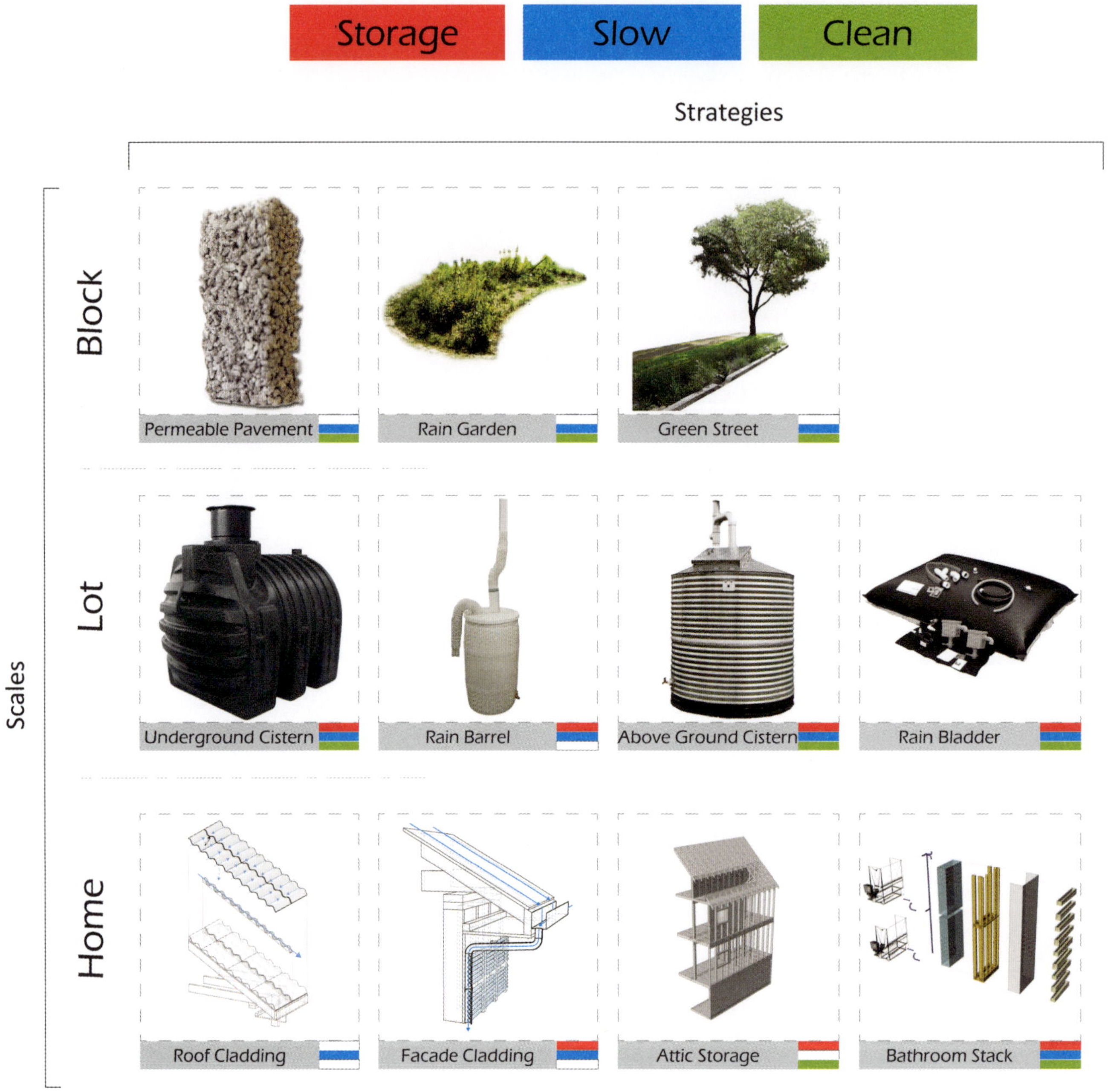

	Storage	Slow	Clean

Strategies

Scales

Block
- Permeable Pavement
- Rain Garden
- Green Street

Lot
- Underground Cistern
- Rain Barrel
- Above Ground Cistern
- Rain Bladder

Home
- Roof Cladding
- Facade Cladding
- Attic Storage
- Bathroom Stack

Above:
Matrix of normative,
EPA-funded, or aggregate
solutions at three identified
scales. These interventions
aim to reduce the impact on
the combined sewer system
while improving the social
and economic environment

Habitation Dimensions

Average American Family:

# of persons	4
Gallons of water per day	300
Loads of Laundry per yr	300

Cost / 5 gallons of water	$0.01

Distribution- gallons [%]

Toilet	80 gallons [26.7%]
Shower	50 gallons [16.8%]
Faucet	47 gallons [15.7%]
Washer	66 gallons [21.7%]
Leaks	42 gallons [13.7%]
Other	15 gallons [05.3%]

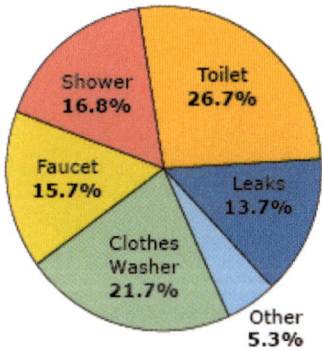

Fixtures / Appliances			Total
Toilet	**Gallons**	**Uses/ Day**	**Gallons**
Poor	3.5	22	77
Good	1	22	22
Washer			
Poor	23	0.8	18.4
Good	15	0.8	12.3

Precipitation Dimensions

A precipitation even is characterized by both its total depth and by the time period over which the rain occurs, the duration. The most severe damage caused by short, deep storms.

Milwaukee WI:

Avg. Yearly Precipitation:	34.8"
Avg. Yearly Snowfall: 52.4"	
Record 24/hr event:	6.8"
(Aug. 6, 1986)	

"Rain events of greater than 1 inch in a 24 hour period are expected to increase by seven events per decade, and rain events greater than 2 inches are expected to increase by three per decade.... Changes in the distribuition of rainfall may result in an increase in the frequency of intense storms that deliver high amounts of precipitation over short periods of time."

Prototypical Conditions:

If existing conditions were calibrated to retain on-site (block and lot combined) the first inch of precipitation, 90% of wet weather events could be prevented from adversely affecting the combined sewer system, thus preventing or reducing combined sewer overflows and improving receiving water quality.

Inches Precipatation	.5	1.0	1.5	2.0	2.5	3.0	3.5	4.0	-	7.0
Volume / Lot (cu. ft.)	228	457	685	913	1,141	1,370	1,598	1,826		3,196
Volume / Lot (gal)	1,689	3,378	5,67	6,756	8,445	10,134	11,823	13,512		23,647
Volume / Block (cu. ft.)	8,217	16,434	24,651	32,868	41,085	49,302	57,519	65,736		115,038
Volume / Block (gal)	608,06	121,612	182,417	243,223	304,029	364,835	425,641	486,446		851,281

Pervious concrete

Component

Type: Pervious Concrete
Lifespan: 20-40 years
Typically 15-20% porosity
Rate: (3-5 gal/min) / sq. ft.
Cost: $2.00-5.00 / sq. ft.
Minimum: None
Maximum: None

Description:
Porous pavement can reduce and infiltrate surface runoff through its permeable surface into a stone or filter media below. Runoff then percolates into the ground, is conveyed offsite as part of a stormwater system, or is collected and contained for future use. Porous pavement can be asphalt, concrete or pavers, but differs from traditional pavement because it excludes fine material and instead provides pore spaces that store and pass water.

Upkeep:
> Maintenance inspection once a year
> Vacuum swept 4 times a year

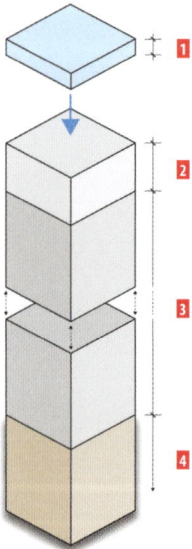

Construction diagram

Construction

1 **Water:** The amount of rainfall or snow-melt to be designed for. This volume helps design the depth of the course aggregate layer.

2 **Porous Concrete/Asphalt [P.C.]:** Course surface layer that allows water to infiltrate into substrate.
Depth: 5-8 inches
Void Percentage: 15-20%
Infiltration Rate: (3-5 gal/min) / sq. ft.

3 **Course Stone Aggregate Substrate [C.A.]:** The primary storage bed for stormwater runoff in the porous concrete paving system. *note: the base of this layer must be a minimum of 24" above the seasonal high water table or bedrock.
Depth: 12-36" (Varies based on design volume)
Void Percentage: 25-40%

4 **Uncompacted Sub-grade:** The earthen soil uncompacted, this includes no machinery allowed over the design area.
Infiltration Rate: Specific to soil type

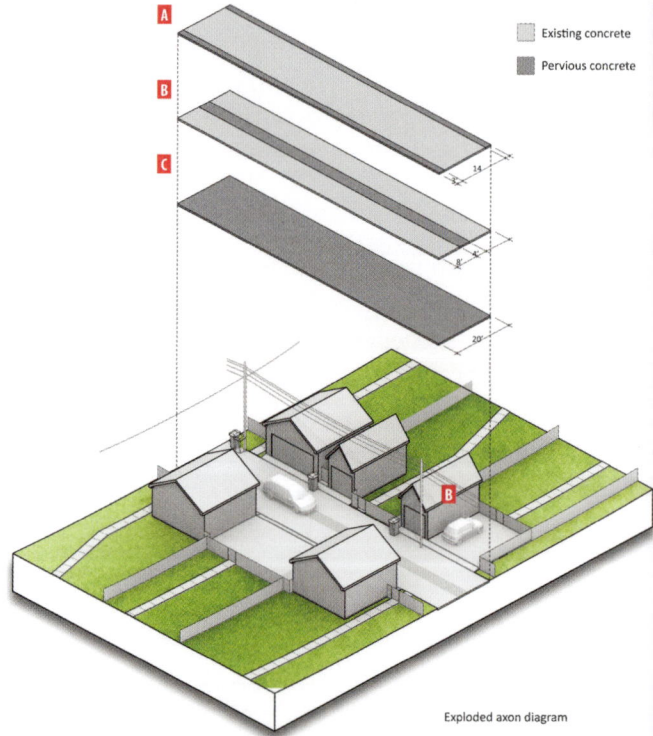

Existing concrete
Pervious concrete

Exploded axon diagram

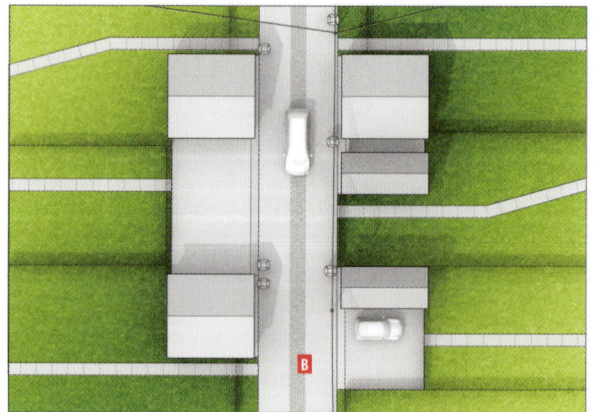

Plan

Permeable Pavement:

Porous concrete, when applied to the entire alley and on-street parking area, can handle almost a capacity of a .75" rainstorm. However, porous concrete is in its infancy and is known to be difficult to install with complete consistency. Continual maintenance and initial installation costs make porous concrete a somewhat effective solution, but financially strenuous.

Isolated

Back to Back Lot Calculations
(33' lot width)

A **Edges:**
width (total): 6 ft
Depth of [P.C.]: 6" (15% void)
Capacity: 14.9 cu.ft. (110 gals)

Depth of [C.A.]: 18" (30% void)
Capacity: 89.1 cu.ft. (659 gals)

Total Capacity: 769 gallons
Cost: $400-1000

B **Middle:**
width (total): 4ft
Depth of [P.C.]: 6" (15% void)
Capacity: 9.9 cu.ft. (73 gals)

Depth of [C.A.]: 18" (30% void)
Capacity: 59.4 cu.ft. (439 gals)

Total Capacity: 512 gallons
Cost: $270-670

C **All:**
width (total): 20ft
Depth of [P.C.]: 6" (15% void)
Capacity: 49.5 cu.ft. (366 gals)

Depth of [C.A.]: 18" (30% void)
Capacity: 297.0 cu.ft. (2200 gals)

Total Capacity: 2566 gallons
Cost: $1,325-3,400

Aggregate

Entire Block Calculations
620 ft. block Length
.5" storm = 60,800 gals

A **Alley All:**
Total Capacity: 48,174 gallons
Cost: $96,200 - 245,000
% of .5" storm: 80%

B **Street Parking:**
Total Capacity: 36,053 gallons
Cost: $72,000 - 182,000
% of .5" storm: 60%

C **Alley All + Street Parking:**
Total Capacity: 84,227 gallons
Cost: $168,200 - 425,400
% of .5" storm: 140%

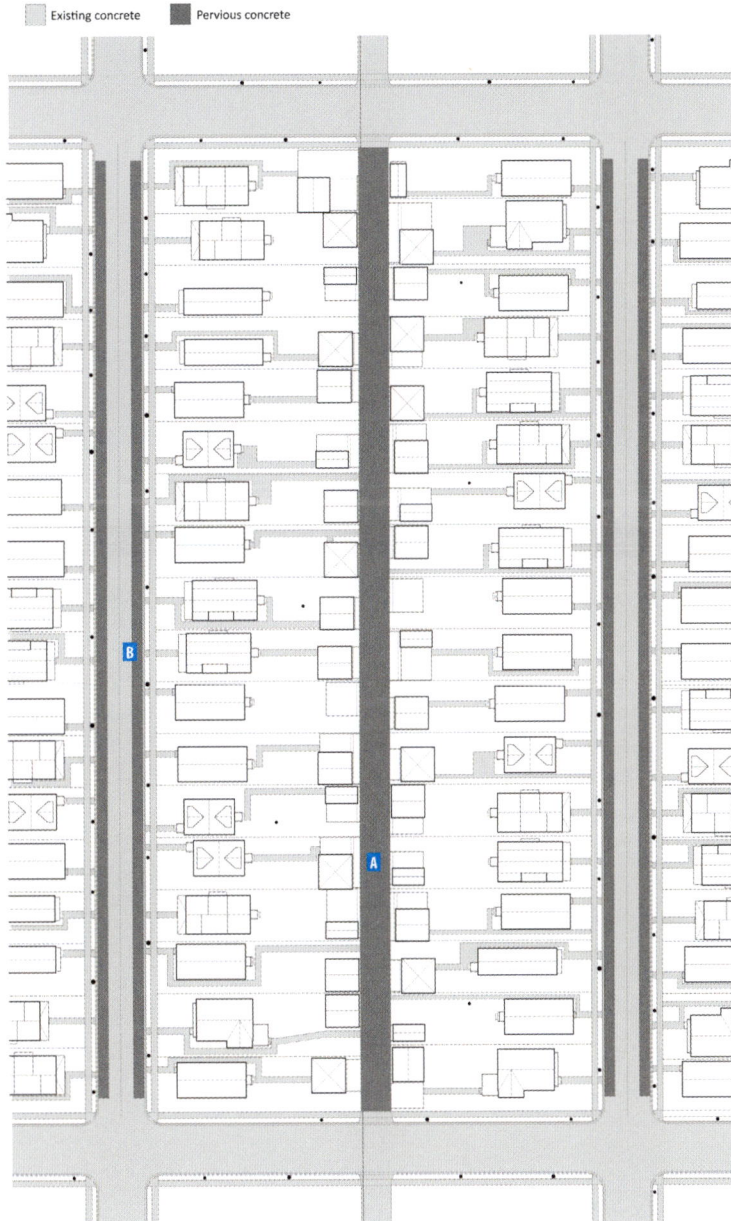

Existing concrete   Pervious concrete

Component

Type: Rain Garden
Lifespan: N/A (requires seasonal upkeep)
Rate: 1-3 gals / sq. ft.
Cost: $3-12 / sq. ft.
Minimum: 10' from structural foundation
Maximum: None

Description:
Rain gardens are gardens that are watered by collected or pooled stormwater runoff, slowly infiltrating it into the ground along root pathways. They are typically planted with wildflowers and deep-rooted native vegetation, which helps infiltrate rain channeled to them from roofs, driveways, yards and other impervious surfaces. They can be placed near downspouts on homes (although away from building foundations and sewer laterals), and are an excellent means of removing pollutants from stormwater runoff. They should be slightly depressed to adequately hold and infiltrate stormwater runof

Upkeep:
> Seasonal weeding
> Seasonal trimming

Construction

1 **Clearence:** There needs to be a minimum of 10 ft between the start of the Rain Garden and any home/ garage structural footing.

2 **Dimensions:** The depth & width of the rain garden depends on designed capacity and slope of existing lawn. Soil characteristics also help define the depth of the garden.

3 **Berm:** A berm is created at the low end of the garden to help contain and store water during a wet wether event. Generally no new earthen material is needed to create this berm.

4 **Vegetation:** Plant selection should be include deep rooted native plants. Plant lists include wildflowers and low shrubs typically.

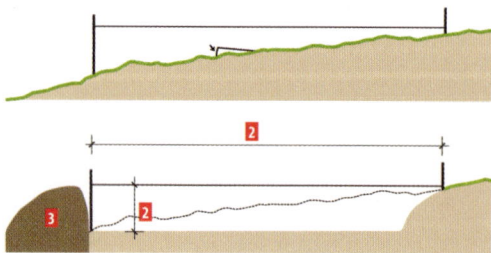

Plan/Section of existing availible area for intervention

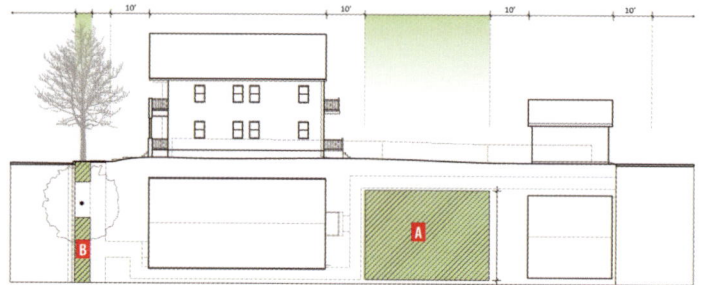

Typical construction section

A public intervention

B Private intervention

Rain Gardens

Rain gardens are extremely effective solutions to clean, slow, and store water. However, they require a substantial amount of space to yield this effectiveness, whereas the lots of the prototypical block are narrow and densely packed. The space needed to handle a 1" storm would require the entirety of the private sector's open space. Furthermore, there is little to no space in the existing public sector of the block to implement rain gardens effectively. Although it's a good intervention strategy, the rain garden does not have the capability to make an effective impact without major spatial changes to the existing conditions.

Isolated

Individual Lot Calculations
(33' lot width) & (2 gals / sq. ft. rate)
(Cost: $8 / sq. ft.)

A **Public**
Avg. available area: 118 sq. ft.
Capcity: 237 gallons
Cost: $946

B **Private**
Avg. available area: 865 sq. ft.
Capcity: 1,730 gallons
Cost: $6,920

C **Public + Private:**
Avg. available area: 983 sq. ft.
Capcity: 1,966 gallons
Cost: $7,860

Aggregate

Entire Block Calculations
(2 gals / sq. ft. rate)
.5" storm = 60,800 gals

A **Public All:**
Avg. available area: 4,260 sq. ft.
Capcity: 8,500 gallons
Cost: $32,000
% of .5" storm: 14%

B **Private All:**
Avg. available area: 31,128 sq. ft.
Capcity: 62,250 gallons
Cost: $233,500
% of .5" storm: 102%

C **Public + Private All:**
Avg. available area: 35,400 sq. ft.
Capcity: 70,750 gallons
Cost: $265,400
% of .5" storm: 116%

Potential rain garden area   Building foundation safe offset

Component

Green Street

Type: Green Street
Lifespan: N/A
Rate: 3-17 gal / sq. ft.
Cost: $250/each
Tree Types:
Minimum:
Maximum:

Description:
Green alleys, streets and parking lots are typically in the public right-of-way and can provide a combination of different benefits designed to channel, infiltrate and evapotranspire rainwater. They include permeable pavement, sidewalk planters, landscaped medians and bio-swales, inlet restrictors, greenways and trees (as described above), and can also take advantage of recycled materials.

Construction

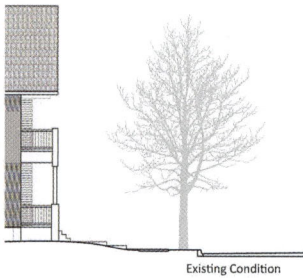

Existing Condition

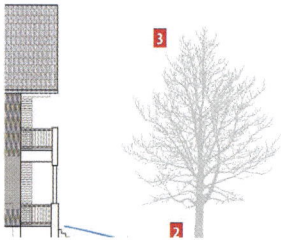

1 **Porus Pavement:** Rate of 4 gallons per square foot at a cost of $0.35 gallon of storage.

2 **Rain Garden:** Rate of 2 gallons per square foot at a cost of $3.75 per gallon of storage.

3 **Stormwater Tree:** Unknown rate when linked to storm event, however holds up to 500 gallons per year.

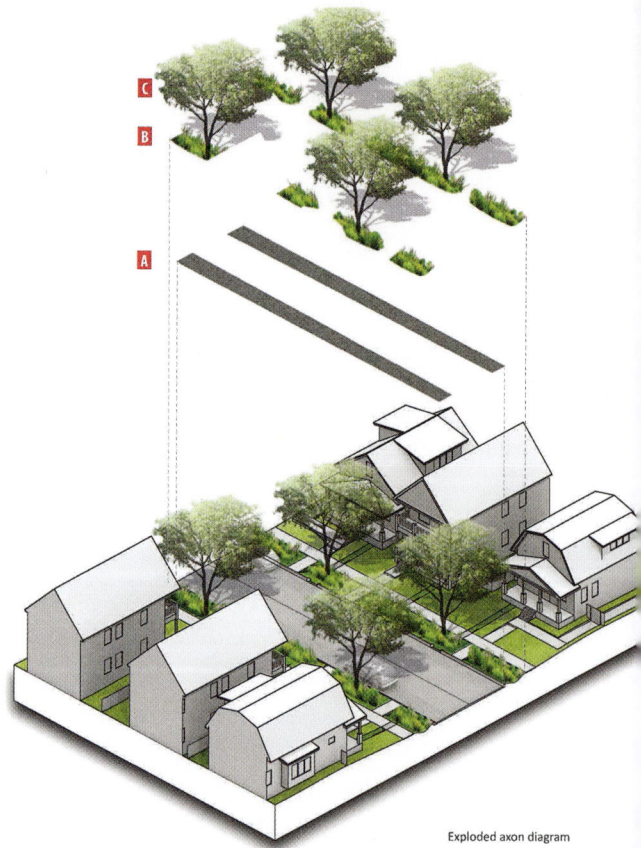

Exploded axon diagram

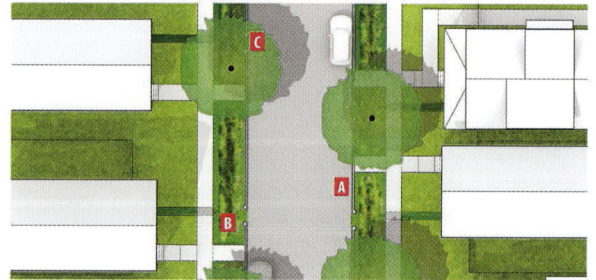

Green Street

The green street interventions could handle up to an entire .5" storm and completely transform the social quality of the existing blocks. However, they are comparable to porous concrete in their financial strain and would reduce the size of the street to make space for the rain gardens and stormwater trees. Potentially a lane of on-street parking would need to be eliminated to have enough space to implement them correctly, while still not being able to handle an entire 1" storm.

Isolated

Front facing lot calculation
(individual house facing street)

A Porus Concrete :
Width: 16 ft (8' both sides)
Area: 528 sq. ft.
Capcity: 2,112 gallons
Cost: $740

B Rain Garden
Width: 16 ft (8' both sides)
Area: 448 sq. ft.
Capcity: 896 gallons
Cost: $3,360

C Stromwater Tree
# of trees: 1.5
Capacity
Cost: $375

TOTAL CAPACITY
Approx. 3,000 gallons
Cost: $4,500

Aggregate

Entire block calculations
(620 ft long block)

A Porus Concrete :
Width: 16 ft (8' both sides)
Area: 9,280 sq. ft.
Capcity: 37,120 gallons
Cost: $13,000

B Rain Garden
Width: 16 ft (8' both sides)
Area: 8,064 sq. ft.
Capcity: 16,128 gallons
Cost: $60,480

C Stromwater Tree
# of trees: 25
Capacity
Cost: $6,250

TOTAL CAPACITY
Approx. 53,250 gallons
Cost: $80,000
% of .5" storm: 87%

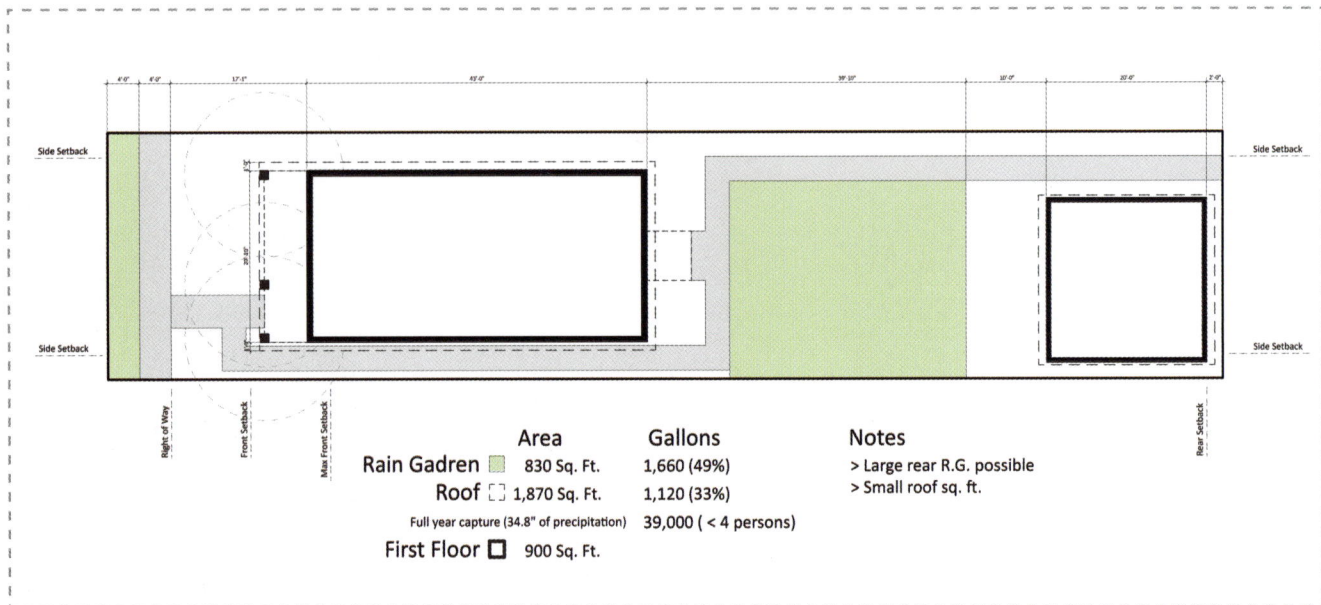

	Area		Gallons
Rain Gadren		850 Sq. Ft.	1,700 (50%)
Roof		2,100 Sq. Ft.	1,120 (40%)
Full year capture (34.8" of precipitation)			50,000 (annual water use family of 4)
First Floor		900-1,200 Sq. Ft.	

	Area		Gallons	Notes
Rain Gadren		830 Sq. Ft.	1,660 (49%)	> Large rear R.G. possible
Roof		1,870 Sq. Ft.	1,120 (33%)	> Small roof sq. ft.
Full year capture (34.8" of precipitation)			39,000 (< 4 persons)	
First Floor		900 Sq. Ft.		

Prototypes

(left top) The prototypical lot needs to address all variables. New construction within the existing
fabric needs to reach these goals in order to make an adequate impact within a decentralized water management system.
(left bottom) The existing condition presents a constant to measure new interventions against. New lot organization needs to respond to existing parameters, as well as to existing adjacent lots.

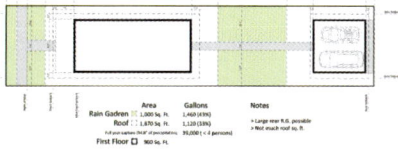

Full Foundation | Front Loaded Aligned | Split Garage

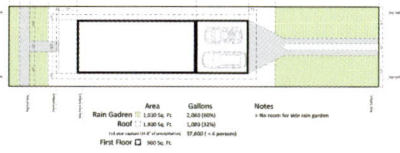

Full Foundation | Front Loaded | Aligned

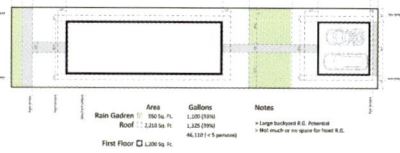

Full Foundation | Split

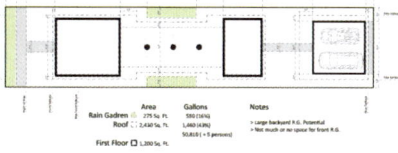

Full Foundation | Split | Bridge Center

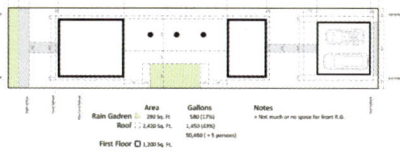

Full Foundation | Split | Bridge Offset

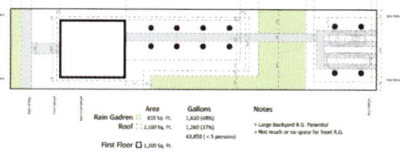

Full Foundation | Split | Cantilever Offset

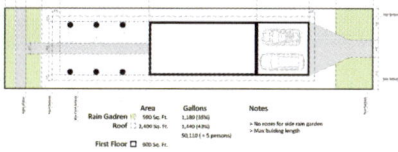

Foundation Back | Piles Front

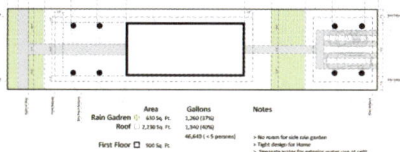

Foundation Back | Piles Front | Split Garage

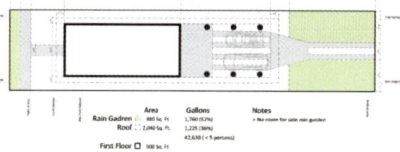

Foundation Front | Piles Back

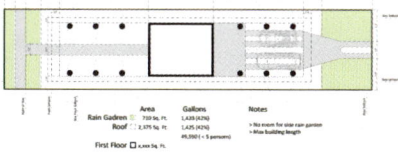

Foundation Middle | Piles Front & Back

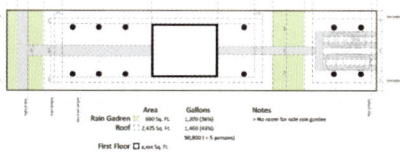

Foundation Middle | Piles Front & Split Back

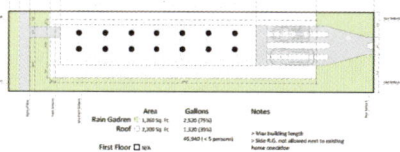

Full Piles | Narrow

Above:
Numerous spatial iterations demonstrate an array of possible lot organizations. Iterations highlighted by a red box reach the defined goals for rain capture, rain garden, and habitation square footage. The highlighted iterations inform the need for an extensive linear rain garden that stretches to both ends of the lot to allow for overflows in both directions. Also, a narrow meeting of the ground in either a foundation wall or pile will result in more space for the necessary rain garden space to exist. Finally, a carport should replace the need for a full garage, which spatially and visually opens up the narrow lot as well as increases the horizontal space for the linear rain garden

The design of this infrastructural landscape—an economic, ecological, and industrial territory extended beyond its location at such a confluence—becomes a delicate negotiation of global and local flows of economy, ecology, and politics. The aerial view is looking east, demonstrating how the breakwater creates a barrier between the interior and exterior conditions while providing a new public beach for residents.

--- ---

Index of tools | infrastructural urbanism, aquacultural industries, hydrological confluence, scalar processes, industrial + ecological hybridization, wetland and aquacultural bars, infrastructural conduits.

DETOXI-CITY

Barra Do Furado, Brazil. Speculative, 2012.

Zuhal Kol, Rodney Bell, and Julia Gamolina

Infrastructural Urbanism

The Campos Basin, an oil-rich offshore territory located along Brazil's southeast coast, is becoming a new frontier of oil exploration. While the discovery of oil here has spawned offshore development, production facilities onshore are currently being planned for the management of these oil reserves. This economic bliss is complicated by the sobering reality that at even optimistic projections, the estimated 850 billion barrels of oil will yield for merely 20 years. As logistical ports and refinement facilities are being planned for the rural coastal villages, most of which thrive on local fishing and tourist industries, questions immediately emerge regarding the ecological reverberations of such rapid urbanization. With more than $220 billion invested by Petrobras, which manages and regulates Brazil's petroleum reserves, the country has exhibited a dedication to becoming a leading global petroleum producer. What is to become of these production facilities after the boom?

Re-Forming the Company Town

The site of this proposal—Barra do Furado—is exemplary of this condition. Situated at the confluence of a man-made canal connecting coastal waters to an inland lake, tidal marshes, seasonal flood zones, and salt-water tributaries, Barra do Furado is not the optimal location—ecologically—for oil refinement. Logistically, however, with its proximity to the offshore facilities,

its dredged canal, and its proximity to infrastructural corridors, it serves as a strategic location for oil receiving, processing, and distribution. The design of this infrastructural landscape—an economic, ecological, and industrial territory extended beyond its location at such a confluence—becomes a delicate negotiation of global and local flows of economy, ecology, and politics. Accepting that Barra do Furado cannot resume as a fishing village once the oil industry has expired, what opportunities exist at this hydrological confluence? With the Brazilian government's investment in its burgeoning aquacultural industries—fish farming, paddy farming, et cetera—can infrastructures supporting the oil industries be reprogrammed through an embedded logic to support a more sustainable aquacultural industry? And can both of these industries thrive on these hydrological flows—both man-made and natural?

Design Strategies: Managing Contingency

Detoxi-city is a project born of both the constraints and opportunities afforded by the confluence of industry and hydrological ecosystems. In the wake of globalization, the market economy, and the spatial fluidity of neoliberalization, environmentalist calls for the suppression or displacement of industry and production away from ecologically critical thresholds has largely fallen on the deaf ears of national environmental regulation. The reality is that industry requires such sites for logistical efficiency and, given its complicity

Campos Site and Oil Issues
Issues to be addressed based on current state of oil extraction

CAMPOS BASIN OIL PRODUCTION

CAMPOS BASIN OIL RESERVES

daily oil production : **1.2 million bopd**
proven oil reserves : **8.5 billion bbl**

19.6 years
production

Process Loop
Oil and Fishing inputs and outputs

CONSTRUCTION   OIL INDUSTRY   +25   FISHING INDUSTRY

with both federal and private economic interests, will indeed occupy such ecologically critical sites. Therefore, the question no longer resides in the idealism of whether we should build on and around estuaries, river deltas, wetlands, and tidal marshes; rather, the question is how can we urbanize such ecologically critical territories utilizing scalar processes and strategies that opportunistically manage risks through a hybridization of industrial and ecological processes? Rather than waiting for catastrophic infrastructural failures that resonate through larger territories, Detoxi-city proposes a co-opting and coupling of hydrological flow to industrial production (in this case, oil refinement and processing) in order to both contain, cleanse, and reintroduce waters that have been used and affected by the industrial processes proposed for the site. Extending from this more pragmatic necessity, these processes, their inputs, and outputs, have been further coupled to the larger economic and social forces that will inevitably engage the site, as described below.

Formatting Urban Growth

Thus, fully utilizing the available inputs of hydrological flow—from tidal flux, seasonal flooding, surrounding estuaries, and tributaries—Detoxi-city anticipates and facilitates the conversion of one industry to another. While these infrastructures of contamination management initially serve the oil industries, they will later form a framework for an emergent aquaculture industry—the beginnings of which exist locally in Barra do Furado and are being further subsidized through the Brazilian government. This transition is managed exclusively on hydrological terms through detoxification—a transcalar process operating through the three main systems, which constitute this proposal: (1) housing islands, (2) wetland and aquacultural "bars," and (3) the infrastructural conduits. These systems mutually reinforce the overall strategies of detoxification, containment, coexistence, and conversion. While aquacultural systems deal with detoxification through production and cleansing, the boundary zone of the city serves to contain toxins within this overall system such that they may be cleaned before reintroduction; in this way, citizens are more than aware of the contingencies driving the economic, social, and ecological vitality of the city—they participate through employment, leisure, and inhabitation in a coexistence with these temporal processes.

Design Proposal

The primary aim of this proposal is to format urban growth and development such that latent processes integrally emerge over time. Thus inputs from one process (such as paddy farming) are fed directly into another process (waste management wetlands). All such infrastructures create a foundation for a new public space. This public space is not the abstract open plane of modernity. Rather, it is a space of productive engagement where citizenry take part in this system at various scales and are able to thus assert themselves within the system. While a thriving local fishing community exists at the site, this industry is far too small to support the influx of a projected 10,000 workers that will inevitably be economically displaced with the expiration of oil operations. Already, Brazil's oil industry constituted the largest share of permanent work visas for migrant workers in 2013. This incoming workforce can thus transfer, over time, into the emergent aquaculture industries, which occupy and replace the oil infrastructure.

Temporality

Detoxi-city does not envision growth in indefinite terms. With the oil- yield projections discussed above, this urbanism operates on a very finite temporality. Thus, there was a need to make an explicit and formally clear boundary of intervention, which is then densified and within which cleansing happens. This boundary, while conceptually and pragmatically determining the interior of the city (housing, industry, infrastructure, production, et cetera), becomes ambiguous over time because of the hydrological processes that shape the site.

Formal Organization: The Re-Formed Company Town

Within this zone of intervention, programs are organized into bars, which maintain certain productive adjacencies with one another. These bars contain aquaculture, farming, housing, wetlands, and port facilities. Each bar's relationship to its neighbor is based on the utilization and reappropriation of processual inputs and outputs. A central spine of infrastructure cuts through the bars. This X—in plan—contains the existing oil lines that connect existing regional infrastructural corridors to offshore facilities. These infrastructural corridors serve to store oil and move resources throughout the site and to the refineries at the periphery. Thus, there is a tangible rendering of more global and national resource movements registered into the

Opposite Page:
Rather than waiting for catastrophic infrastructural failures that resonate through larger territories, Detoxi-city proposes a co-opting and coupling of hydrological flow to industrial production. The graph demonstrates the flows of production, starting with the oil industry, and how that may evolve into the economic and social needs of the future city.

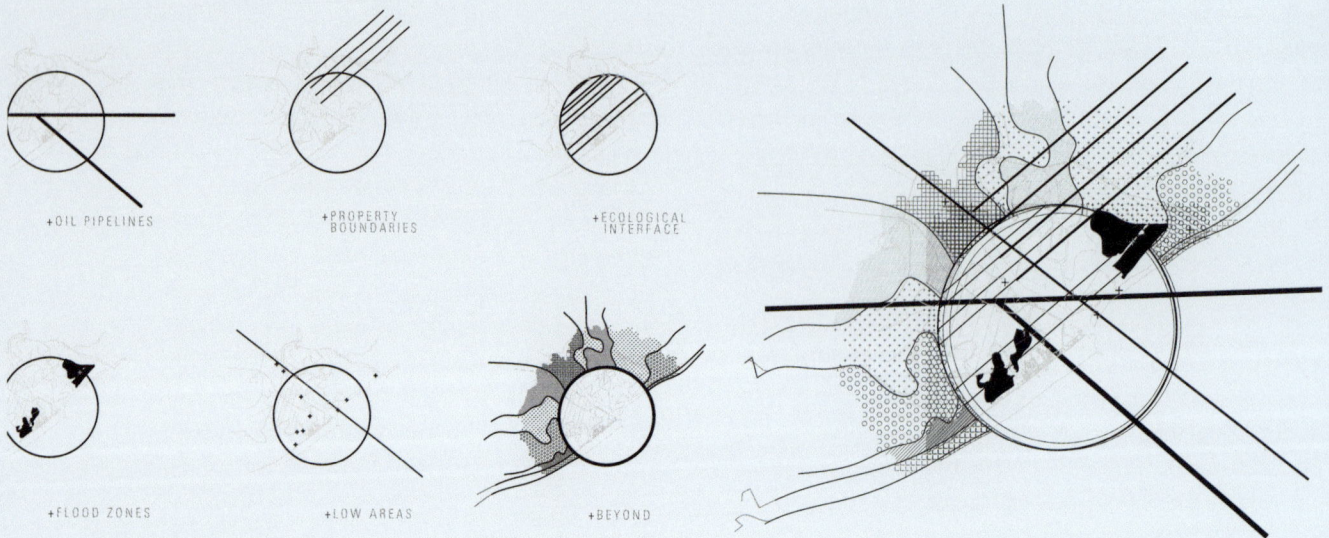

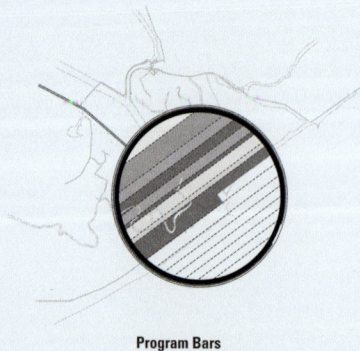

+OIL PIPELINES

+PROPERTY BOUNDARIES

+ECOLOGICAL INTERFACE

+FLOOD ZONES

+LOW AREAS

+BEYOND

Program Bars

Program and functions are organized across the site in bars of performative adjacency. Each bar spans the diameter of the site connecting across the ring and interacting performatively with the bar next to it.

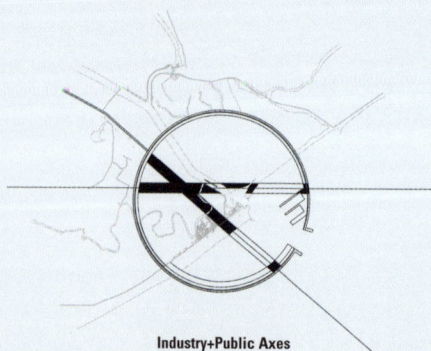

Industry+Public Axes

The oil pipelines—already planned for the site—provide an axis for the organization of public functions. This axis houses oil bulks which inhabit the public program on their floating roof and are later repurposed for the emergent fishing industry.

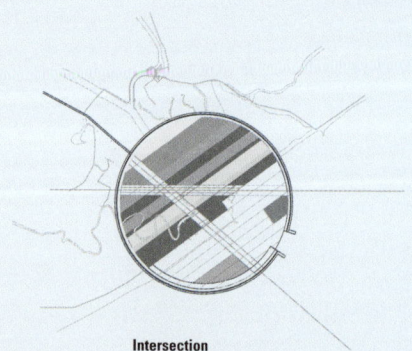

Intersection

The intersection of the public axes with programmatic bars provide opportunities for short-circuiting the linearity of the bars both physically and programmatically.

urban morphology of Detoxi-city. Mapped onto these signifiers of the industry which predicates the city are public spaces of gathering and cultural production. Additionally, each each bar of the program engages this public and infrastructural corridor in various ways. Unlike traditional company towns, which employ a strict dichotomy between industry and housing, Detoxi-city makes explicit the industrial infrastructure and renders it public over time. Thus, this space becomes a place for the production of industry and the production of culture. A constant negotiation exists between the industrial and cultural use of this urban connector. Urbanistically, these corridors are envisioned as "urban connectors" where circulation cross-cuts the bar's logic and people can gather as they move through the city.

Territorial Implications

While the interior of the city is well defined in terms of its processes and their relationship to the public and private domains, the boundary of the city becomes more ambiguous and, in fact, begins to reorganize and rehabilitate ecological flow at a more territorial scale. Because of Barra do Furado's location at the confluence of multiple hydrological systems, the containment of the city's refuse is necessary. However, as water and toxic refuse is cleansed through various processes on the bars, its percolates to holding ponds at the periphery of the city. With the seasonal flooding of the farmlands and tidal marshes, water is allowed to flush into the city but is cleansed before being reintroduced into the surrounding territory. Additionally the two rivers, which cut through the city, are lined with stepped-wetlands in order to further filter toxins carried from the surrounding farmlands (fertilizers/pesticide). This more mechanized interior reorganization resonates into the exterior territory of the city.

Detoxification

The strategies of decontamination and detoxification are operated through hydrological flow in various forms. The processes on the site largely depend on the flow of water into, out of, and within the city. Toxins brought in from the surrounding farmlands and the refuse of the oil industry (runoff pollution, oil leaks, and ballast water contamination) introduce harmful toxins into this critical area. However, these same processes of hydrological flux, which disperse such toxins, are re-deployed within the city towards cleansing waters for reuse and reintroduction. The aquacultural systems operate on two fronts: (1) as actively productive agents within the city—processing

water for reuse, cleaning water as it enters from the surrounding rivers, canals and estuaries, processing and producing dredged fill material for earthworks, producing protein in the form of hatchery-grown fish, and producing food in the farm plots; and (2) as passive receptacles of hydrological flux—absorbing rainwater and runoff flooding. Using strategies of productive adjacency, the bars move, store, contain and process water throughout the site. The rational and mechanized mediation of hydrological flow within the site prevents the dispersal of toxins into the surrounding territory and conversely cleanses the toxi- rich tributaries, which enter the site from the surrounding farmlands. Along with the contaminants produced by industry and the biological waste produced by the inhabitants, the port must constantly be dredged to maintain water depths in the canal and port and to replenish the breakwater which will inevitably be eroded by ocean currents. Thus, the wetland bar adjacent to the interior port on the site is dedicated to the cleansing of dredge material such that it may be redeployed throughout the site for various uses.

Infrastructural Public

The programs and their adjacencies were chosen in order to maximize the cleansing of water at this ecologically critical juncture. Beyond these more pragmatic concerns, however, these programs provide the foundation for a new public space and more sustainable industry once the oil production is phased out. The wetlands serve to cleanse both toxins and biological waste for the city. The aquaculture bars—for both fish farming and paddy farming—lay the foundation for a new more sustainable industry while also cleaning and preparing water for reuse.

Detoxi-city proposes more productive modes of urbanization that deal with the material realities of contemporary urbanism. However, as these realities are a starting point, Detoxi-city projects the possibility for a new public, emerging productively from such private developments as oil refinery towns and fully engaging with the infrastructure which constitutes the urban realm. Far from the first generation of hard infrastructure constructed during the New Deal era, these infrastructures are responsive to ecological, environmental, economic and social variation over time. Moreover, beyond merely being sublime impositions on the landscape—the taming of nature—these new infrastructures provide a space of cultural production—a space for citizen engagement and interaction on productive terms.

Above & Opposite:
The various layers of the city.

Conversion of Industry
Below: Oil Industry
Above: Fishing Industry

- labor
- wastewater
- cleaned water
- re-introduced water
- contaminated water
- agricultural output
- aquacultural output
- oil exports
- oil imports
- maritime shipping
- automotive shipping
- locomotive shipping
- power production

+ WATER INPUTS & OUTPUTS

+ WATER MOVEMENT

+ SEDIMENT PROCESSING & REDEPLOYMENT

[2] Dredge Material Processed
Processed water percolates through material before being sent to adjacent wetlands.

[1] Dredge Material Arrives
Dredge boats offload material at the shipyard which is placed into wetland here.

[3] Process Cycle
The dredge material goes through several cycles of percolation and aeration before further reuse for biofarming.

[1] Wastewater Input
Wastewater is pre-processed at the treatment facility before being discharged into the filtration system.

[2] Wastewater Input
After percolating through sediment, water is filtered and passed to the next stage.

[3] Filtered Water Output
After several stages of filtration surrounding wetlands.

+ CANAL

+ SYSTEMS AXONOMETRIC

+ FARMING PLOTS

Ecological Interface

Ecological Interface

Waste Processing Surfaces

Productive Surface System

tidal flux wetlands / shrimp farming

paddy fields

small farm plots

growing ponds

nursing ponds

fish hatching ponds

reservoir

hatching ponds

aqueduct

water transfer node

water transfer node

fish hatchery - adult growing pond

paddy field farming

absorption wetlands

catchment reservoir

outer wetlands

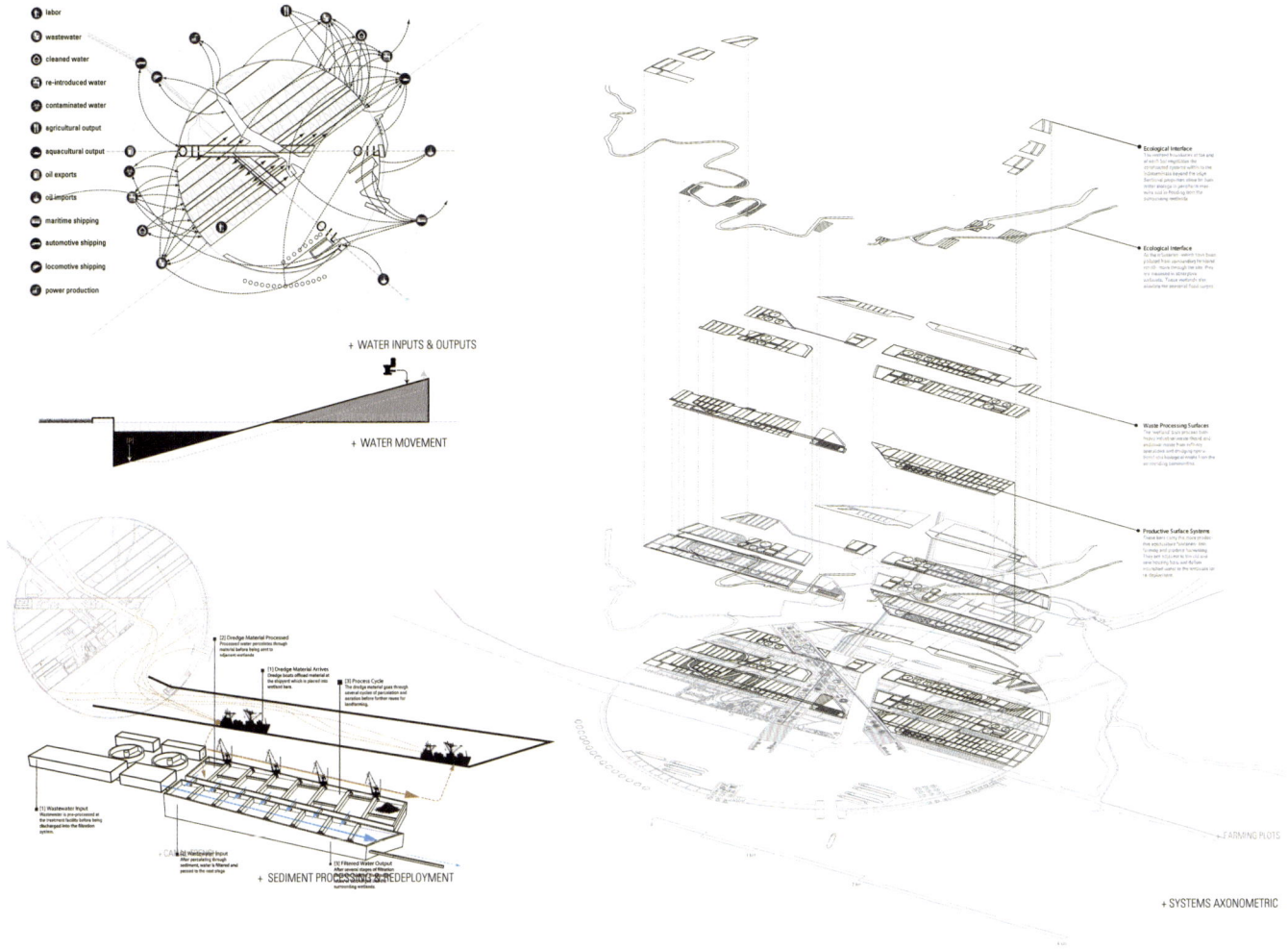

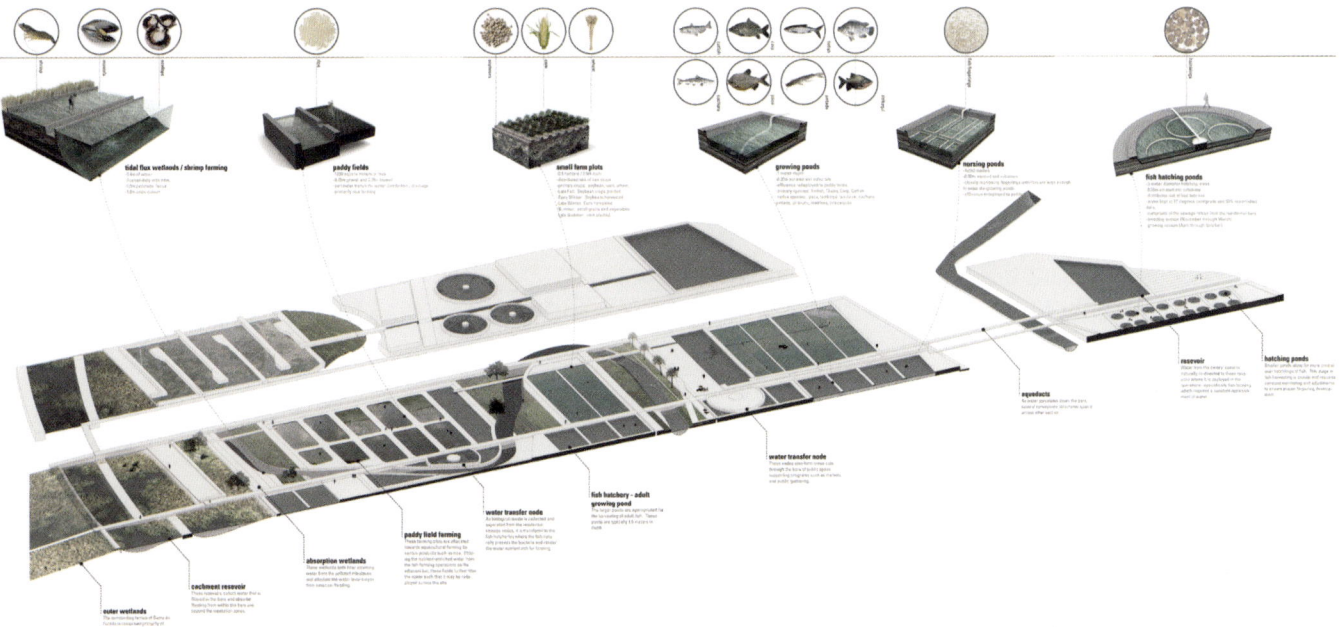

The strategies of decontamination and d
hydrological flow in various forms.

Connectors // Coexistence+Conversion

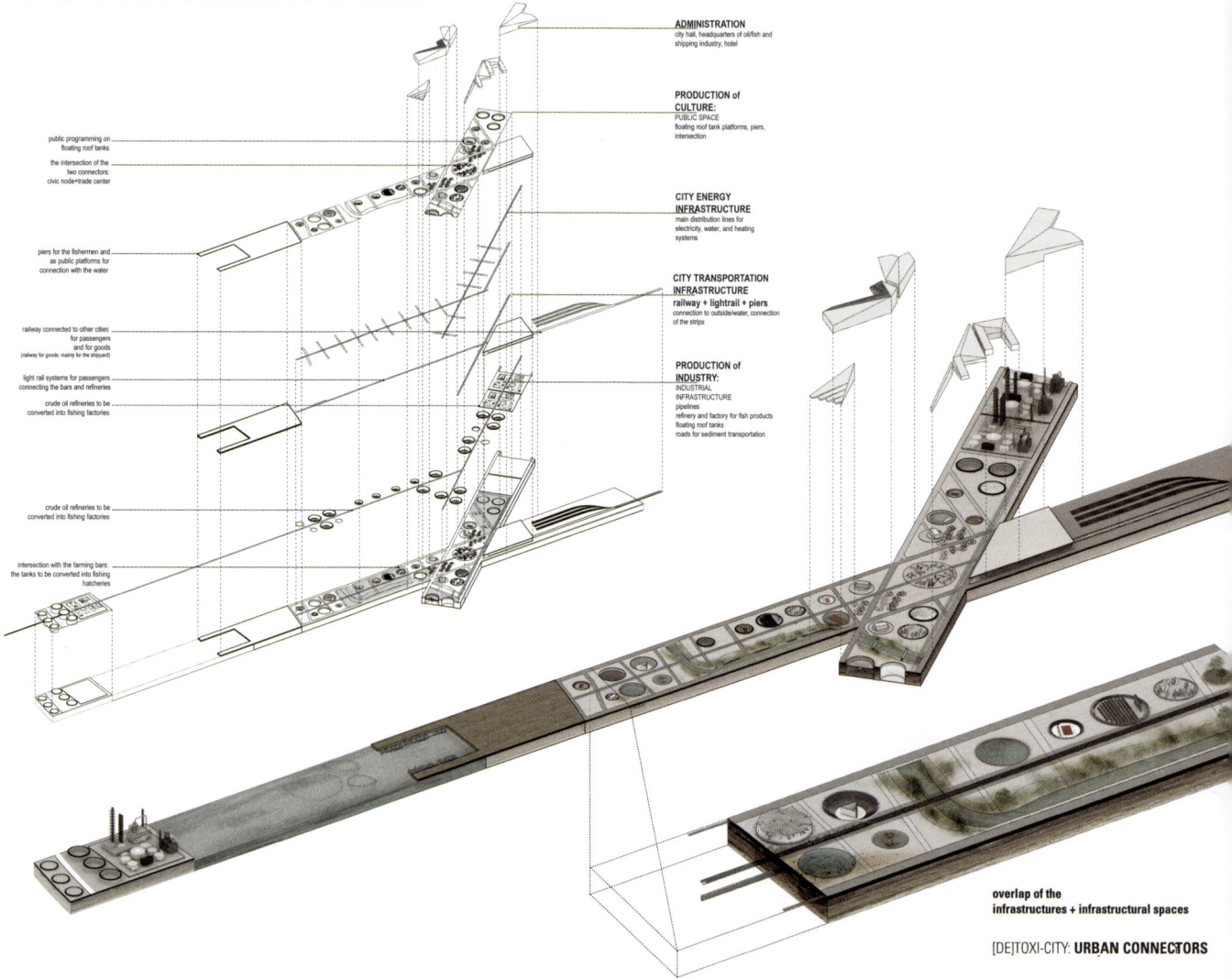

ADMINISTRATION
city hall, headquarters of oil/fish and shipping industry, hotel

PRODUCTION of CULTURE:
PUBLIC SPACE
floating roof tank platforms, piers, intersection

public programming on floating roof tanks

the intersection of the two connectors:
civic node+trade center

CITY ENERGY INFRASTRUCTURE
main distribution lines for electricity, water, and heating systems

piers for the fishermen and as public platforms for connection with the water

CITY TRANSPORTATION INFRASTRUCTURE
railway + lightrail + piers
connection to outside/water, connection of the strips

railway connected to other cities for passengers and for goods
(railway for goods: mainly for the shipyard)

light rail systems for passengers connecting the bars and refineries

crude oil refineries to be converted into fishing factories

PRODUCTION of INDUSTRY:
INDUSTRIAL INFRASTRUCTURE
pipelines
refinery and factory for fish products
floating roof tanks
roads for sediment transportation

crude oil refineries to be converted into fishing factories

intersection with the farming bars: the tanks to be converted into fishing hatcheries

overlap of the
infrastructures + infrastructural spaces

[DE]TOXI-CITY: URBAN CONNECTORS

As an anomaly throughout the city, two bars are diagonally placed on the rest of the bars. The Xs house the oil industry and are based on the existing logic of the oil-carrying pipelines. The bars help to connect the program bars and create moments of intersection that serve as key public realms.
The bars provide continuity, unity, and control. The X that runs through the project uses the existing logic to make the canals applying it to connect all of the bars, a public space connector (providing a corridor through the bars), and a host of industry. Key industry buildings help to reinforce the spine, through their permanence (since just transportation, or just spaces of flux in the spine don't suggest that it is a temporary barrier).
X as the landscape of industry nourishes and intersects all the strips; however, as it intersects the strips and connects the spine, it also creates the landscape of the public realm through the negotiation of programmatic strips. Unlike the classical company towns where industry and town kept away from each other, X unites them and becomes a place designed both for the production of industry and the production of culture. The public space shares the industry's infrastructure and directly negotiates with industry while experiencing the processes in town through the strips and the industry. Rather than a 'public' space designed explicitly for consumption, X creates and produces a culture out of the processes of a company town. Using its infrastructure also to serve for the later industry, X experiences the transformation of the oil i ndustry infrastructure into fishing infrastructure. Allowing this transition smoothly from one to the other, X inhabits the conversion and coexistence through adapting its infrastructure.

CONVERSION :
Oil Industry—Fishing Industry
+
COEXISTENCE:
Landscape of production—Landscape of public realm

CRUDE OIL/BALLAST WATER
STORAGE

CLEANING
MECHANICAL SCRAPER & SOLVENT FREE
POLYURETHANE COATING

FISHING HATCHERY
FISH FARMING, WATER STORAGE,
PUBLIC POOL

EXTERNAL FLOATING ROOF TANK

An external floating roof tank is a storage tank commonly used to store large quantities of petroleum products such as crude oil or conden- sate. It comprises an open- topped cylindrical steel shell equipped with a roof that floats on the surface of the stored liquid. The roof rises and falls with the liquid level in the tank. As opposed to a fixed roof tank, there is no vapor space (ullage) in the floating roof tank (except for very low liquid-level situations). In principle, this eliminates breathing losses and greatly reduces the evaporative loss of the stored liquid. There is a rim seal system between the tank shell and roof to reduce rim evaporation.

gram

l system:
aircase/membrane covering layer

ection system

l slab for the public program

oof+base of the tank

open-air auditorium

market place

fishing pond**

sports field

skating park

playground

open-air auditorium/screen

open-air theatre

ice-skating rink

installations

swimming pool**

rain water storage**

COMMUNITY
COURT YARDS

HOUSING ISLANDS
5 distinct districts with distinct unit
typologies

EXISTING HOUSING

WETLANDS

FARMING

WETLAND ORGANIZATION
Wetlands,farming, and hatcheries &
their interaction with the housing bars

CONVERSION INTO
FISHING

PUBLIC REALM

OIL INFRASTRUCTURE

INFRASTRUCTURAL : PUBLIC X
Conversion from oil to fishig and the
coexistence of oil with the public realm

SHIPYARD

CITY BOUNDARY

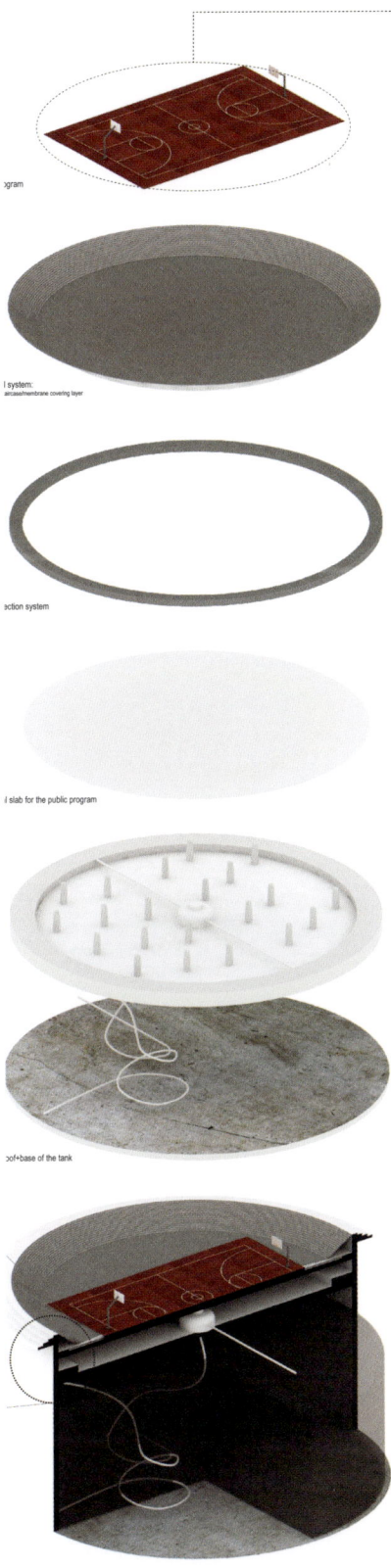

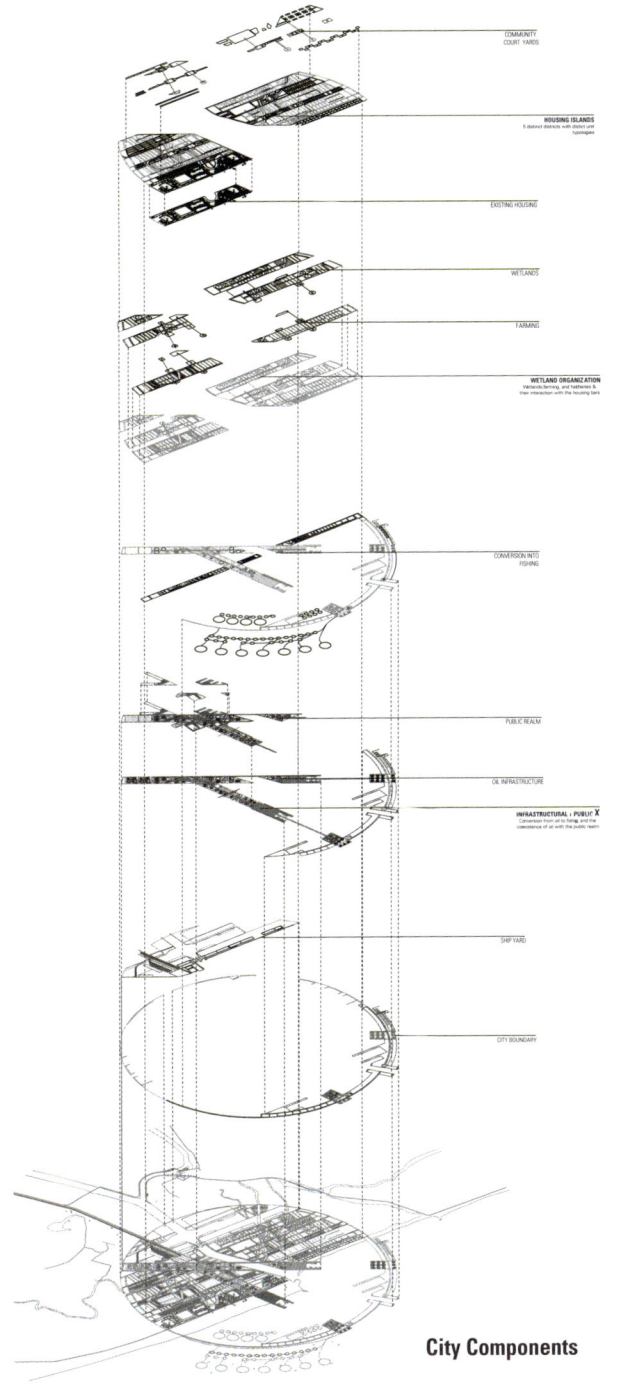

City Components

ACT THREE
THE DISAPPEARING

The same quantity of water exists on the planet now as existed on the earth 5 billion years ago. Water is not disappearing from the planet, as all water is recycled. However, water *is* disappearing in key locations around the globe—locations where it is greatly needed for agriculture, manufacturing, energy, recreation, and human consumption. Increasing populations in environments that are already water-stressed instigate the disappearance of water. Each year the earth's population increases by 83 million more people, and water demand continues to rise. According to a 2010 National Geographic article, "Nearly 70% of the world's fresh water is locked in ice and most of the rest is in aquifers that we are draining much more quickly than the natural recharge rate."[1] In 15 years, 1.8 billion people will live in regions of severe water scarcity. [2]

This is the situation of "The Disappearing," an investigation of water's departure from populated or ecologically sensitive regions. The projects and essays presented here examine how cities, architecture, and landscapes work to counteract the depletion of water sources. Traditionally, distance has been a major obstacle in connecting thirsty populations to water sources. Development in the American West, in cities like Los Angeles or Las Vegas, depends upon the rerouting of distant water from the Colorado River. In undeveloped regions in the world, human feet traverse this distance, not networks of infrastructure. Women in developing countries walk an average of 3.7 miles to get water.[3] Thus, the critical element to examine in the "disappearing" of water is the relationship between population growth, resource depletion or recovery, and distance to an available water source. The proposals in this act generally take a position within the system of hydrological distribution: the point of intake, the line of transport, the pool of collection, or the point of outtake. They offer possibilities for living in arid, water-stressed situations.

Two Call to Action texts set the tone for this act of water depletion. William Morrish's essay "The Urban Spring: Formalizing the Water System of Los Angeles," first published in 1984 in **Modulus** volume 17, examines how a desert city is formulated by the parameters of water distribution and collection. In this early survey of Los Angeles's water system, Morrish highlights how the extreme separation between point of intake and point of use instigates irresponsible water consumption habits. Morrish notes, "to the average person's perception of the city, this labyrinth [of water infrastructure] remains hidden from view," until they confront the water bill. Thus, "The Urban Spring" calls for a visual acknowledgment and celebration of water distribution systems in order to promote conservation efforts within the city.

Morrish's analysis of Los Angeles is followed by Barry Lehrman's essay "Reconstructing the Void: Owens Lake," which was first published in Kazys Varnelis's 2008 book *The Infrastructural City: Networked Ecologies In Los Angeles*. Here, Lehrman examines Owens Lake, "a silent victim of the city's destructive thirst." Lehrman has provided new graphics, revised text, and a postscript to his original essay in *The Infrastructural City* for this publication.

These Call to Action texts examine the specifics of Los Angeles and that city's continuous struggle to import water. They reveal that urbanization in arid environments is an extremely fragile operation that prompts a major rerouting of water sources. These texts present a "this will kill that" situation that is all too common in efforts to provide water to the masses, in which supplying water to a city often provokes the disappearance of aquatic ecologies for the rural population.

To defend against the disappearance of water, mechanisms are constructed that bring water from distant lands to people, buildings, landscapes, and cities. Key tools that enable this importation are the reservoir, the aqueduct, and the pump. Remote access to water sources is overcome by engineering solutions that import water, such as South-to-North Water Diversion Project in China or The New Croton aqueduct that supplies New York city with water. In addition to the case study of Los Angeles provided by Morrish and Lehrman, this section includes an analysis by Benjamin Gregory of Las Vegas. Vegas is also a city in the desert playing a vicious game of importing water and fighting over ground rights for artesian wells. As this analysis shows, vast hydrological extensions support this oasis in the desert.

To retreat from the effects of the disappearing entails moving people, wildlife, and the built environment away from stressed water environments and to new water-rich environments. This is the strategy of Sea Tree by Waterstudio, a project that imagines the sea as a new frontier. Sea Tree offers an alternate habitat for flora and fauna that must escape urbanization and desertification. While the project is dedicated to nonhumans, it could be imagined as a solution for human retreat, recalling Kiyonori Kikutake's *Marine City*.

To adapt to the condition of the disappearing is to harvest, store, and restore healthy water within the spaces of cities, landscapes, and buildings. This is the contemporary strategy—if not obsession—where control of water distribution lies in the hands of the designer. The examples are plentiful for this strategy and we have provided a wide range of both realized and speculative projects of fluctuating scales.

The Winton Wetland Restoration project by Taylor Cullity Lethlean presents an adaptive strategy within the water-stressed Australian frontier. This project of national scientific, cultural, and environmental significance is the largest wetland restoration project in the southern hemisphere. The project aims to create a major national facility for wetland education and research, as well as to demonstrate best-practice natural resource management and to develop nature-based tourism activities and recreation. The public becomes a key ingredient in the restoration of water and ecology within a fragile nature.

"Lima Beyond the Park" by Antje Stokman, Bernd Eisenberg, Eva Nemcova, and Rossana Poblet examines the Peruvian capital of metropolitan Lima, located on a desert coast overlooking the Pacific Ocean. With its more than 9.5 million inhabitants, Lima is considered the most extensive desert city in the world after Cairo. The project takes a multi-scalar approach to addressing the vast expansion of informal settlements and their lack of basic urban water services. The proposal considers urban, open space as a key ingredient to save water, purify water, treat wastewater, and recycle nutrients, or even to harvest water. Public space is coupled with water infrastructure, creating a spatial hybrid, a common form for adaptation.

"WASH: Urban Hydrological Networks for Resilient Cultural Ecologies" by Aja Bulla-Richards takes us back to Los Angeles and the problems outlined by Morrish and Lehrman. This project proposes prototypical interventions that reconfigure stormwater and greywater infrastructure to initiate layered social and ecological structures in a typical Los Angeles neighborhood. Here, water shortage is addressed within the domestic space: the home, the yard, and the street become key components for a decentralized water-recycling intuitive.

A similar approach to decentralizing water distribution and harvesting can be seen in Izaskun Gonzalez Barredo and Oriol Valls Guinovart's "Immaterial Water," a speculative project that imagines a time when water can be extracted from the air. Here, a refrigerated ceiling becomes the answer to collecting water vapor from the atmosphere. Water is harvested within the boundaries of architecture and distributed on-site.

The projects of "The Disappearing" address drought and water shortages resulting from climate change, desertification, and rapid urbanization. They range from fantastic speculations of water-harvesting gadgets to realized parks of reclaimed wetlands. While some provide very specific, technical solutions, others are included to generate a larger, conceptual discussion. The hope is that this index of possibilities for the future can catalyze a discourse about the disappearing waters and design methodologies to deal with them.

Notes ---

1. "Water: Our Thirsty World," *National Geographic* (April 2010) 217, no.4: 52.

2. Ibid, 54.

3. Ibid.

ACT THREE: THE DISAPPEARING

CALL TO ACTION

The following text is reprinted from the essay "The Urban Spring: Formalizing the Water System of Los Angeles" by William Morrish, first published in *Modulus* volume 17, 1985. Reproduced by permission of William Morrish.

THE URBAN SPRING: FORMALIZING THE WATER SYSTEM OF LOS ANGELES

from Modulus, 1985.

William Morrish

Western Civics
"To a considerable extent, the problem of water in Southern California is a cultural problem. By this I mean that newcomers to the region, who have always made up a majority of the population, have never understood the crucial importance of water. Crossing the desert, they arrive in an irrigated paradise in which almost anything can be grown with a quickness and abundance that cannot be equaled by any other region in America. There does not seem to be a water problem. Nor are they told there is such a problem, for Southern California has always been extremely reluctant to discuss its basic weakness."
—Carey McWilliams (1946)[1]

Los Angeles is a city finally awakening to its basic weaknesses. With more frequency than ever before, the issues of water quality and supply are making front-page news. Developers interested in expanding their projects, as well as residents who want to insure the environmental quality of their homes and neighborhoods, are beginning to realize that the issue of water is synonymous with city building in Los Angeles. The city of the angels is now entering a phase of maturity. Its history of speculative growth is being left behind; developers, residents, and city architects must now answer questions regarding the city's image and identity. These questions cannot be addressed apart from a consideration of the water supply that makes the city possible. To many, Los Angeles and its environs are one large, sprawling, homogeneous city. Upon closer view, one can begin to see a diverse set of neighborhoods which are defined by geographic, microclimatic or ethnic features. The existing fabric of the city, however, does little to support this diversity. As Los Angeles evolves into a mature city, identification and orientation of its separate neighborhoods will become more critical.

The water aqueduct system creates the basic pattern of the sprawling Southern California gridiron. The gigantic scale of the three-water aqueduct system has combined with the availability of flat cheap land to create a city with a disparate hierarchy of spatial orders. The city is separated into two scales: the individual home and the freeway. These two scales are mediated by linear strip commercial zones, office centers, and neighborhoods, but the intermediate zones tend to relate either to the residential scale or the freeway scale.

The freeway, the most famous symbol of Los Angeles, hovers over the landscape like the aqueduct system of ancient Rome. In contrast to the freeway, the water system of Los Angeles, which is much older and fundamentally more critical to its existence, is not as well known. One key reason is that the freeway is a public space and the water aqueduct is not. Daily, thousands of drivers keep their eye on the road and their ear to the radio, listening for SIG ALERTS, warnings of accidents ahead on the freeway. The ritual of the freeway is an everyday activity for residents of Los Angeles. The consumption of water is also an everyday ritual, but one which has been removed from our daily consciousness.

This loss of consciousness is primarily due to the removal of the aqueduct from public sight. The ritual of water is no longer a public activity like commuting.

Los Angeles is an excellent example of a man-made desert oasis. Its present-day physical form, however—like that of Phoenix, San Diego, or other cities in the American Southwest—does not effectively celebrate the water system that nurtures its existence. Most residents thoughtlessly assume that their garden paradise merely comes from "turning on the tap." In reality, a gigantic system of aqueducts, pumps, reservoirs, canals, and pipes delivers water from 500 miles away. To the average person's perception of the city, this labyrinth remains hidden from view, except when he receives the monthly water bill or when he has to vote on water-related bond issues. Here we offer a design exercise with two purposes in mind. The first, and more specific purpose is to reintroduce the water system to public view. To do this, we will explore the possibilities of externalizing the hidden water aqueduct system into a set of public spaces, activities, and monuments. Potentially, these new public spaces could be the articulated intermediate scale of urban spaces now missing from the Los Angeles landscape. New and existing developments can begin to infill and reorient themselves to the water places, rather than to the scale of the home or freeway.

Our second and more general purpose is to identify city design principles, which are inherent in the formal structure of the western American city. Intuition suggests that cities like Los Angeles might be translating the universal principles of city building through their own vocabulary, what I call the Western Dialect. This is more than an expression of regionalism; it is a specific set of city design principles which are inherent to the formal construct of the western city typology. Because Los Angeles's form is so fundamentally tied to the transportation of water, it provides a unique opportunity to understand the factors that control the translation of universal design principles into the formal spatial structure called Los Angeles.

Introverted City: 1781–1888

In the late 18th century, water management was a principal concern of Southern Californians. From San Diego to Santa Barbara, the Spanish Franciscan padres employed Indian labor to build systems, sometimes elaborate ones, for the conservation and distribution of water. The availability of arable land and water was a basic requirement for successful settlement of the missions. So it was that when the Spanish authorities determined to establish a pueblo in the south, they chose a low-lying alluvial terrace adjacent to that portion of the Los Angeles River through which the water flowed year-round. With its founding on 4 September 1781, the *Pueblo de Los Angeles* began its enduring relationship with the river.

The pueblo was an introverted community centered on the mission and plaza. The water was distributed from the center of the city to the outlying agriculture. The first settlers erected a brush diversion dam and excavated a *zanja madre* (main ditch) along the base of the hills, past the northeast corner of the plaza. A ditch master was appointed, and rules established for the operation of the system, which supplied water to the residents and surrounding agriculture. In 1868, the Spanish water distribution system was leased by a private company for a period of 30 years, during which the company constructed a system of supply lines, cast-iron and steel water mains, and storage reservoirs. By 1888, almost 3,000 acres of irrigated farmland lay within the town's borders. The adobe Pueblo de Los Angeles had grown to a city of almost 50,000 persons, the state's second-largest urban place. The water systems established by other missions in California had been dismantled or had fallen into disrepair by the 1830s. Various urban observers have put forward the idea that Los Angeles was, in its pueblo days, a dispersed but homogeneous environment. I propose that before the large water projects of 1913, Los Angeles and other communities in Southern California were a set of quite separate and different ranchos. Each community attempted to create its own urbanity in the desert, but the links which fused Anaheim (a German colony in late 1800s), San Bernardino (a Mormon settlement of 1857), and Los Angeles were the water systems of 1913 and 1938. The super-grid of cheap water and its subsequent standardization connected the communities in a way not seen again until the coming of the freeway in the 1950s when the grid of auto transportation would further tie the pueblos, colonies, and communities into an irrigated megalopolis

The Boom City: 1888–1968

The two forces which changed the pattern of water distribution and, in turn, the urban fabric of Los Angeles, were the acquisition of the water system from the private water company by the city of Los Angeles, and the cheap and efficient transportation provided by the railroad, and eventually by the automobile. Cheap land with access to transportation lines, and irrigated by an abundant source of water, supplied the formula for success.

Due to widespread disenchantment with private water system operators, the city, under public pressure, acquired the system for $2 million on 2 February 1901. Three days later the city's first Board of Water Commissioners was established to manage that system. William Mulholland, a talented, self-trained engineer, retained his position as system superintendent, and would later become the chief engineer and visionary of the future aqueduct system from Owens Valley and the Colorado River.

During the boom years, the growth of Los Angeles was constrained by the dwindling supply of artesian wells and river water. California land speculators needed more irrigation water to transform the desert into the vision of lush gardens advertised in magazines in the Midwest and on the East Coast. Immigrants dreamed of their own orange trees and gardens in the sun; syndicated real estate developers envisioned the lucrative possibilities of a moist San Fernando Valley.

Both dreams required quantities of water which far surpassed the limited supply of the existing water system. In a series of unprecedented political and legal maneuvers, the city and land speculators gained access to waters from the Owens Valley, located 300 miles to the north. At the turn of the century, the Owens Valley was a flourishing cattle and agricultural community, nurtured by the waters of the Inyo and Mono basins of the eastern slopes of the Sierra Nevadas. Before their intentions were publicized, land speculators from Los Angeles acquired much of the land surrounding the Owens Valley from the farmers, with the intention of channeling the water to their land holdings in the parched San Fernando Valley. Following a vicious court fight and actual violence over the rights to the water resources, the syndicate and other politicians put the city of Los Angeles in a position to pay for the land acquisition and the construction of the system through a public bond of $24 million in order to supplement the existing city water supply.

In 1905, the city of Los Angeles hired 5,000 men to build the Owens Valley Aqueduct, with the purpose of bringing water not to the city, but to the edge of the San Fernando Valley. People still argue about whether the water brought by the residents of Los Angeles from the Owens Valley was for their use or was intended only to turn the dry land of the San Fernando Valley into a speculator's dream. On 5 November 1913, the aqueduct opened, and immediately began delivering four times as much water as the city of Los Angeles was then capable of consuming for domestic purposes.

The city's ability to dispose of this water surplus was severely restricted by federal law. In response to allegations concerning the land syndicate's role in planning the project development, President Theodore Roosevelt attempted to ensure that water from the public enterprise would not be used to benefit the syndicate's holdings in the San Fernando Valley. As a condition for his approval of the aqueduct's right-of-way in 1906, Roosevelt stipulated that no water from the aqueduct could be offered to a private interest for resale as irrigation water outside the city limits.

City officials, encouraged by land speculators, responded to the restrictions by rapidly extending Los Angeles's political boundaries. Between 1914 and 1923, Los Angeles initiated a series of annexations which nearly quadrupled its land area and eventually embraced all of the syndicate's holdings. Once annexed by water-rich Los Angeles, the San Fernando Valley blossomed into fields of beans and groves of citrus. The aqueduct, as an urban water development, initially operated for the principal benefit of agriculture. With the opening of the Panama Canal in 1914, the city of Los Angeles became a major seaport for agricultural exports. But, in time, the notion of a strictly agricultural valley gave way to a more domesticated agrarian vision—that of homes within a garden city. At the end of World War I, Los Angeles was flooded with immigrants. The water system and irrigated infrastructure of the future city was waiting for the 100,000 immigrants per year to create their own gardens in the desert.

Los Angeles is equipped with an elaborate and expensive infrastructure. Because the city requires large areas of land to irrigate before the expense of a system is feasible, each new system is over-scaled to anticipate future growth. This usually means that the new system is in place long before urban growth catches up. Unlike a typical city where urban density radiates outward, Los Angeles grew along points on the irrigated distribution pattern. Later, the introduction of the automobile would enable more and more people to live in a dispersed pattern along the water system.

In the 1930s, the collective community spirit of getting America back on the road to recovery provided the basis for building the Colorado River Aqueduct, an idea originated by engineer William Mulholland of the successful Owens Valley Project The Colorado River Aqueduct would provide water not only to back up Los

Angeles's Owens Valley water system, but also to supply water to other Southern California cities, which were rapidly growing even in the depression years. Since the Department of Water and Power served only the city of Los Angeles, a new agency was established: The Metropolitan Water District (MWD). The Colorado River Aqueduct was part of an elaborate water, power, and flood-control plan. One of the major dams constructed, Hoover Dam, was proposed to provide cheap electrical power to the growing cities of the Pacific Southwest. Down river, Parker Dam was built to create a lake from which water for Southern California would be drawn. Originally the water was used for agriculture, but again, as in the Owens Valley Aqueduct, the result was to spread urban growth out across the now-fertile land.

In the late 1950s, the state of California, responding to arguments over the adequacy of water supplies to booming postwar agribusiness and urban expansion, decided to create a comprehensive statewide water plan which would manage the limited resources. This project, dubbed the California Water Plan, was developed to build dams and control the rivers, to store water for use during dry months, to generate electric power, and to move the surplus water to the San Francisco Bay area, the San Joaquin Valley, and Southern California. A proposal favoring Southern California industry, homes, and agriculture with a majority of the water resources put the MWD in the middle of a heated debate. Many people in the south saw the water as fuel for yet another real-estate boom. Others saw water availability as a threat to delicate negotiations between neighboring states over Los Angeles's rights to the limited water resources of the Colorado River. Then, as now, this debate reflected a divided state. Northern Californians view Los Angeles and the expanding Southern California communities as wrongly placed development. They often claim that sending "their water" to the south (which claims two-thirds of the state population) not only encourages a poor pattern of development, but also threatens the environment and the general quality of life.

Worried about its position in the negotiations over water resources from the Colorado River, the MWD wanted to accept water supplied by Northern California, but such a decision could hold the MWD hostage to the anger of the Northern Californians. To get the support of the MWD, the state legislature passed the Burns-Potter Act, guaranteeing proposed water-delivery contracts subject to ratification by voters in the 1960 general election. This, and other assurances, finally turned the tide: several days before the election, the MWD

came out in support of the project. On 8 November, the Burns-Potter Act and the construction bonds for the project were ratified by fewer than 200,000 votes out of a total of almost 6 million votes cast. The vote, carried by the larger population in the southern part of the state, was approved in only one county in the north. In 1968 the California Water Project began delivering water to the south San Joaquin Valley and Southern California. To this day, the project, only half completed, is still the focus of emotional arguments between environmentalists—who are concerned about the water equality of the Sacramento/San Joaquin River Delta—and the developers of the homes, businesses, industry, and agribusinesses of the south.

"Boom City" now extended beyond the 1781 pueblo ordinance boundary and even beyond the dry mountain ranges. Through its water system, Los Angeles was tied to the politics of distant states across their valleys and delta regions. As growth patterns developed in Southern California, they became issues of discussion in San Francisco, Denver, and Phoenix. Extending urban fingers out into the surrounding wilderness, the garden city carried the identity of Los Angeles far beyond its topographic and political boundaries. Aqueducts now tie Los Angeles into a network that flows in both directions. The irrigated city of Los Angeles has become the irrigated megalopolis of Southern California in the irrigated region of the American Southwest.

Paris in the Desert: Present and Future Prospects
Los Angeles, like its sister cities, has evolved from an introverted settlement to an extroverted, sprawling conglomeration. Many architects of the city's future now wish to repent for past sins and transform this conglomeration into a real city, like Paris, or maybe more realistically, Barcelona. Whatever model prevails, the future shape and form of that city will necessarily be closely tied to its water system. Juxtaposing the geometry of the water system against the order of the city is a steppingstone which can begin to establish an urban order that is not just a memory of past heroic cities, but is rather a blend of future excitement and past traditions of city building.

The "City of Gardens" seems assured of adequate water supplies and a long-term flow of national and international capital. This new, mature growth cycle allows the city to become wiser in its development strategies. This cycle still responds to short-term problems, but more importantly, it provides the time to explore the long-term ramifications of its actions.

Unfortunately, anti-tax legislation like California's Proposition 13 has curtailed city government involvement in the public sector. Therefore, the burden of this exploration is transferred to individual designers. The city's future will be based on whether or not individual developers question the history of urban evolution in Southern California. Any design solution must conform to the specific site and, one hopes, connect with the whole of the city. This sort of design reflection will, hopefully, replace the generalized speculative answer with one which translates and shapes general typologies of architecture and city building into the specifics of the environments called Los Angeles and Southern California.

At this point it might be useful to embark on a tour through a series of design exercises which explore the possibilities of this new urban outlook. Using the water aqueduct system as a test case, the goal of our design is to create a set of specific urban sources in and around Los Angeles, which will simultaneously provide utilitarian service, spatial clarity, and ritual places which celebrate a city created from water and sand. The method for this search we can call the design scenario—a process by which statements of policy are translated into three-dimensional architectural or city-building programs. Our method takes the statements of policy and poses the question: What if the information were expressed as architectural spaces or public monuments? The results of this process will generate two types of information. From the first, we can begin to see the effect a policy has upon the physical structure of a place. From the second, we can identify a spatial vocabulary. The following design scenarios look at the potential ritual places which can be created to celebrate both the spiritual and the utilitarian relationship of the city to its water system.

The First Ritual: The Point of Intake
The aqueduct begins hundreds of miles away from the city boundaries. At the Point of Intake, water is pooled from the natural water courses into holding channels. At one end, the large pumps of the aqueduct's lift station draw water up out of the pool, into the pipes of the aqueduct, and on to the distant city. At this point of transference, the water leaves the wilderness, or rural state, and enters the geometry of the city. To many, the lift station can be seen as the gateway to the city. To others, it is the outermost tentacle of the city as it stretches into the countryside.

The lift station, or Point of Intake, also symbolizes the battle for control of water resources, in which there are two participating parties. The first is composed of those who feel they have control over the water because of riparian rights. Since they own land from which the water originates, they feel that they should be in control of its future. The other party usually lives outside the area of the water origins and argues that an area's water resources should not be limited and controlled by the few who own the land at its source; the water should be put to maximum use. They claim the need for appropriation rights. Two hundred years of litigation, legislation, and emotional arguments have been generated by this conflict over the control of limited water resources. This argument is rooted in the historic American conflict between rural virtue and urban intellectualism.

In order to ensure that no other remote region would face the fate of the Owens Valley, the state legislature passed the County of Origin law in 1931, prohibiting the draining of one area's water in order to supply other areas. This law helps small counties stop larger municipalities from looting local water resources.

In interpreting the law, the Point of Intake can be seen as the middle ground of the debate. It is proposed that a line be drawn between the intake lift station and the water pool of the natural water system, on which a building called the Basilica of Origin will reside. From this point, the basilica mediates between the values of the rural and wilderness landscapes and the geometric aqueduct lines of the city, which terminate here.

In the Basilica of Origin there would be two icons representing the two sides of the water debate: those of the city and those of the county of origin. The basilica would create a place for the debates about balancing water supplies. It would be the formal space where the process of deciding the amount of water entitlement would take place annually.

Each year, lawyers, officials, and citizens from both sides would gather at the Basilica of Origin and act out the ritual of balancing the area's water resources. This Act of Entitlement would be debated and recorded within the Basilica of Origin at each aqueduct. These basilicas would be created at the delta, the Colorado River, and the Owens Valley, and each would represent the debate particular to that area of origin.

The Second Ritual: Lines of Transport
As it leaves the Point of Intake, tunnels, canals, conduits, and siphons carry the water across the dry landscape of the Southwest. These lines of transport tell the story

of the land they traverse—a dry landscape marked by broad, open valleys which lie between the high Rocky Mountains. The lines of transport zigzag across the desert floor, and at times lift their cargo up and over rocky routes. These are the same routes taken by early settlers; today they are followed by travelers on the freeways which parallel the water system. These lines of transport act as ritual passageways from the open land and its ridges to the garden cities of Southern California.

The lines of transport unleash their power onto the landscape, a power which has been contained and withheld from the parched land it has just passed over. Each line is unique in its technology, its historical moment of construction, and the terrain it traverses. With its own rite of passage, each is perceived differently by the participants of the passage. To some, the Owens Valley Aqueduct represents a period of ruthless political exploitation. To some, the Colorado River Aqueduct represents the collective work spirit of the WPA. Finally, there is the California Aqueduct, which, to those of Northern California, represents the power and the insatiable thirst of the southern part of the state. Whatever the image, lines of transport act collectively as fingers extending the city into the distance, carrying with it the image and characteristics of that city. The symbolic functions of the "city fingers" are to demarcate the distance and passage of time across the landscape, and to inform the traveler of the past and present effect of the city upon the land, by creating a three-dimensional timeline.

This scheme can be realized by visually externalizing the system on the land. During the day, at the bases of these ridges, the lift station can be landscaped with compact stands of trees, creating an oasis that demonstrates the life-giving power of the cargo carried in the lines. At night, when the drive across the landscape can be quite monotonous, the lift station can be lighted to create a focal point in the darkened landscape. The traveler counts off the illuminated ridges, assuring himself, "Only a few more before I get home." It is a point on the horizon, marking time and distance, and extending a fragment of the city into the desert. Thus the monotonous landscape takes on meaning and texture.

The Third Ritual: Pool of Collection
Each aqueduct delivers its water to a reservoir. Like the water cisterns and fountains of Rome which collected, stored, and distributed the water from the aqueducts, the Los Angeles reservoirs can represent both urban-

entry landmarks and neighborhood fountains.

Located at the outer edge of the city, reservoir pools perform a utilitarian function by distributing the aqueduct's water to the homes and gardens of the city. They also represent a transition from the linear aqueduct axis of the lines of transport, to the spreading grid of the distribution system: a transition from the open, expansive scale of the surrounding mountains and desert to the more articulated individual scale of the irrigated city—from wilderness to civilization. Paralleling the terminus reservoir, the major interstate freeways breach the surrounding mountain walls of Southern California and Los Angeles. At this point, where water and traveler pass into the garden, the terminus pool can be developed into a formal entry space. This pool would be emblematic of both the land it is entering and the journey taken to get there.

Like the previous two rituals, these junctures can celebrate each aqueduct differently by representing the unique qualities of the particular system they serve—for instance, their geographic and historical origins. To the east of the city, the Colorado River Aqueduct greets those who have just crossed the desert. To the north, the terminus pool can be formed to greet the traveler who has traversed the mountain pass from the agricultural grid of the San Joaquin Valley. Finally, the terminus pool represents a potential gateway to the presently inarticulate sprawl of Southern California cities.

Spread out over the landscape of the city are Pools of Collection which could articulate distinctly different areas in the environment. As part of the distribution system, each terminus pool passes water into a series of smaller distribution reservoirs. These Pools of Collection are interrelated as parts of a larger distribution system, yet each should be distinct. Physically, they could be seen as landmarks, perhaps as super- scaled fountains like their antecedents in ancient Rome.

The Fourth Ritual: The Grid of Distribution
Fed by the Pools of Collection, the Grid of Distribution transports water to the individual consumer. It further reduces the scale, breaking down into a fine-grained complex of pipes and pumping stations which bring water to each house and garden.

Los Angeles and its environs are created by three overlapping grid systems: one from the Owens Valley Aqueduct, one from the Colorado River Aqueduct, and another from the California Aqueduct. Each is operated by a separate agency, but they are tied together to provide supplementary water as needed. Historically, city development has responded to the grid pattern of each system. At the smaller scales, growth has clustered around the major supply lines of the distribution system. Field patterns of agriculture have become large blocks of residential neighborhoods. At a larger scale, the shape of the cities of Southern California have followed each aqueduct system. The Owens Valley Aqueduct system caused the city of Los Angeles to extend northward from the original pueblo site rather than to the coastline in the west. The Colorado River Aqueduct allowed development to fill in the valley extending from the coastline on the western edge and eastward to Riverside. Rather than follow the Jeffersonian or Spanish grid, the city of Los Angeles, and other cities of Southern California, follow the Grid of Distribution pattern of the irrigation and water distribution.

The Grid of Distribution is the lifeblood of the city. It could be said to represent the dialogue between the natural environment and its man-made settlements. To make its importance evident, Water Parks could be placed throughout Los Angeles and other communities to commemorate the rapport between man's irrigation system and the ecology of Southern California. Each park would have three functions. The first would be to exhibit the wise utilization of water in a dry climate. The second would be to commemorate the bringing of water to the specific neighborhood. The design of each water park would reflect the origin of its water, such as the Colorado River, for example. The third function of the Water Park would be that of civic landmark. Each park would be site-specific and at the same time regionally tied, thus giving further definition of space to the Southern California plain, devoted not just to the domestic landscape, but to one of community.

The Fifth Ritual: The Private Spring

The homes and gardens on the grid plan of the San Fernando Valley sit like private oases. Faucets, sprinklers, appliances, and other fixtures provide pleasure, life-sustaining fluid, and cleanliness, with minimum inconvenience to the individual. Even in the arid climate, water to quench one's thirst is never far away. The city is made up of millions of these private springs, each catering to individual ritual patterns.

While bathing in a household spring, there is little to remind one of the water's sources. Actually, the faucet and water fixtures can be seen not only as utilitarian conveniences, but as connections to the community and to the distant landscapes at the end of the water aqueduct. Water in Beverly Hills is actually drawn from the Colorado River or the San Joaquin Delta. Across the street in West Hollywood, the tap water comes from the Owens Valley.

Domestic habits tie in to the whole system of water rituals from the Point of Intake down to the individual faucet tap. Therefore, the design of the individual spring could reflect, through its image and usage patterns, the form and significance of the larger aqueduct system. The citizen is reminded daily of his debt to the entire water system. The private spring can achieve these ends by:

1. Shaping the home and garden into patterns reminiscent of the components of the water system.
2. Redesigning water usage-fixtures to recall the origin of the water sources, such as a sink shaped like the delta reservoir of the Colorado River.
3. Readapting the garden to plant material and patterns, which utilize and represent irrigation techniques.

The private spring terminates a long line of water transportation and thus, in many ways, is a representation of all the issues and physical patterns of the water system. If the private spring is designed properly, it can be a source within the city from which residents can reflect on the balance of water usage in their city in relation to other rural and urban areas.

Western Dialect

The Pools of Collection, the Basilica of Origin, and other points of ritual along the water aqueduct system provide just a few examples of possible public activities that can be associated with the water system in Los Angeles. It is hoped that the design alternatives in this article will stimulate interest in the potential of developing urban places in the arid western city. These sketches have illustrated a few of the many prototypes found in history and in other cultures that can be used to house civic and private rituals.

These design exercises emphasize that each aqueduct is part of a unique system, constructed to carry out the same tasks: the transportation and distribution of water. Each system must convey water a long distance from its source and also represent its historical and geographical origins. It is this collision between the utility of water transport and its contextual response that creates a set of structures which are simultaneously universal in principle and specific in response to locale. For example, each aqueduct might have a Basilica of Origin, but the articulation of that building would be different for the California aqueduct than for the Colorado aqueduct, since the former has its source in the lush river delta, and the latter is located on the edge of the desert.

The final step of the design exercise was to build a scale model. The model reassembles the aqueducts of the Los Angeles water system in relation to the elements of its landscape. It illustrates, in abstract geometric forms and patterns, how the city is structured by three similar aqueduct systems that have been constructed at different times over different terrains, and which flow from different sources. Their combination results in the distinctive pattern of Los Angeles. The model is composed of three formal elements: the three aqueducts, the mountains which surround the city of Los Angeles, and the freeway system entering the city. The model illustrates how circumstances of time, terrain, and social interaction can translate similarly functioning systems into radically different spatial structures. For example, the zigzag pattern of the Colorado River Aqueduct is very different from the long line of the Los Angeles Aqueduct from Owens Valley. The forms in this model illustrate the concept of the Western Dialect: the aqueduct model is a vernacular expression of a universal principle of transporting water. It should be noted that the term "Western Dialect" is used to emphasize that there are very specific factors at work in the West, factors that usually are bypassed by using the words "regional" and "contextual."

In thinking about the design of a western city like Los Angeles, three issues must be addressed. The first issue is that of context. The dominant contextual elements of Los Angeles are not architectural, but natural. In the model, the mountains surrounding Los Angeles are viewed as part of the formal vocabulary of the urban landscape. The mountains protect the city, create its major gateways, and provide a backdrop to its architecture. In many ways they suggest elements of the ancient walls of a glorious past civilization. The mountains are a resource for a literal translation of new forms, and provide a metaphorical basis from which to translate prototypes from other places.

Second is the issue of scale. In the western city, time, distance, and proportion exist on an extremely large scale. Los Angeles and its sister cities cover 5,000 square miles of land. The land is defined by cloudless blue skies combined with the broad horizon of the city grid, receding to the 3,000- foot-high mountains in the distance. Within this colossal environment, the placement of a single office tower or house implies that

Los Angeles is described by the dynamics of its setting and not by the particular style of its architecture, unlike New York City, where the setting is described by its architecture. This is not to say that architecture can never comfortably fit in the Los Angeles landscape. Rather, I propose that a building in Los Angeles relies on the issues of site planning more than on its architecture. In Los Angeles, the unique positioning of the building in terms of the site—simultaneously tying it to the distant mountains, to the horizon, or to the sprawling grid of the city—becomes the dominant issue of the building's form. The formal structure of the aqueduct system illustrates this point. Since each ritual has a corresponding function in each of the three different aqueducts, there is a basic set of organizing principles which ties the distant parts together. There is a sense of the whole over their 300- to 500-mile lengths; the unique formal identity of each is retained. All respond to the intersection of a particular water system crossing a specific piece of terrain

The third issue is the dialogue between technology and art in the landscape. The engineering structures of the aqueduct system provide an excellent design laboratory for studying the interrelationship of natural and man-made forms found in the western landscape. The juxtaposition of the earthen dams, the sinuous lines of the gravity-flow canals, and the geometric forms of the lift stations create an incredible environmental sculpture from which to draw architectural inspiration.

The aqueduct system of Los Angeles, the principles of a western dialect, and the five ritual sections of the water celebration are design elements that provide inspiration for the future planning and shaping of the city and its architecture in the western oasis of Southern California. This exploration, which is not typically part of the architect's repertoire, redirects traditional elements of architecture into new relationships. The West is a gigantic unyielding landscape: it should be used as an architectural context from which to develop the future shape of the city.

Notes ---

1.   Carey McWilliams, *Southern California: An Island on the Land* (Layton, UT: Gibbs Smith, 1946), p. 192.

ACT THREE: THE DISAPPEARING

CALL TO ACTION

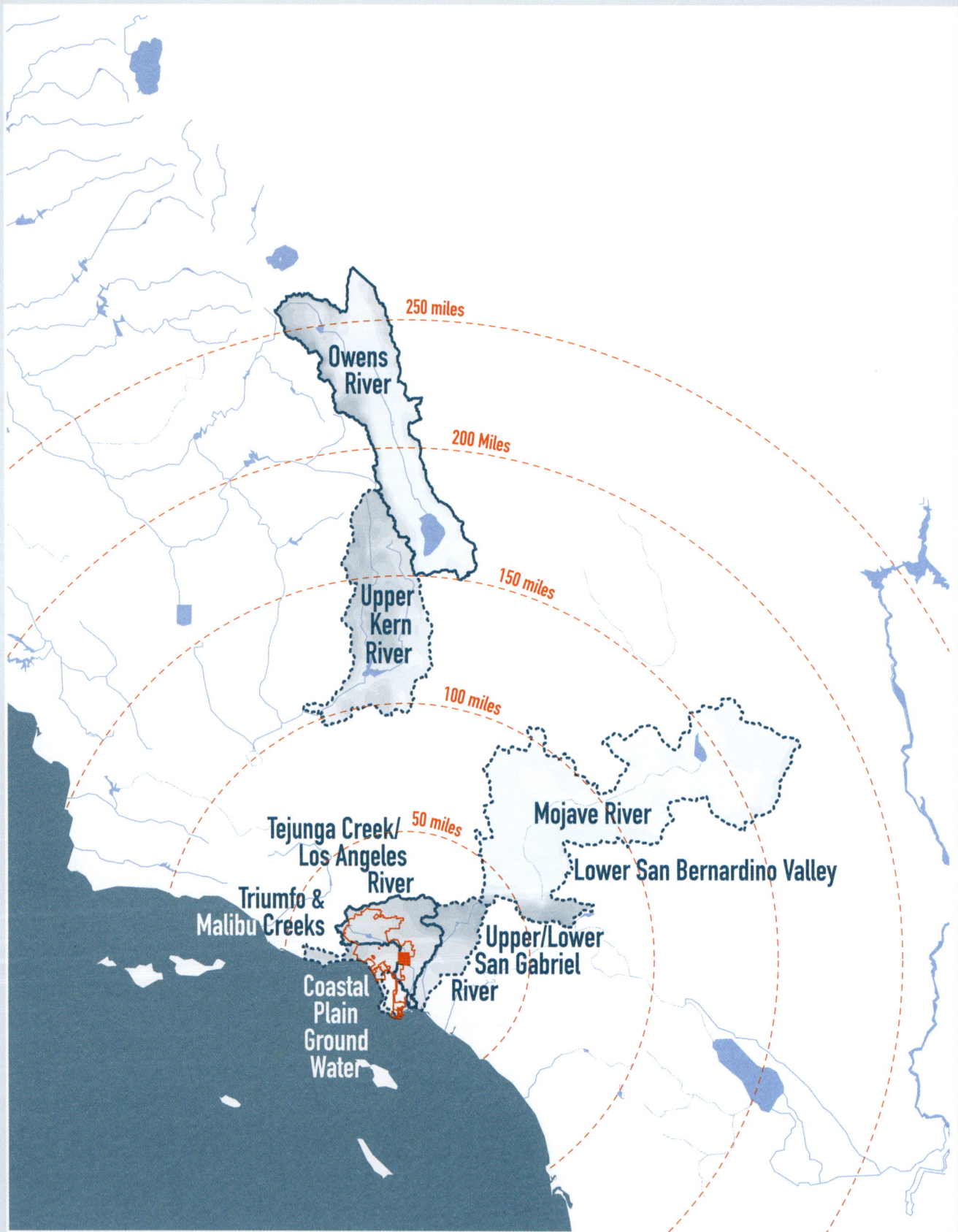

250 miles

Owens
River

200 Miles

150 miles

Upper
Kern
River

100 miles

Mojave River

Tejunga Creek/
Los Angeles
River

50 miles

Lower San Bernardino Valley

Triumfo &
Malibu Creeks

Upper/Lower
San Gabriel
River

Coastal
Plain
Ground
Water

In 1905, Los Angeles considered several other sources of water to enable the continued expansion of the city. The Owens River watershed was chosen over the alternatives, as much for plentiful water (the gradient shows precipitation), the cost of the water rights (compared to that of the Kern River), as for lining the pocket of Fred Eaton—the LA mayor who was a major landholder in the valley and proponent of an Owens River Aqueduct.

(Sources: Los Angeles Water Commissioners' Report for the Year Ending November 30, 1905; LADWP, USGS, US National Atlas, California Water Resources, and California Biodiversity Council/University of California.)

RECONSTRUCTING THE VOID: OWENS LAKE

From *The Infrastructural City: Networked Ecologies in Los Angeles*, 2008.

Edited and updated for *Water Index* **in 2014.**
Barry Lehrman

Two hundred miles due north of Los Angeles is a 108-square-mile playa, the abandoned corpse of Owens Lake, a silent victim of the city's destructive thirst. A century ago, Los Angeles became dependent on this distant watershed, funneling its life-giving liquid into a vast aqueduct to nurture its delirious growth. But this history of water, politics, and exploitation has grown ever more complicated and inextricable, reshaped by networks of negotiation, litigation, and politics. As much artificial as natural, the result is a second nature, a wild, uncontrollable condition created by social, infrastructural, and organic ecologies interacting with the environment and with each other.

Los Angeles is now a thoroughly urbanized landscape, while the Owens Valley faces stagnation, desiccation, and toxic dust storms. But in taking the water from the Owens Valley (see map 1), Los Angeles ensured that agribusiness and exurban sprawl would never take over. Thus, a unique fragment of the American frontier was preserved as a permanent rural antipode to the sprawling metropolis to the south.

Owens Valley is a land of extremes. Running for 75 miles along the eastern slope of the Sierra Nevada Mountains, the valley floor sits at 4,000 feet above sea level. To the west, the Sierra Crest rises up above 14,000 feet, while to the east, the Inyo and White Mountains reach nearly the same height. This is one of the deepest valleys in the world. Mount Whitney, the tallest peak in the continental United States, looms above Owens Lake. The Methuselah tree, one of the world's oldest living things, lives in the White Mountains.

Relic Landscape

This relic landscape stands in stark contrast to the San Joaquin Valley, a mere 50 miles to the west on the other side of the Sierras (but there are no roads connecting these two worlds). In the Central Valley, farmers turned once-verdant wetlands and sloughs into farmland, creating the nation's vegetable garden. Then the millennium's housing bubble turned vegetable fields into subdivisions, which were then economically devastated in the 2008 crash and have yet to recover—a fate that Owens Valley escaped.

The tangled relationship between the desiccated Owens Valley and the city that drained it first emerged a century and a half ago. Long before the water wars, Owens Valley capitalized Los Angeles's rise from a sleepy Spanish cattle town into a vibrant global metropolis. The 1865 discovery of silver and lead at Cerro Gordo above Owens Lake funneled the first flush of wealth through Los Angeles. Cerro Gordo bullion helped bring the Southern Pacific Railroad to the city in 1876, in turn ensuring the success and growth of San Pedro harbor, and contributing to the real estate boom of the 1890s. This first significant wave of population and economic growth was sustained until Los Angeles started to exceed the available local water resources (see map 2).[1]

Los Angeles Growth
1781–1935
the first 80 annexations = 442 square miles

1940–2013
added only 27.1 square miles more

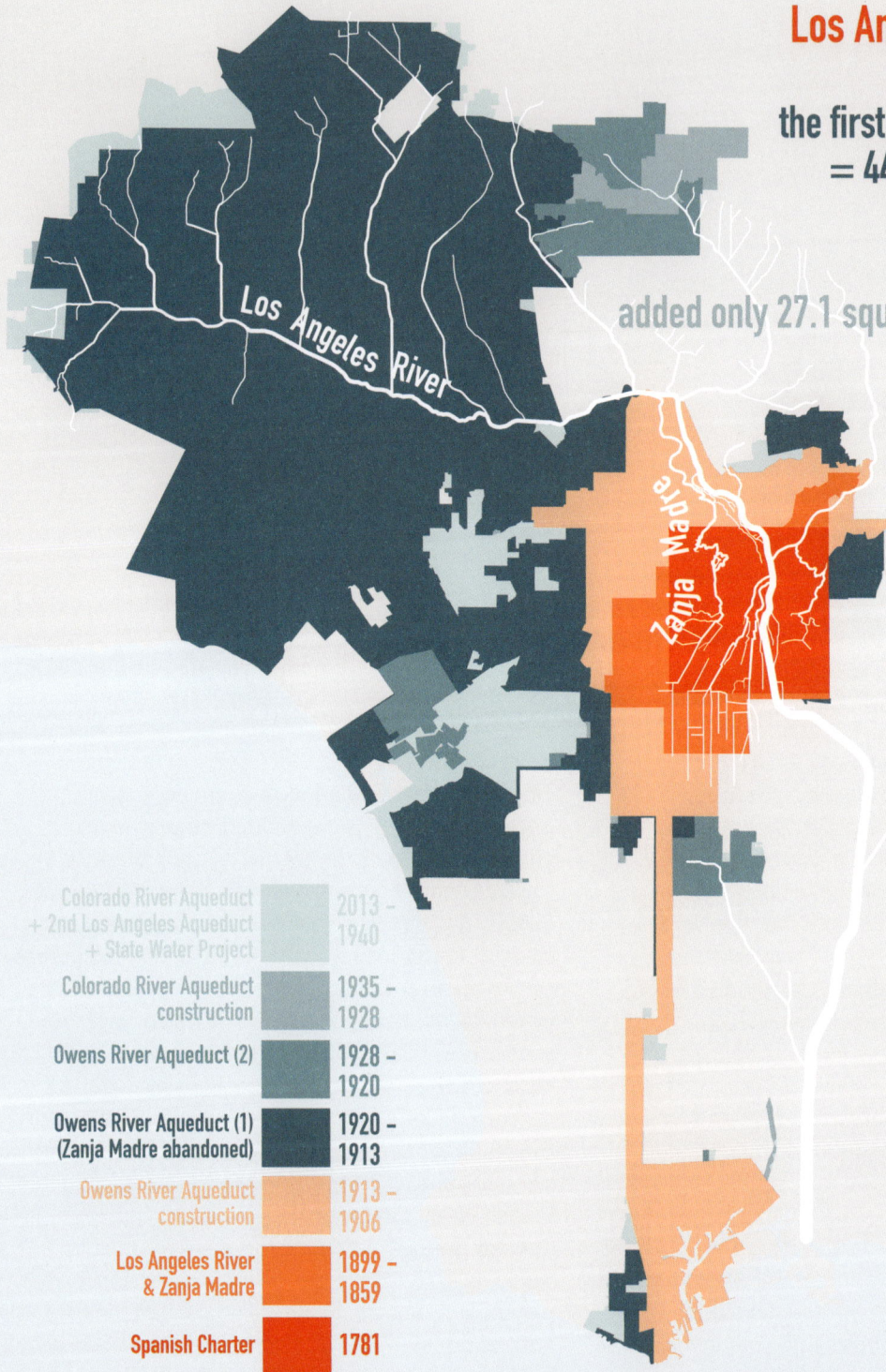

Los Angeles River

Zanja Madre

Colorado River Aqueduct + 2nd Los Angeles Aqueduct + State Water Project		2013 – 1940
Colorado River Aqueduct construction		1935 – 1928
Owens River Aqueduct (2)		1928 – 1920
Owens River Aqueduct (1) (Zanja Madre abandoned)		1920 – 1913
Owens River Aqueduct construction		1913 – 1906
Los Angeles River & Zanja Madre		1899 – 1859
Spanish Charter		1781

The physical size of Los Angeles is tied to imported water. It scarcely grew in area before 1900, and then grew eightfold between 1906 and 1917, with a jump from 43 to 351 square miles. Following the 1936 opening of the Colorado River Aqueduct, growth of Los Angeles slowed considerably, when adjacent communities started getting water from the Metropolitan Water District and didn't need to be annexed into LA to grow. City ownership of Owens Valley parallels the growth of Los Angeles, with 492 square miles owned by the LADWP in 2014 exceeding the city's 462 square miles. (Source: Los Angeles Bureau of Engineering).

By diverting the Owens River 200 miles south across the Mojave Desert to slake Los Angeles's thirst, chief city engineer <u>William Mulholland established the precedent for urban growth to dominate water "reclamation" policy, thereby changing the future of the West.</u> When the Los Angeles Aqueduct opened in 1913, a major wave of urban growth followed (see map 2), creating the city we recognize today. The aqueduct could deliver ten times as much water as local sources to the city. Yet the phenomenal growth that ensued meant that within a decade, Southern California once again was searching for even more water—a quest that continues today. Now the focus is on finding water through reuse and conservation, not diversion from distant watersheds. The aqueduct became the new Los Angeles River, extending the city's watershed some 350 miles north to the Mono Lake Basin (see map 3). With the arrival of the aqueduct, the eponymous river became superfluous, allowing engineers to funnel its unpredictable flows down to the ocean as rapidly as possible, through a system of concrete channels not unlike those that hijack the Owens River on its epic journey south.

Despite the destruction wrought to Owens Valley agriculture by the aqueduct's arrival, Los Angeles saved the rural character of the valley. To protect the city's water rights, a series of legislative acts at the local, state, and federal levels effectively prevented development in the Owens Valley, as the political clout of the adolescent city steamrolled over the sparsely populated countryside. Adding to this regulatory colonization, Congress established the Inyo National Forest at the city's behest to preserve the watershed in the mountains surrounding the valley, while the Los Angeles Department of Water and Power (LADWP) became the second-largest landholder in the county. Because of these external limits, the Owens Valley and surrounding Inyo County have remained rural, one of the few places in California with no projected growth.[2]

The irrigation network that crisscrosses Southern California predates the Spanish missions, and originates with the Tsonga and Paiute tribes. Long before California was home to Europeans, a Tongva village was located by the banks of the Los Angeles River, near where Pueblo was established. These first Angelenos created a modest network of ditches to irrigate the crops that supplemented their hunting and gathering. The Spanish took over the network, expanding the ditches and instituting rules and regulations for access to the water of the Zanja Madre, or <u>"mother ditch."</u> The

colonists moved the diversion point up into the Elysian Valley, and the city of Los Angeles was officially born.[3]

Once California became an American state, the growing population and thirst for water continued to extend the ditches further from the Los Angeles River and further upstream. Eventually the diversion points reached nine miles north of the plaza to the southern edge of Burbank. By 1886, so little water remained in the river below the Zanja's intakes that the city tunneled below the riverbed to capture the remnant subsurface flow.[4] With such a precedent of engineering exploits, looking north to the Owens Valley for more water was not such a great leap of faith for William Mulholland in 1904. His challenge was just <u>to extend the ditch digging by an order of magnitude.</u>

When the Los Angeles Aqueduct diverted the Owens River's water to the city, the desiccation of Owens Lake was already well underway. <u>After the 1913 opening of the aqueduct, the lake's water level continued its precipitous drop.</u> On the dying lakebed, a new industry harvesting the precipitating salt was born. Salt works sprung up on the eastern and western shores, but soon the falling water level left the lagoons on the east dry. After an effort to pump the brine into these saltpans failed, they were abandoned. By 1923, <u>only a dusty brine pool covered the lowest part of the once- mighty lake.</u> In 1926, the dust storms became a regular occurrence.[5]

Owens Lake was destined for this fate, even if it had remained free of the city's thirst. Water, a scarce commodity in the American West, is subject to <u>the "Colorado Doctrine," which states that the first person to divert and use a water source acquires future rights to it.</u> Well aware of this, farmers established extensive system irrigation ditches along the Owens River in 1872, starting the lake's decline. Thirty years later, the Federal Bureau of Reclamation proposed to divert water to a vast irrigation network to serve the area's agricultural interests. Owens Valley farmers were destined without the federal project to desiccate the lake by 1929, so <u>the lake was doomed</u>, no matter who controlled the water.[6]

With Los Angeles's rapidly growing population exceeding projections and the city suffering from drought in the 1930s, the LADWP began drilling hundreds of wells to tap the Owens Valley's groundwater to bolster the flow down the aqueduct. Until the 1970s, these wells had a relatively minor impact on life in the valley—at least compared to the draining of the Owens River—but with the 1970 completion of the second Los Angeles Aqueduct (south of

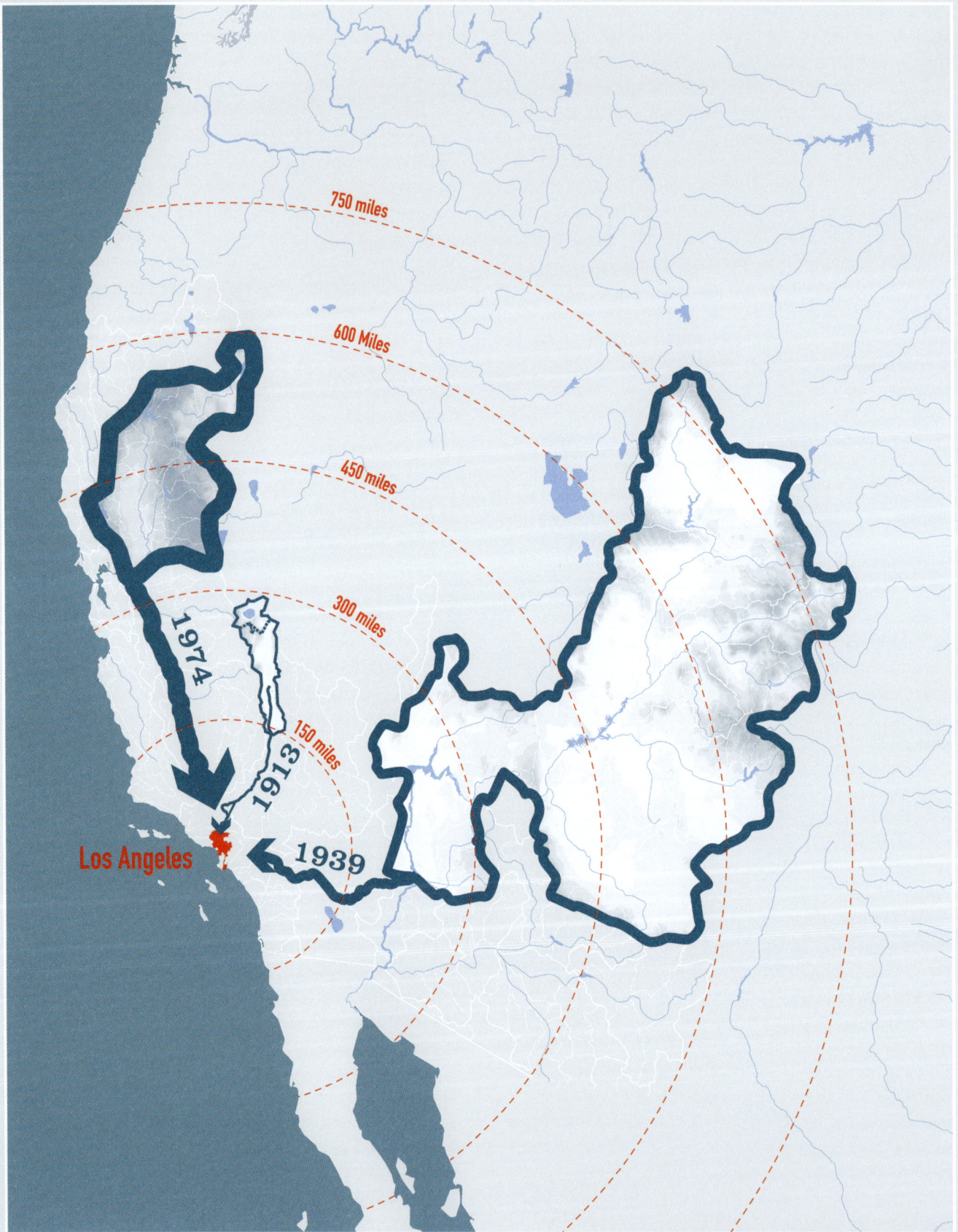

750 miles

600 Miles

450 miles

300 miles

1974

150 miles

1913

Los Angeles

1939

Los Angeles's aqueduct shed now extends to the peaks of the Rockies via the Colorado River Aqueduct (completed in 1939) and the headwaters of the Feather River via the California Aqueduct (begun in 1960). This vast infrastructural sprawl is exceeded by few other cities, perhaps most notably by the South-North Water Project to supply Beijing, which was supposed to be completed by 2014.
(Sources: LADWP, Metropolitan Water District, USGS, US National Atlas, and California Water Resources)

Haiwee Reservoir), Los Angeles greatly increased the pumping. Predictably, groundwater level dropped and the meadows of verdant phreatophytes plants fed by a myriad of sparkling springs were replaced by a xeric landscape of sagebrush and sand dunes.[7]

Contested Landscape

The Owens Valley remains a contested landscape, with disenfranchised residents chafing against the colonizing infrastructure serving the distant metropolis. With a population less than 1% of Los Angeles's, Owens Valley holds little sway in legislation from Sacramento or Washington compared to the city, and has no means to engage the LA City Council or the mayor, who appointed the LADWP senior staff.

Groundwater pumping in the 1970s sparked the "Second Owens Valley Water War," as residents sought means to protest the death of their land. This round, unlike the infamous water war of the 1920s and 1930s when valley residents dynamited the aqueduct to little avail, Owens Valley had a new weapon—the California Environmental Quality Act (CEQA). CEQA provided Inyo County the legal leverage and venue to challenge the environmental impacts associated with the Second Aqueduct on equal footing with the city. Back in 1925, California's legislature had passed a law requiring compensation to all business and property owners for financial losses due to water diversions. It wasn't enforced until the California Supreme Court ordered Los Angeles to comply in 1929, but this injunction only compensated folks for direct financial losses and did not require any remedy for the environmental damage. Half a century later, CEQA gave ecological needs for water a legal standing. When the original Los Angeles Aqueduct was built at the turn of the century, there were no environmental review laws anywhere in the United States. By the time of the Second Aqueduct, the world was in the midst of the burgeoning environmental movement.

Litigation against the city to rectify environmental damages in Owens Valley started in 1970, with the first successful case being the 1972 Inyo County suit to limit groundwater pumping. Only in 1990, once extensive scientific studies and groundwater modeling defined a sustainable level of pumping that would preserve plant life above, was this 1972 lawsuit completely settled. In response to legal pressure, Los Angeles undertook production of an Environmental Impact Report (EIR), but it took several lawsuits for a rigorous and scientifically sound document to be produced and for the city to follow up by mitigating the reported effects. CEQA had leveled the battlefield for the disenfranchised residents of the Owens Valley by taking the fight out of the legislature and into the courts. At least to a degree, CEQA made it possible to assess the environmental costs of the city's vast infrastructure. The EIR for the Second Aqueduct discovered several conditions required urgent mitigation: dropping groundwater levels from the increased pumping (see photo 1), the death of Owens River below the aqueduct intake, the lowering of Mono Lake's water level, and the toxic dust storms from Owens Lake Playa.[8]

The second successful fight under CEQA was over water levels in Mono Lake, a unique ecology of two million migratory water birds nesting among dramatic tufa towers formed by the lake's unique saline chemistry. In 1940, LADWP bored 11 miles through solid rock to create the Mono Craters Tunnel, and began diverting water into the Owens River for the aqueduct. Like Owens Lake, Mono Lake's water level dropped. With CEQA, Mono County residents and scientists were able to get the courts to intervene. With the lowered water level, coyotes and raccoons could walk across the lakebed to reach the formerly isolated island on which endangered California gulls made their nests. In 1979, with the leadership of biologist David Gaines, the Mono Lake Committee launched lawsuits against the LADWP to protect the bird nesting sites. The 1994 Mono Lake Accord established a minimum water level for Mono Lake to protect the rookery and established a schedule of allowable water diversions that would gradually increase water in the lake to a sustainable level.

Even with reforms upstream, Owens Lake remained an ecological disaster. Any wind gust on the lakebed above 20 miles an hour lofted more than 50 tons per second of "Keeler Fog" from the playa. These dust storms sent 130 times the United States Environmental Protection Agency's limits two miles or more up into the atmosphere, where dust could travel 250 miles. Such storms occurred two dozen or more times each year, generally in the spring and fall. Composed of microscopic particles smaller than 10 microns (PM10), Owens Lake dust contains efflorescent salts plus significant levels of toxic metals, including selenium, arsenic, and lead. As the largest single source of PM10 pollution in the country, these dust storms were a health threat to the 40,000 people in the immediate region (most of whom live south of the Coso Range in the Indian Wells Valley). Even the US Department of Defense suffered, as dust reduced

0   2.5   5   10 miles

Owens Lake 3600' contour ———
Highways ———
Roads ———
1781 Spanish Grant ———
City Boundaries ———

Juxtaposed onto Los Angeles, Owens
Lake stretches from the terminus of
the aqueduct in Sylmar almost to the
Getty Center in Pacific Palisades.
(Sources: LADWP, Los Angeles
Bureau of Engineering)

visibility so much that China Lake Naval Air Station to the south had to stop flight operations five to ten days each year—costing the Navy over $5 million annually. Physicians at China Lake linked the dust to significant health problems in the region, including higher rates of cancer, lung disease, and eye problems.

Control Mechanism

Finally, in 1998, the City of Los Angeles and the Great Basin Unified Air Pollution Control District (GBUAPCD) reached a court-negotiated settlement to abate the dust blowing off the lakebed. The dust mitigation process initiated by the Memorandum of Agreement (MOA) focused on a few specific and tangible results: the reduction of dust being blown off the dry lake and the preservation of the nesting habitat of the snowy plover, an endangered wading bird. By 2005, Los Angeles had created 30 square miles of dust control with 300 miles of mainline pipe (some as large as five feet in diameter), thousands of miles of "whiplines" supporting 5,000 irrigation bubblers spaced every hundred feet (see photo 2), and hundreds of miles of fiber optic control cables. In comparison, Los Angeles has 7,225 miles of distribution mains for its 462 square miles (see map 5).[9]

But these initial phases of dust control dictated in the MOA still failed to limit dust emissions to safe levels. Every request by the GBUAPCD to achieve compliance by expanding the dust control areas was vigorously litigated by Los Angeles. Then in November 2014, the city changed tactics and signed the Final Stipulated Judgment (under negotiation since 2012) that limited dust control to a total of 53.4 square miles. The judgment allows LA to convert the water-hungry shallow flooding into corduroy fields of rows of tillage (see photo 3) that only require turning on the water every few years after the clay clods disintegrate when the trenches and berms need to be re-excavated.[10]

Los Angeles's dust control efforts on Owens Lake have been equivalent to building waterworks for a city of 350,000 from scratch in just 10 years. By 2013, dust control costs had exceeded $1.2 billion. Converting much of the existing shallow flooding into low-water tillage is estimated at another $1 billion. The future of the dust control project will have many fewer pipes to leak, valves to stick, or controls to break, while using less water and requiring almost no energy to operate. Owens Lake will be transformed over 20 years into zigzagging corduroy of berms and trenches sheltering instead of the sparkling braids of water over bright red salt flats (see photo 5) created by the bubblers.[11]

Half the water once bound for Los Angeles via the aqueduct is being diverted back into Owens Lake. This will never refill the lake. For that to happen, the Owens River would need to flow unimpeded back onto the playa for seven years. Still, this bounty of water has been the genesis of a vast habitat consisting of mudflats and brine pools teeming with brine flies and microbes, creating a vital oasis on the Pacific Flyway for the migratory shore birds that inspired the 1998 MOA (see photo 4). Birds and birdwatchers are visiting Owens Valley in increasing numbers, enhancing the local ecology and economy. There are even plans to create a road loop for tourists to venture onto the northern end of the lake. The sky may never be completely darkened by millions of ducks, as reported by early pioneers, but at least the city is finally allocating a fraction of the water that is needed by nature and the residents of the Eastern Sierra.[12]

The future of Los Angeles starting to look dry. Very dry.

Los Angeles is adjusting to the imposed restrictions on water imports via the aqueduct, though most of the impetus to conserve has been from the drought. In the days of unlimited water exports, the Los Angeles Aqueduct provided 70% of the water used by the city. For the rest of the 21st century, at least half of the aqueduct's historic flow will remain in the Eastern Sierra.[13]

The days of building massive aqueducts in the United States are over. There are no distant watersheds remaining for the city to tap. The last frontier for importing water is draining the fossil groundwater pockets under the Mojave (many containing rainfall from a million years ago), as the Cadiz Water Project is doing. With the 2014 water bill, California finally established groundwater regulations—so the Wild West days of rampant pumping and exports are fading into the sunset. To the east of Los Angeles, the over-allocated Colorado River remains a vicious legal battleground, pitting state against state, the upper basin versus the lower basin, and US interests against Mexico. Moreover, the Colorado River's water quality is threatened by acid mine drainage, excess salinity and pesticide runoff from farms, and piles of radioactive tailing on its banks. To the west and north, the massive California Aqueduct is grappling with preventing the extinction of several fish species, saltwater intrusion into the Sacramento Delta caused by collapsing levees from earthquakes or neglect, sea level change, ground

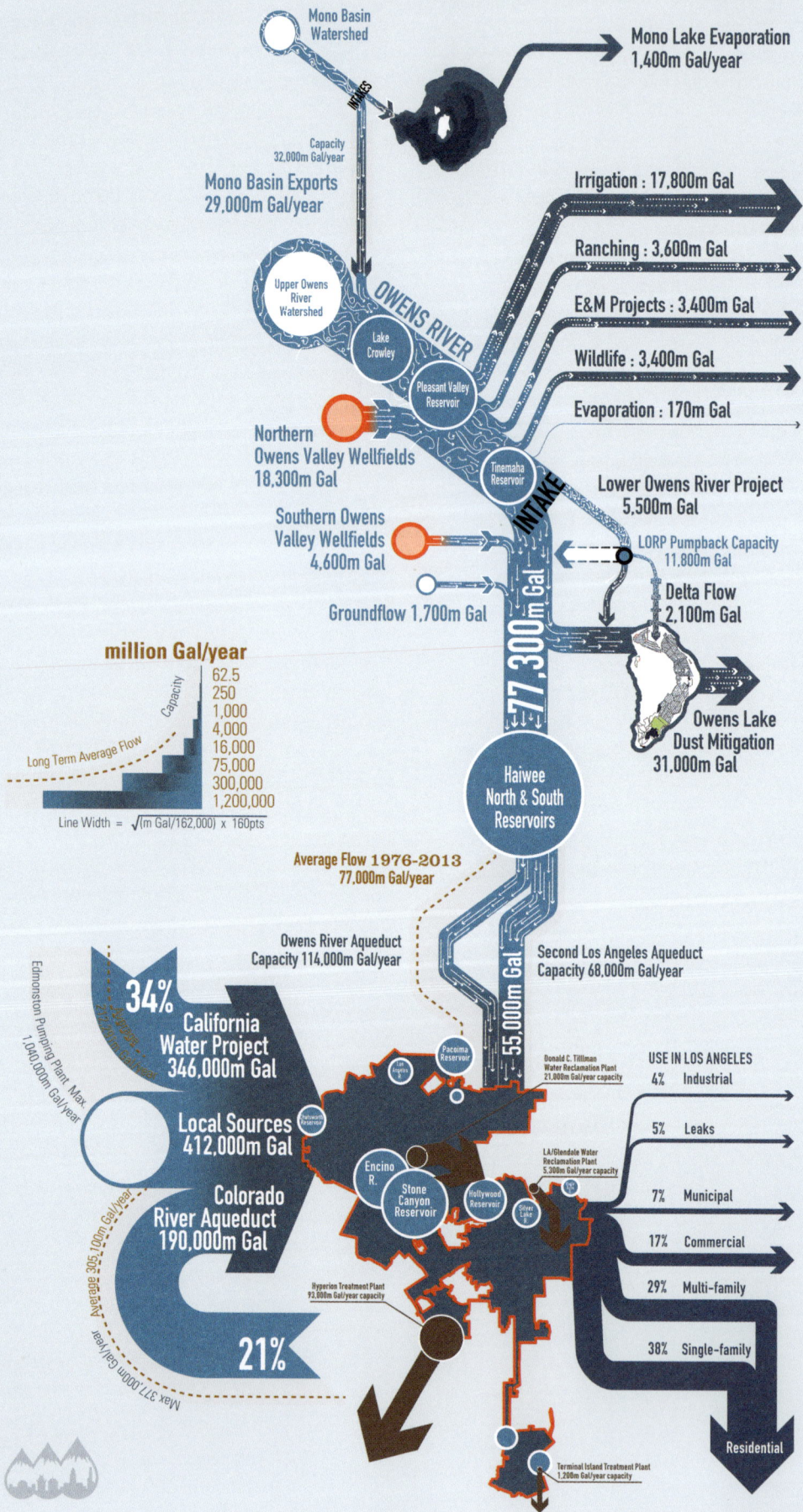

SOURCES

SINKS

Mono Basin
Watershed

Mono Lake Evaporation
1,400m Gal/year

INTAKES

Capacity
32,000m Gal/year
Mono Basin Exports
29,000m Gal/year

Irrigation : 17,800m Gal

Ranching : 3,600m Gal

E&M Projects : 3,400m Gal

Wildlife : 3,400m Gal

Evaporation : 170m Gal

OWENS RIVER

Upper Owens
River
Watershed

Lake
Crowley

Pleasant Valley
Reservoir

Northern
Owens Valley Wellfields
18,300m Gal

Tinemaha
Reservoir

INTAKE

Lower Owens River Project
5,500m Gal

Southern Owens
Valley Wellfields
4,600m Gal

LORP Pumpback Capacity
11,800m Gal

Groundflow 1,700m Gal

Delta Flow
2,100m Gal

million Gal/year

62.5
250
1,000
4,000
16,000
75,000
300,000
1,200,000

Capacity

Long Term Average Flow

Line Width = $\sqrt{(m\ Gal/162,000)}$ x 160pts

77,300m Gal

Owens Lake
Dust Mitigation
31,000m Gal

Haiwee
North & South
Reservoirs

Average Flow 1976-2013
77,000m Gal/year

Owens River Aqueduct
Capacity 114,000m Gal/year

Second Los Angeles Aqueduct
Capacity 68,000m Gal/year

55,000m Gal

34%
California
Water Project
346,000m Gal

Edmonston Pumping Plant Max.
1,040,000m Gal/year

Pacoima
Reservoir

Donald C. Tillman
Water Reclamation Plant
21,000m Gal/year capacity

USE IN LOS ANGELES

4%   Industrial

Los
Angeles

Local Sources
412,000m Gal

Chatsworth
Reservoir

5%   Leaks

LA/Glendale Water
Reclamation Plant
5,300m Gal/year capacity

7%   Municipal

Encino
R

Stone
Canyon
Reservoir

Hollywood
Reservoir

Silver
Lake

17%   Commercial

Colorado
River Aqueduct
190,000m Gal

Average 305,100m Gal/year

29%   Multi-family

Hyperion Treatment Plant
93,000m Gal/year capacity

38%   Single-family

21%

Max 377,000m Gal/year

Terminal Island Treatment Plant
1,200m Gal/year capacity

Residential

subsidence, shrinking snow packs in the Sierra, and the recalcitrant and belligerent agricultural interests that consume 80% of water in California. Towing icebergs from Alaska or floating giant bags of fresh water from the Columbia River are just pipe dreams that are not economically or technically feasible. All forecasts point to less water flowing into Southern California.

In response, Los Angeles (and the Metropolitan Water District) is aggressively developing local water resources. Unlike water from the Owens Valley, which generates a surplus of hydropower on its journey down to Los Angeles, all local sources are energy-intensive. In 2013, decades after plumes of TCE and other industrial solvents rendered 50% of the groundwater in the San Fernando Valley unfit for human consumption, LADWP announced it was building two groundwater treatment plants (one of them reported to be the largest in the world) for $800 million that might provide up to 25% of the city's water. While the San Gabriel River east of the city captures 92% of the rainfall for groundwater recharge, Los Angeles is far behind capturing both stormwater and other urban runoff, with just 27,000 acre-feet/year infiltrated.

The city's main wastewater facility, Hyperion Sewage Treatment Plant in El Segundo, sends over 362 million gallons of secondary treated effluent into the Santa Monica Bay every day. Reclaiming this discharged water with tertiary treatment and additional filtration would provide enough water for all the homes in Los Angeles (based on the 2014 per capita water residential use rate). In the San Fernando Valley, the Tillman and Glendale Reclamation Plants contribute 89 million gallons per day (77% of the total) to Los Angeles River's base flow. Los Angeles River restoration plans depend on this increased flow and have nurtured a growing constituency that may limit diversion of this water back into the supply system. The "ewwww" factor that has limited deployment of "toilet to tap" water recycling is slowly fading as the reality of persistent drought changes attitudes. For now, the accepted method of psychologically "sanitizing" this connection is to recharge groundwater with treated effluent, then later pump it out for use. For now, however, most of this valuable water is dumped into the Los Angeles River or Santa Monica Bay.[14]

Los Angeles has a long tradition of pursuing conservation of water, dating back to Mulholland's era, when the first water meters were installed in the city. Even with the rapidly growing population, the total amount of water used in Southern California has stayed almost steady since 1990. Indeed, the entire state of California has risen to the challenge of using less water. Post-suburban 21st-century California is shifting from the earlier ideal of a verdant, irrigated Eden of swimming pools and lush lawns, to embracing the xeriscape landscape of native chaparral and oak savannahs. Within the culture of conservation lies the new water source for continued urban growth. Water recycling, off-stream reservoirs, and in-ground storage are a few proven solutions to creating more available water with the existing supply.

Antipode

As the antipode to sprawling Los Angeles, the artificial emptiness of Owens Lake is now an analogue of Turner's 19th-century frontier—a place for aspirations and heartbreak. As you stand at the edge of the lake, the stark flatness of the playa recedes into the distance, the monumental dust control project barely visible, comprising only a subtle shading of greys, reds, and greens against the white salt. Only as you become immersed in the dust infrastructure do you notice the bubblers, rising like alien plants on the terraformed lakebed . This vast infrastructure, dedicated to preserving the integrity of the void, undoes any notion that this is a pristine wilderness. When you look back out towards the surrounding mountains, the other subtle traces of human activity jump into stark relief—the road snaking up the sheer wall of the mountains, the gossamer threads of the power lines, and the absolute horizontal trace of the aqueduct cutting into the foothills.

Once a natural Eden, 21st-century California is thoroughly transformed into an anthropogenic cyborg. Perversely, only in places as heavily regulated and mechanized as Owens Lake are there palimpsests of frontier. Ironically, Los Angeles preserved the 19th-century rural landscape in Owens Valley—re-creating the void where, by all rights, we shouldn't expect to find it.

2015 Afterword

This narrative about reconstructing the void didn't originally delve deeply into the future of Owens Valley or the aqueduct. However, significant political, regulatory, and climate shifts occurred (in addition to those mentioned earlier) leading up to this 2015 revision that open up new possibilities to this saga. California State Lands Commission (SLC) shifted from a peripheral role in the dust control project to actively regulating habitat and aesthetics in 2009. SLC established the Owens Lake Master Planning

Opposite:
A Sankey diagram of the Los Angeles Aqueduct system reveals quantities of water flowing from the sources to their destinations (sinks).

(Data Sources: LADWP; City of Los Angeles; MWD; Inyo Water; Mono Lake Committee; State of California; USGS; US National Map; Vincent Ostrom's dissertation, UCLA, 1950;

Armstrong, History of Public Works in the United States, 1776-1976; and Kahrl et al, California Water Atlas, 1979)

Above:
LADWP Well 104 captures water from 260 feet below the Owens Valley surface. 36°53'12"N, -118°14'19"W near Independence, CA. Photo by author.

Below:
The northern end of Owens Lake is a mosaic of shallow flooding reflecting the sky and the Inyo Mountains, pale green corduroy of managed vegetation, and flat grey polygons of gravel. 36°30'27"N, -118°4'29"W west of Owens Lake, CA at +4800'. Photo by author.

Project, which is looking holistically at the landscape of the playa, and brought landscape architects into the process. Provoked by the politically insensitive siting of the Southern Owens Valley Solar Farm just a few miles from Manzanar, a multicultural coalition emerged in Los Angeles to advocate for Owens Valley issues. Los Angeles elected Mayor Eric Garcetti in 2013, while the candidate backed by the powerful LADWP union lost. Marcie Edwards was appointed as the LADWP General Manager—the first woman to hold this post.

In this changing context, 137 landscape architecture students from the California State Polytechnic University Pomona participated in the courses and activities sponsored by the Aqueduct Futures Project. Established in early 2012 to commemorate the aqueduct's 2013 centennial, Aqueduct Futures continues to map changes wrought by the aqueduct, and works to create a framework for peace between the Owens Valley and Los Angeles. From this framework, several scenarios have emerged that explore potential changes in land ownership and water exports from Owens Valley.

Resolving the liabilities related to LADWP's property ownership and management is a pressing issue, as this is the most significant source of anger for residents in the Eastern Sierra. There is an inherent conflict of interest between the DWP's mission of supplying water to Los Angeles and managing their 492 square miles. There is no justification for the LADWP to continue owning any property in Owens Valley towns and communities beyond their own facilities—so they must divest. Better management of the vast valley floor is required to enable local prosperity and to provide stronger protections for the fragile ecosystems. A radical solution would be transferring nonessential LADWP holdings to another agency better suited for managing recreation and natural resources. Perhaps Los Angeles Parks and Recreation will step in or will establish an Owens Valley National Monument or other park to be run by the state or county.

In Los Angeles, the Los Angeles River Revitalization Master Plan refocused the city back on riverbanks and nature it had long ignored. A similar process could shift the role the aqueduct plays for the communities and desert valleys it traverses. Perhaps a living aqueduct would be the result, as the confined waters are freed to create a linear oasis alive with fish, birds, and invertebrates as the water flows to Los Angeles.

We can also consider a future when the aqueduct is no longer needed. Local Southern California water sources (from reuse, conservation, and decontamination) will soon exceed the volume of water supplied by the aqueduct, so perhaps this day is not too far off. The Lower Owens River Restoration project hints at the verdant transformation possible when sufficient water flows to the valley floor. If all the water remains in the valley, then the shell-game of pumping to supply water to restore habitat elsewhere will end. With all the water, Owens Lake can refill. With all the water, justice will be served as Los Angeles renews the void.

Notes ---

1.   Jeff Putnam and Genny Smith et al., *Deepest Valley: A Guide to Owens Valley, Its Roadside and Mountain Trails,* 2nd ed. (Palo Alto: Genny Smith Books/Live Oak Press, 1995), 145-149. Cerro Gordo produced $17 million of silver and lead between 1865 and 1879. While this seems small compared to other bonanzas, it was enough to catapult Los Angeles from being a town of less than 2,000 people in 1850 to over 50,000 in 1890. Between 1880 and 1890 alone, the city experienced 450% population growth, yet its area barely expanded. Remi Nadeau, *The Water Seekers* (Santa Barbara: Crest Publishers 1997), pp. 11-15, calculated that water from the Los Angeles River could support a population of 200,000 people (less than Los Angeles's 1904 population).

2.   See Norris Hundley, *The Great Thirst: California and Water—A History* (Berkeley: University of California Press, 2001), 141-166, on the politics behind the Los Angeles Aqueduct. For an economic analysis, see Gary Libecap, "Chinatown Revisited: Owens Valley and Los Angeles— Bargaining Costs and Fairness Perceptions of the First Major Water Rights Exchange" (working paper, University of Arizona, Tucson, 2004), www.international. ucla.edu/cms/files/Liibecap.pdf. Owens Valley/Inyo County has projected population growth at less than 10% over the next 50 years and may even lose population. See State of California, Department of Finance, "Population Projections for California and Its Counties 2000-2050, by Age, Gender, and Race/Ethnicity" (Sacramento, 2007).

3.   Establishing the Zanja system in 1781 on the bones of the Tsonga irrigation system was paralleled by the American settlers in the Owens Valley in the 1850s who commandeered the Paiute's extensive irrigation system containing over 60 miles of ditches. See Harry Lawton, Philip Wilke, Mary DeDecker, and William Mason, "Agriculture Among the Paiute of Owens Valley," *Journal of California Anthropology* 3 (1), 1976, UC Merced; and seek out Jenna Cavelle's documentary *PAYA: A Documentary About California Water Wars*; partial transcripts are available in *Arid Journal,* Fall 2013, aridjournal. com/a-paiute-perspective-owens-valley-water-jenna-cavelle/.

4.   Blake Gumbrecht, *The Los Angeles River: Its Life, Death, and Possible Rebirth* (Baltimore: Johns Hopkins University Press, 1999).

5.   Marith C. Reheis, "Owens (Dry) Lake, California: A Human-Induced Dust Problem," United States Geological Survey, geochange.er. usgs.gov/sw/impacts/geology/ owen. The *Inyo Independent* first reported about a dust storm back in 1871 (LADWP). Hoyt S. Gale and F. L. Ransome, "Salines in the Owens, Searles, and Panamint Basins, Southeastern California," *United States Geological Survey Bulletin* 580 (1913) describes Owens Lake as being 270' deep at the end of Pleistocene Ice Age and the first in a chain of lakes, with the Owens River flowing over Fossil Falls and through Rose Valley into China Lake, filling Death Valley with Lake Manley, before cascading into Lake Lahontan north of Lake Tahoe. In 1872, the lake was 49' in depth; by 1876 it had dropped to 38', and in 1913 it was only about 29' deep. See also the map of Owens Lake in 1888 reproduced in Putnam and Smith, *Deepest Valley,* 47.

6.   Todd M. Mihevc, Gilbert F. Cochran, and Mary Hall, "Simulation of Owens Lake Water Levels," Publication No. 41155, Great Basin Unified Air Pollution Control District, June 1997, ftp://

Above:
36°32.460' N
118°3.007' W

Opposite page, above:
Halobacteria & Salt on Owens Lake.
Opposite page, below:
2nd LAA in Antelope Valley.

gbuapcd.org/HydroReports/ Mihevc,%20et%20al.,%20 1997.pdf.

7. Droughts in California tend to occur every 30 years. Significant ones since statehood were 1863-64, 1887-88, 1912-13, 1922-24, 1928-34, 1947-50, 1959-61, 1976-77, 1987-92, and as this chapter is being revised in late 2014, Los Angeles and most of its aqueduct shed is in an "exceptional drought" that started in 2013. See "Climate of Los Angeles," National Weather Service Forecast Office (ND), wrh.noaa.gov/lox/climate/ climate_intro.php; and A. S. Jayko and C. I. Miller, *Preparing for California's Next Drought: Changes Since 1987–92* (Sacramento: Department of Water Resources, State of California, 2000).

8. Out of the litigation over groundwater pumping, Los Angeles and Inyo County jointly developed the *Green Book for the Long-Term Groundwater Management Plan for the Owens Valley and Inyo County* (Los Angeles: City of Los Angeles and Inyo County, 2000; revised in 2009). Additional Technical Memorandums have been issued as addenda to the Green Book that address specific subjects such as refining the management and monitoring of groundwater pumping.

9. The 1998 Memorandum of Agreement approved just three "Best Available Control Measures" (BACM) for dust control: shallow flooding (either ponds or from sprinklers), planting, and gravel cover. It is also specified that all berms have "snowy plover crossings incorporated every 500 feet."

10. The main points of the 2014 Judgment were: (1) Acknowledging that controlling dust outside the historic shoreline (such as from the Keeler Dunes) is the GBAQMD's responsibility—not LADWP's. 2) The extent of mandatory LADWP dust control (to be operational by the end of 2017) is for 48.6 square miles. 3) Designates 4.8 square miles as contingency dust control areas, so 53.4 square miles is the absolute upper limit for LADWP dust control. 4) Additionally allows the study of other waterless dust-control measures for possible use.

11. In 2013, LADWP reported that the dust control project was using 95,000 acre-feet of water annually. This is three times the water used by a comparable area of Los Angeles.

12. Wayne Bamossy (project manager at LADWP), conversations with the author, 5 January 2005. Richard Cervantes, Inyo County Supervisor, conversation with author, 7 January 2005. LADWP, Policy for Public Access to LADWP Facilities at Owens Lake (Los Angeles, 20 May 2004).

13. Dan Cayan et al, *Climate Change Scenarios for California: An Overview* (Sacramento: California Climate Change Center 2006), www.energy.ca. gov/2005publications/ CEC-500-2005-186/ CEC- 500-2005-186-SF.pdf.

14. City of Los Angeles Department of Public Works, Bureau of Sanitation, "About Wastewater," www.lacity.org/ san/wastewater/factsfigures. htm.

Right, above:
Cheaper to construct and operate, the new dust control zones on Owens Lake will be more of these 4' deep salt-encrusted ditches that zigzag across the playa. Many of the existing zones of shallow flooding will be converted to tillage. 36°23'33.73"N, -117°55'28"W on Owens Lake, CA. Photo by author.

Right, below:
Shallow flooding has generated a mosaic of pools that shore birds and ducks now forage. 36°21'38"N, -117°56'21"W on Owens Lake, CA. Photo by author.

ACT THREE: THE DISAPPEARING
DEFEND

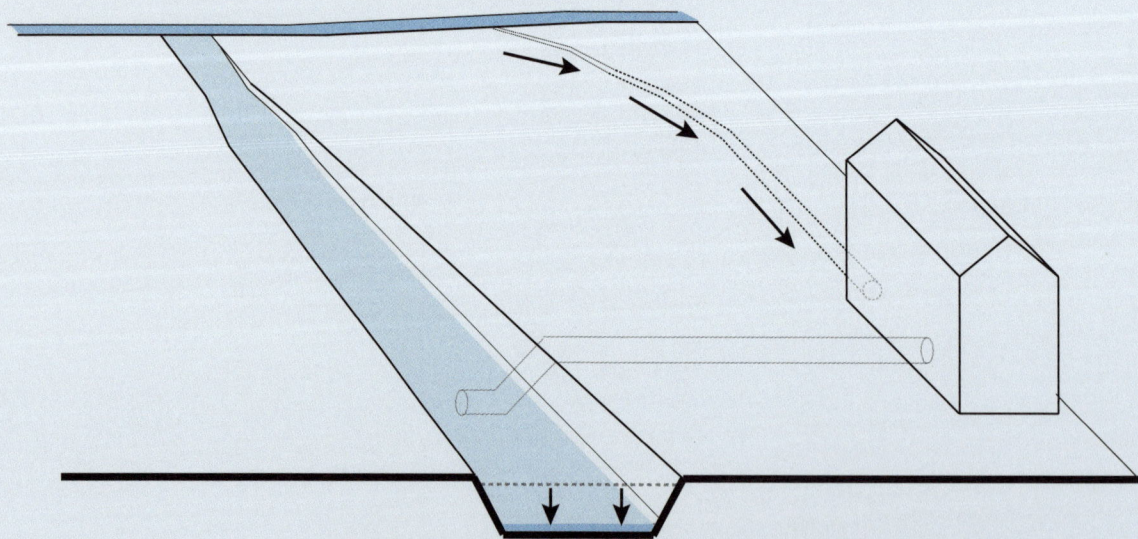

To construct a mechanism to bring water from distant lands to people, buildings and cities.

--- ---

Index of tools | reservoir, pipeline, groundwater rights, drainage, flow-return credits, interstate water banking, aquifer recharge, flow surplus, water rights Transfer/Exchanges.

DRAINING THE OASIS

Las Vegas, Nevada. 2015.

Benjamin Gregory

In 2013, the water level in Lake Mead reached its lowest point since it was filled in 1936. Even the largest reservoir in the United States is at risk of falling short of the water demands of the American Southwest. Since the Colorado River Compact was signed by seven states and Mexico in 1920, there has been a constant search for water sources to "augment" the supply of the Colorado River Basin. As early as the 1950s, major pipeline projects were proposed that would bring water from as far away as Alaska to increase the river's flow. The latest big infrastructural idea: tap the Mississippi, and channel its excess waters across Iowa and Nebraska and into Colorado.

Of course, there are alternatives to laying hundreds of miles of pipe. Few people have had to consider them as much as Pat Mulroy. As the recently retired manager of the Southern Nevada Water Association, she has fought to stretch Nevada's share of the Colorado River Compact, a paltry 350,000 acre-feet (as compared to California's 4.2 million a-f) as far as it will go. Water is a right, she says. There's plenty of it on this earth, it's just a matter of getting it to where we need it. During her tenure, she diversified the "water portfolio" of Las Vegas Valley, gaining groundwater rights from other parts of the state, brokering deals with neighboring states to "bank" their water, and with the federal government to take advantage of flow "surplus." The terminology surrounding water has moved from one of natural resource management to one of financial management. This thinking brings with it all the corollary tools to "end up in the black" that exist in the financial sector. Some of these tools are beneficial, as they increase the efficiency of management; drainage flow-return credits, interstate water banking and aquifer recharge, Colorado River Surplus, and water rights transfer/exchanges all help make the numbers work while carefully accounting for every drop of water. But as in the financial sector, the short-term game often doesn't pan out in the long run.

The Las Vegas Valley is a sediment-filled structural trough that was formed over many millions of years through compression, extension, and faulting of the original flat-lying marine sediments that form the bedrock. Though the valley itself gets little rain, 3 to 4 inches per year, the surrounding Spring Mountains get upwards of 24 inches. This water drains to the valley and seeps into the layers of sediment and clay, forming large aquifers which are then forced upward along faults and fissures via hydrostatic pressure, creating surface springs and artesian wells. Long a place known for its abundant water supply, Las Vegas Valley (meaning "the meadows") attracted Native Americans, Spanish explorers, and westward travelers alike. Las Vegas was established as a waypoint between Flagstaff and Los Angeles, and began to grow with the introduction of the railroad in the late 19th century. Copious land and water made the area favorable to agriculture and industry. In the 1930s, the Hoover dam brought workers from across the country. The population exploded as real estate interests soared, and casinos took root in the desert. But by the 1960s the flowing springs, which had attracted settlers for thousands of years, had gone dry.

The first well in the valley was drilled in 1907, and by 1912, 125 wells were extracting 15,000 a-f of water per year. By 1942, 40,000 a-f of water were being pumped from the ground each year, as the first pipeline from Lake Mead was extended to Henderson. A second pipeline was built in 1971 to the city of Las Vegas. By 1996, the Valley was using nearly its entire allotment of Colorado River Compact water, 356,000 a-f, as well as pumping 56,000 a-f from the underground aquifers, nearly three times their sustainable yield of 20,000 a-f/yr. That year, it was predicted the valley had only enough water to last until the year 2025. Las Vegas has grown over 40% since and, in its greatest year of tourism in 2013, nearly 40 million people visited the city.

Already stretched to its breaking point, the water supply system of Las Vegas is at a crossroads. The unpredictability of Colorado River water availability puts them at risk. As a result, negotiations within the closed system of the Colorado River Compact will provide only so much leverage. As Lake Mead shrinks and Las Vegas grows, it's only a matter of time until a pipeline from the Mississippi, or some other distant source of water, becomes an urgent, and expensive, reality.

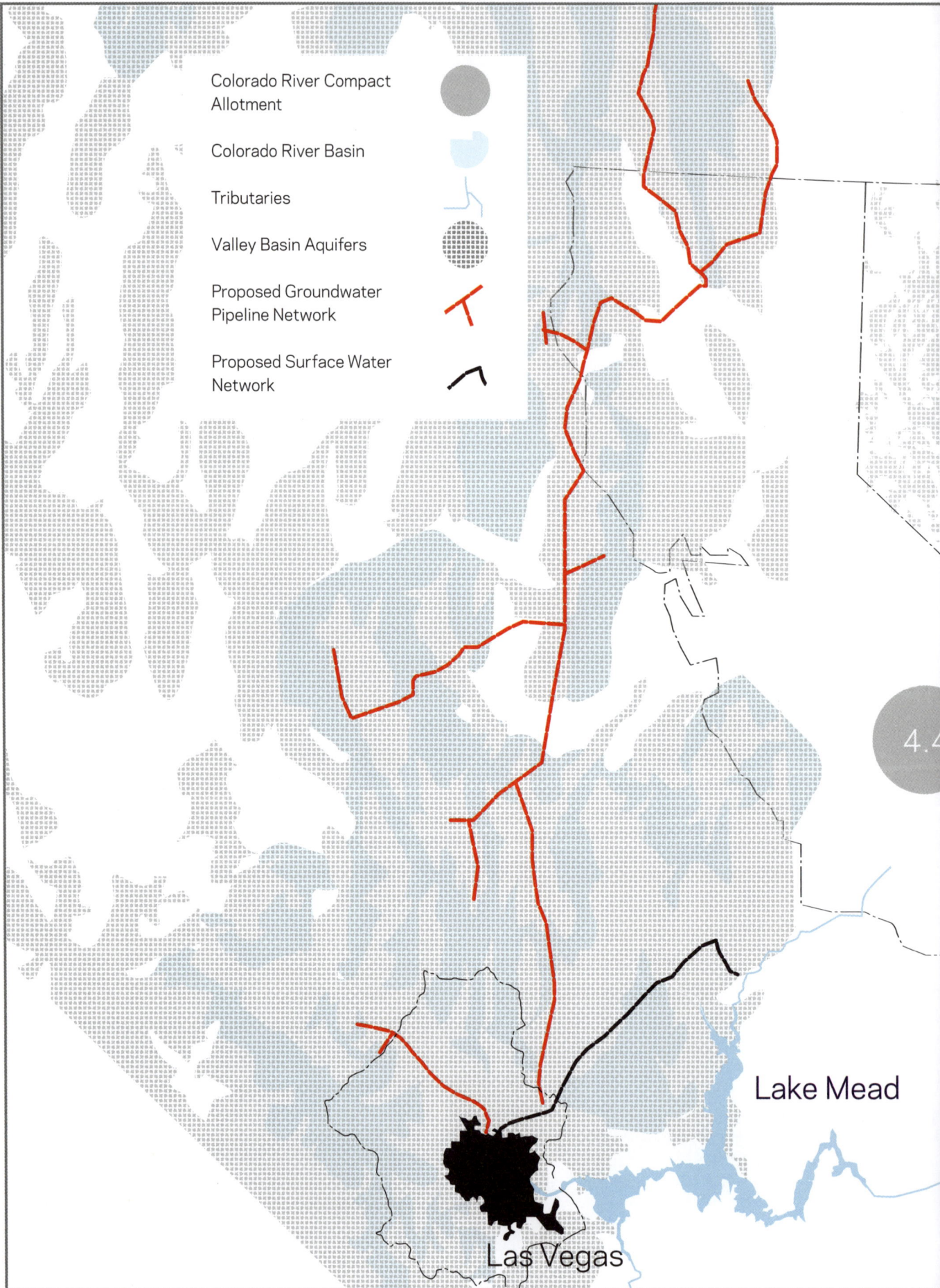

Colorado River Compact
Allotment

Colorado River Basin

Tributaries

Valley Basin Aquifers

Proposed Groundwater
Pipeline Network

Proposed Surface Water
Network

4.4

Lake Mead

Las Vegas

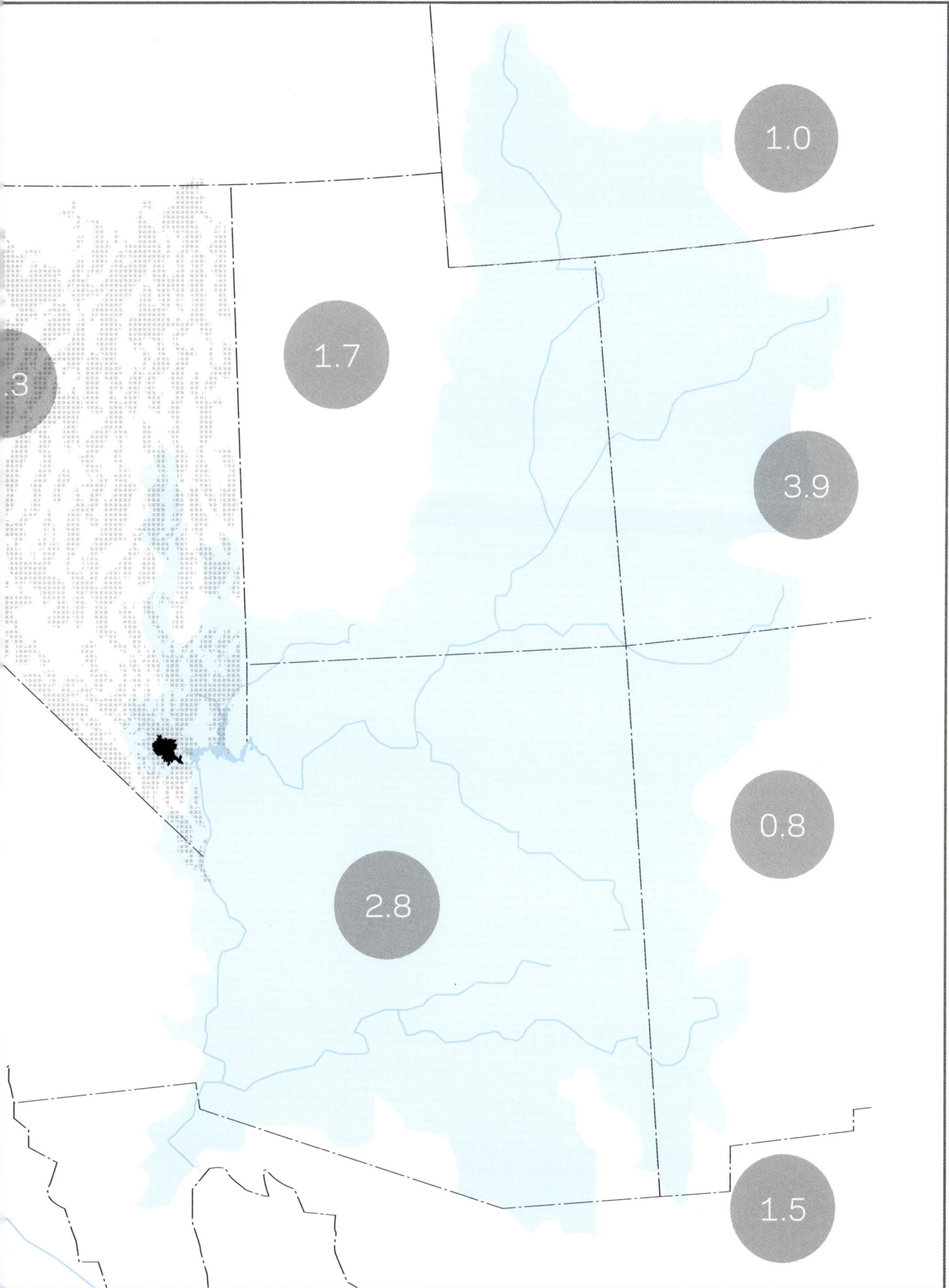

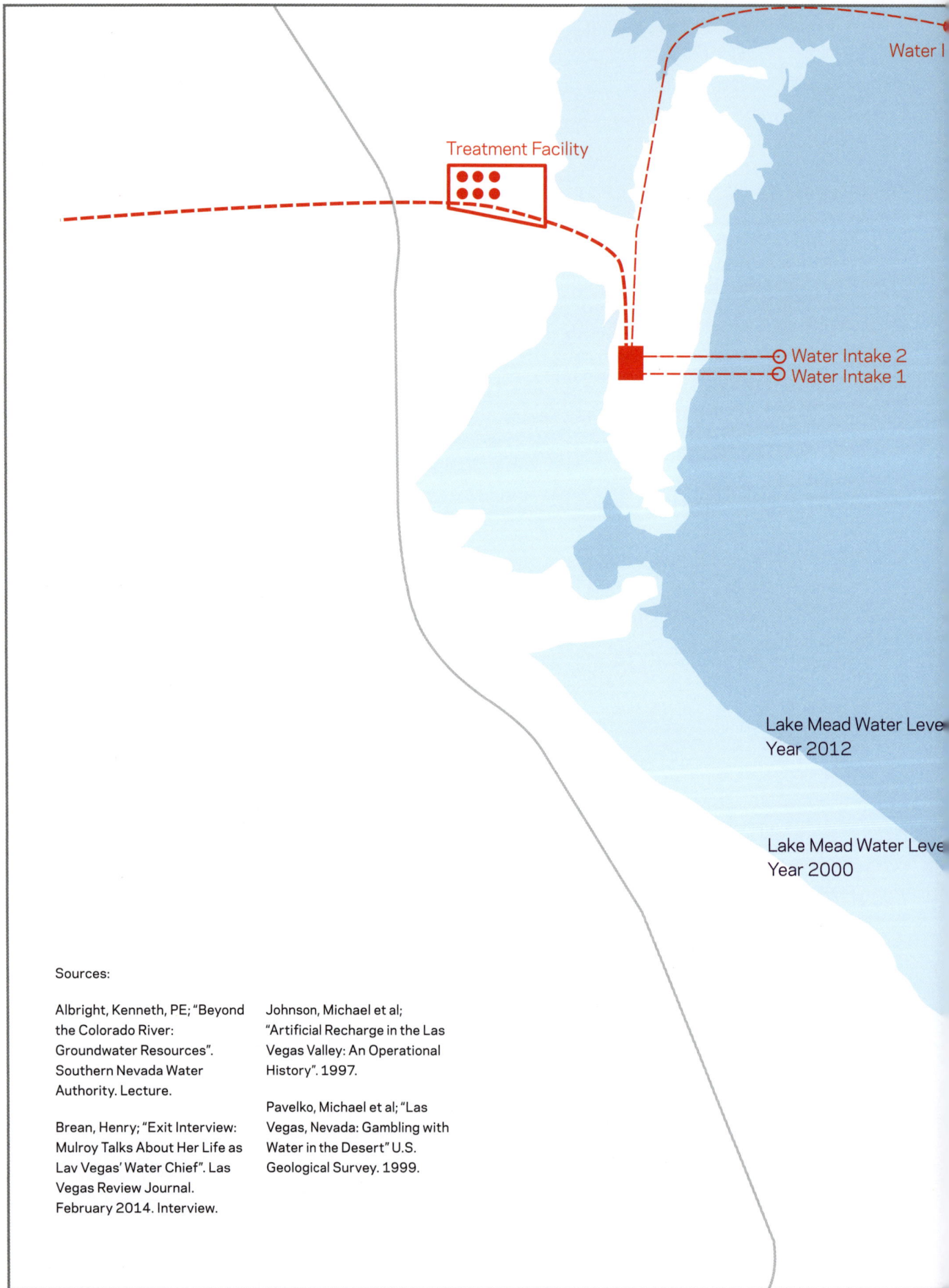

Water I

Treatment Facility

○ Water Intake 2
○ Water Intake 1

Lake Mead Water Leve
Year 2012

Lake Mead Water Leve
Year 2000

Sources:

Albright, Kenneth, PE; "Beyond the Colorado River: Groundwater Resources". Southern Nevada Water Authority. Lecture.

Brean, Henry; "Exit Interview: Mulroy Talks About Her Life as Lav Vegas' Water Chief". Las Vegas Review Journal. February 2014. Interview.

Johnson, Michael et al; "Artificial Recharge in the Las Vegas Valley: An Operational History". 1997.

Pavelko, Michael et al; "Las Vegas, Nevada: Gambling with Water in the Desert" U.S. Geological Survey. 1999.

Hoover Dam

ACT THREE: THE DISAPPEARING RETREAT

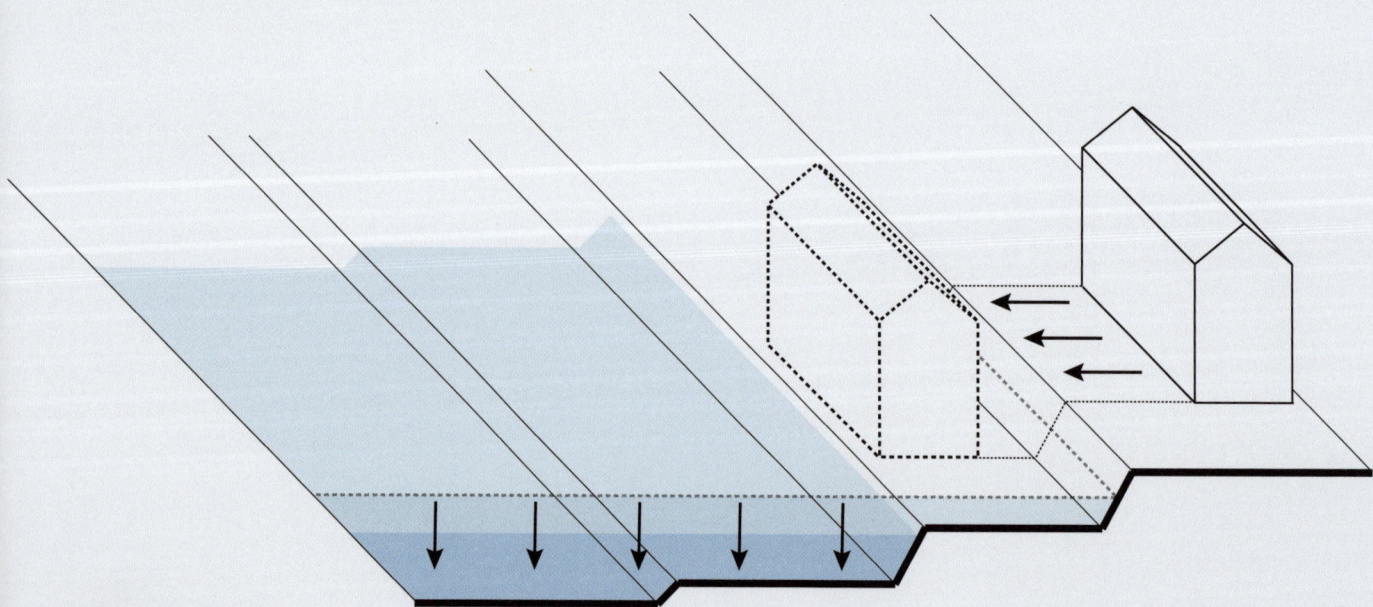

To move people, wildlife, buildings and cities away from stressed water environments and to new water frontiers.

--- ---

Index of tools | offshore technology, vegetated habitats, floating tower.

SEA TREE

Various Locations. Speculative, 2014.

Waterstudio

"The waterfront no longer places a limitation on city expansion; in fact it is the new frontier!"

Urbanization and climate change continue to hold pressure over densely populated areas with a scarcity of land available for developments such as parks or natural reservations. It is becoming ever more difficult to allot an appropriate amount of land for the conservation of wildlife habitats within city limits. The increase of difficulty in promoting park developments has set back several preservation initiatives that were eager to find a way to protect animal habitats. Conserving habitats for birds, bees, bats, and other small animals not only would increase biodiversity, but also would provide any city with a multitude of positive environmental effects, promoting the general health and well-being of the community at large.

In response to an increase in the demand and urgency for wildlife protection, Waterstudio has designed a new concept, the Sea Tree®, in order to create the necessary green spots within a city. Sea Tree is a floating, steel structure made up of vegetated, layered habitats designed exclusively for flora and fauna. As a way of giving back to nature, Sea Tree is designed and constructed taking into consideration every element essential to fostering an abundant habitat for species both above and below the water table. Under water, the Sea Tree provides a habitat for small water creatures or even, when the climate allows, for artificial coral reefs.

Sea Tree is built using the latest offshore technology, similar to the existing and proven technology required for the construction of oil-storage towers found in the open seas. Large oil companies will have the opportunity to give back by using their own intellectual property and resources to donate Sea Trees to a community in need, showing their concern and interest in preserving the distressed wildlife. Sea Tree would provide the ideal environment for a multitude of species, not to mention a significant reduction in CO_2 emissions.

Sea Trees can be located along riverbanks, oceans, lakes, harbors, and even near industrial zones and islands. The height and depth of the Sea Tree can be adapted according to the physical conditions such as water depth, waves, tides, and currents of any location. Much like a land tree, the Sea Tree will gently move along with the wind and waves, and it will be secured to the seabed with a cable and anchoring system.

This design alleviates the conflict of allocating land within a city; simultaneously, the Sea Tree introduces an innovative avenue to protect our wildlife, providing countless benefits and a slew of possibilities. One Sea Tree has the potential to affect a zone of several miles around the moored location by providing habitat to many plants and animals in search of refuge. For all we know, the Sea Tree will be the first floating object that is 100% built and designed for flora and fauna.

The Dutch heritage, consisting of hundreds of years fighting against the water, has provided its citizens with a propensity to design innovative solutions that use the water as an ally to combat

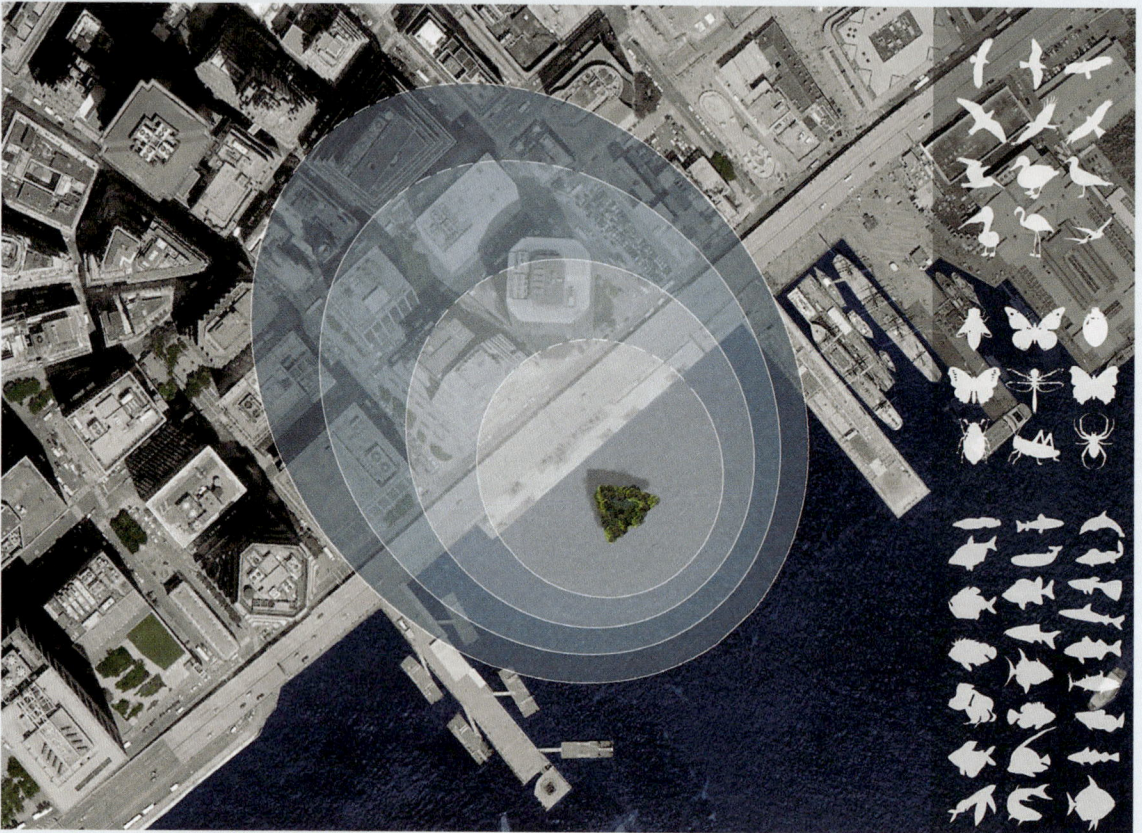

Above:
The Sea Tree has a
large effect on its direct
environment. Opposite

Below:
The Sea Tree, a floating
instant green solution.

modern challenges. Oil companies have used these floating storage towers for years; Waterstudio only gave them a new shape and function. They have consulted experts from the most respected institutes in Holland, who have provided them with the latest research to optimize the design of these structures so they will provide an ideal atmosphere to create catalysts for the growth of flora and fauna habitats.

Companies adopting a Sea Tree will demonstrate their positive attitude towards citizens, the environment, and biodiversity in metropolises. The Sea Tree is considered a "City App," a floating product that can be added to any city, similar to adding an app on your smart phone. The sponsoring company holds ownership of the City App, while the city provides a location.

The inspiration for creating Sea Tree came from a project in Holland where ecologists challenged designers to design a habitat for fauna which could not be disturbed by human beings. Water is, of course, a perfect way to keep people away. Other sources of inspiration were the shapes of floating oil storage structures in Norway and the shapes of land trees with large crowns. Lastly, the concept was developed from park zones in urban areas. Waterstudio divided these areas into sections and placed them vertically on top of each other. In the end, it has become a vertical hangout for wildlife!

Above:
The layered habitats are designed exclusively for flora and fauna.

Below:
The floating foundation is designed to give shelter to underwater life such as plants and fish.

ACT THREE: THE DISAPPEARING

ADAPT

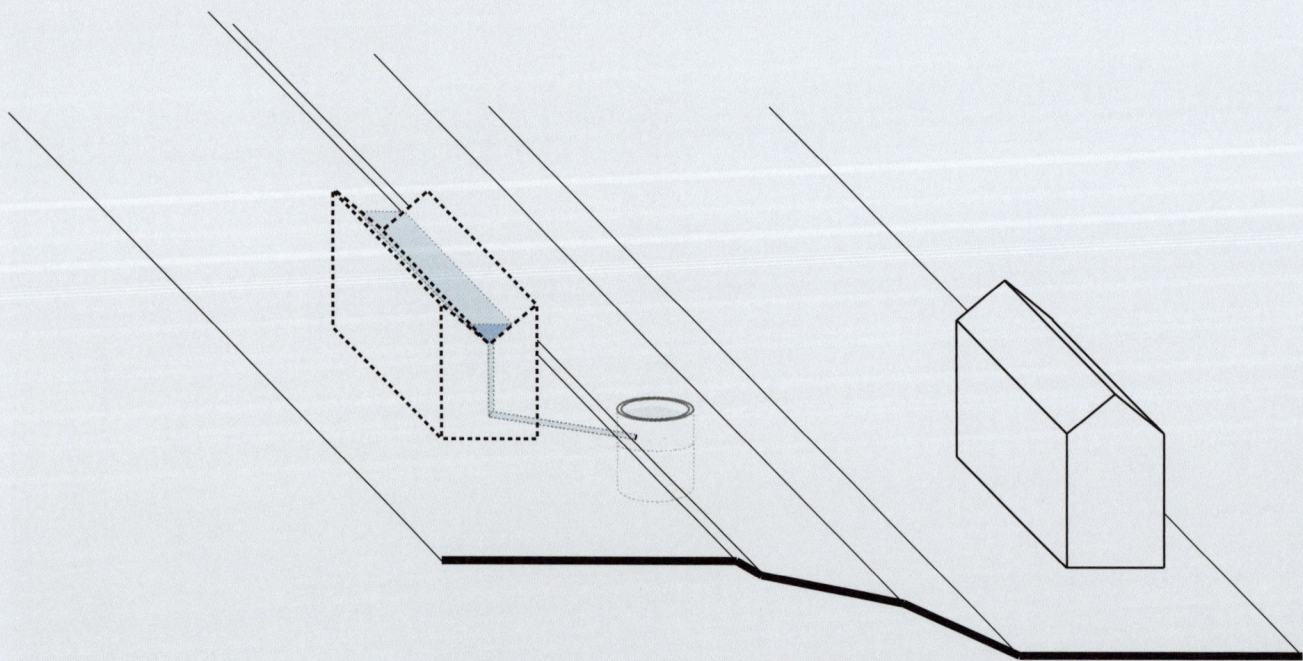

To harvest, store, and restore water within the spaces of cities, landscapes, and buildings.

WINTON WETLAND RESTORATION

Northeastern Victoria, Australia. Constructed 2012.

Taylor Cullity Lethlean

"This site is not a natural environment, it is a 'nature environment'—it is a work-in-progress, carrying the marks of our rich history into a new future."

Taylor Cullity Lethlean (TCL) won the commission for the master plan restoration of 3,000 hectares of the 8,750-hectare Winton Wetlands site. The site, formerly known as Lake Mokoan, has rich Indigenous importance (covering three language groups), fertile agricultural land, and thousands of hectares of River Red Gum forests.

This project of national scientific, cultural, and environmental significance is the largest wetland restoration project in the southern hemisphere. The project aims to create a major national facility for wetland education and research, as well as to demonstrate best-practice natural resource management and develop nature-based tourism activities and recreation.

Work began on the decommissioning of Lake Mokoan in 2009. The re-establishment of the wetlands has allowed for the return of 44,000 megaliters of water per year to the Broken, Goulburn, Snowy, and Murray Rivers, with environmental and economic benefits to both upstream and downstream.

The master plan involves a detailed business model, seeking to provide the project with an economic footing to assist the wetlands in becoming an "eco-tourism" attraction and provide the region with a stimulus, offsetting any negative effects surrounding the decommissioning of Lake Mokoan.

KEY DESTINATIONS

1. **Visitor Arrival Point**
   - Shelter Structure with Wayfinding & Interpretive signage
   - Toilet Block
   - Car Access & Car Park

2. **Yacht Club & Regional Playground**
   - Refurbish Yacht Club as Community Shop / Cafe
   - Regional Playground
   - Gravel Car Park
   - Landscape Planting

3. **Robinson's Hill**
   - Vechicle Access (Gravel)
   - Carpark
   - Toilet
   - Lookout (Deck and Path)

4. **Boat Launch Facility**
   - Vechicle Access
   - Small Carpark
   - Boat Ramp

5. **Duck Pond**
   - Fishing
   - Water Activities
   - Vehicle Access
   - BBQ/Table/ Shelter
   - Jetty
   - Camp ground
   - Toilet Block

6. **Bill Friday's**
   - Canoe activity with associated infrastructure
   - Layby parking off road.
   - Gravel walking track
   - Boadwalk/ Jetty over water.

7. **11 Mile Creek**
   - Bird Watching Site
   - Gravel Walking Path Circuit
   - Small Boardwalk
   - Vehicle layby to Road.

8. **School Group Camping**
   - Fishing

9. **Bike Circuit**
   - BMX X-Trail
   - Mountain Bike Circuit

- **Camping - Walk In**
- **Camping - Drive In**
- **Glamping**

L E G E N D

- Entry into Site
- Access to Site
- Grazing Land Use
- Education Precinct
- Key Destination
- Main Entry for Cyclists
- Future Bike Trail
- Existing Bike Trail
- Future Walking Trail
- Site Boundary
- Railway Line
- Level Crossing
- RV / Camping

Preview Page: Water is the focal point of the site. The project acknowledges the water as the central aspect of the site and as a precious resource, and acknowledges the value of sustainability. Photographer: John Mitchell

Above:
Winton Wetlands Masterplan Stage 1. Credit: TCL

The Winton Wetlands project demonstrates the importance of assessing how particular land uses perform across a range of scales, and how working with, rather than against, ecological processes can benefit the economy as well as the environment.

TCL Master Plan Guiding Principles

Economic driver
It will have a business model which ensures triple bottom-line optimization and will be a driver for the local economy by encouraging partnerships with private and public investors.

Restoration
It will consider the restoration efforts implicit in the Restoration and Monitoring Strategic Plan, while at the same time making it a beautiful landscape to visit in either drought or flood, with sound management of pests, weeds, and fire hazards.

Experiential
It will provide a range of unique nature-based experiences, both water- and land-based, catering for varying visitor needs and market opportunities.

Experience backdrop
The site is to function as a "backdrop" for nature-based experience and events, showcasing the natural beauty of the wetlands, both on water and land.

Complementary intervention
Interventions in the space will balance creative man-made installations with the natural environment. It is a place where we can be with and experience nature.

Education and Research
It will function as a living laboratory and classroom for students of any age, demonstrating the wonders of a wetland and associated ecosystems. The hub can also become a platform for research on wetlands restoration.

By the community, for the community
It will be a place that provides meeting places, recreational assets, and jobs for the community, and which achieves a high level of community ownership, engagement, and pride.

Connecting with Benalla
It will provide a physical and emotional link to the Benalla community.

Personality
Its own brand identity will evolve to define the experience and has the potential, in time, to develop as a regional icon.

Marking history
It will acknowledge the Indigenous heritage and pioneer stories hidden within the site and the historical figures who played the leading roles.

Water
Water is the focal point of the site. The site will give people a water-based experience at Winton Wetlands, maximizing opportunities for water-based activities and acknowledgement of the water as the central aspect of the site and as a precious resource that must be sustained.

The wetlands provide a place where Victoria's native flora and fauna can flourish.
Photographer: John Mitchell

Children of any age can explore the wonders of a wetland and associated ecosystems.
Photographer: Scott Hartvigsen

Above:
Water trucks providing drinking water to informal settlements. Photo credit: Andrea Balestrini

Below:
Aerial view of informal settlements in Ate District. Photo credit: Evelyn Merino-Reyna

--- ---

Index of tools | multifunctional open space system, hydro-urban units, design manual, water demand calculation tool, performance-based open space design.

LIMA BEYOND THE PARK

Lima, Peru. 2012.

Antje Stokman, Bernd Eisenberg, Eva Nemcova, and Rossana Poblet

The Peruvian capital of Metropolitan Lima is located on a desert coast overlooking the Pacific Ocean. With its more than 9.5 million inhabitants, Lima is considered the most extensive desert city in the world after Cairo. It has an average of only 9 millimeters of rainfall per year, the glaciers feeding its rivers are melting, and the groundwater table has already reached critical levels. At the same time, Lima is facing a vast expansion of its informal settlements. These lack many basic urban services including water supply and wastewater infrastructure, which has caused severe environmental degradation. Around 20% of Lima's population, mainly living in the hilly and peri-urban areas, are not connected to the public water supply. They receive drinking water, often of very bad quality, from private water vendors at high prices.

At the same time, most parks and road greenery are irrigated with drinking water. Water-intensive lawns and artificial ponds in parks put even more stress on the limited and inefficiently managed water resources. The increasing water demand of green areas and the high cost of drinking water put pressure on the wastewater infrastructure, which is informally misused for irrigation purposes, with bad hygienic consequences. At the same time, polluted rivers and irrigation channels are concreted and covered. The agencies responsible for different water-related issues are neither sharing their data nor coordinating their actions. Therefore, Lima's hydrological systems and its urban landscape need radical rethinking to make urban and natural systems perform in concert with one another and keep up with the increasing water demand for a growing, more livable, and green city.

The German-Peruvian research project "Sustainable Water Management in Urban Growth Centers Coping with Climate Change: Concepts for Metropolitan Lima, Perú" —known as LiWa—has initiated dialogues and developed scenarios, tools, technologies, integrated planning strategies, and pilot projects for Lima's water-sensitive future urban development. A new approach, combining infrastructure design and spatial design, acts as a catalyst for landscape transformation and assists in developing an alternative water culture in Lima. The "Lima Ecological Infrastructure Strategy" (LEIS) integrates the hydrological cycle into a multifunctional open space system, which is strengthened to guide future urban development. The concept of "Ecological Infrastructure" was adapted to use in the arid climatic conditions of Metropolitan Lima: it consists of natural and man-made ecosystems, considers the hydrological cycles, readapts ecological processes and increases essential environmental services in the city, processes and increases essential environmental services in the city, improves the urban environment, and guides the urban development in a sustainable way. Based on an overall assessment of the city as a system of different "hydro-urban units" using GIS and NDVI analysis, a design manual as well as a simple design-testing and

Construction of water-sensitive
urban design prototypes during a
summer school program in 2012.
Photo credits: Maximilian Mehlhorn

water-demand calculation tool were developed. They describe how to develop a water-sensitive design that takes into consideration differences in the urban structure and open-space typologies (including level of consolidation and condition), the geomorphology (slope and soil), and the hydrological aspects (including the availability of water sources and the current state of water infrastructure).

In order to demonstrate water-sensitive urban development in practice, the lower Chillon River watershed in the north of Lima and Callao was chosen as a demonstration area. Here, the Chillon River does not carry any water between May and December, but becomes a torrent and causes floods in urban areas the rest of the year. Several pollutants affect the water quality in different sections of the river, including the discharge of raw domestic wastewater and industrial wastewater, drainage from agricultural areas with high concentrations of fertilizers, and insufficiently treated effluent from the wastewater treatment plant. Irrigation channels divert water from the river to irrigate the remaining agricultural fields in the valley. At the same time, valuable cultural heritage sites from pre-Inca times, like the temple El Paraiso (2000 BC) can be found within the area.

A Strategic Landscape Framework Plan for the lower Chillon River watershed was created to demonstrate possibilities for a water-sensitive urban development that can serve as a model for the entire watershed through an integrative approach. This plan integrates water management and landscape planning with social, cultural, and economic aspects, thus guiding the implementation of a water- sensitive demonstration area. It was presented to the local planning authorities of Metropolitan Lima with the aim of being considered in actual planning projects such as the Land Management Plan and the Urban Development Plan for Metropolitan Lima.

Additionally, different prototype solutions were designed, built, and tested in an effort to promote water-sensitive solutions at a smaller scale. These on-site solutions were developed within two interdisciplinary summer schools attended by German and Peruvian students of architecture, engineering, or social science. By designing and constructing a series of temporary installations relating to the different water sources in the area, the students in the program and experts from different institutions discussed the viability of different concepts with the local community. The students tested minimal strategic interventions

that focused attention on the topics and the necessity of water-sensitive urban development in this zone.

In parallel, conceptual designs were developed for a lomas park as well as for a river park and agricultural park in the floodplain of the river, to demonstrate future possibilities of modern urban development, while improving the natural environment, cultural landscape, and archaeological heritage. The river park project was developed in detail, in order to trigger the implementation of this and other projects in the area. It was accepted by the municipal administration Services for Parks in Lima (SERPAR), which has allocated a budget for the construction of the river park. This project is currently in the process of technical and legal resolution.

Overall, the project "Lima beyond the Park" shifts the focus from the current practice of "image-based" open space design to "performance-based" open space design. It no longer considers urban open space an expensive luxury but considers it as a necessity forsaving water, purifying water, treating wastewater, and recycling nutrients or even harvesting water.

Map and diagrams of Lima: urban structure + informal settlements + lomas ecosystem.
Credit: Marius Ege

Design Proposal Parque Lomas, plan/
Author: Marius Ege

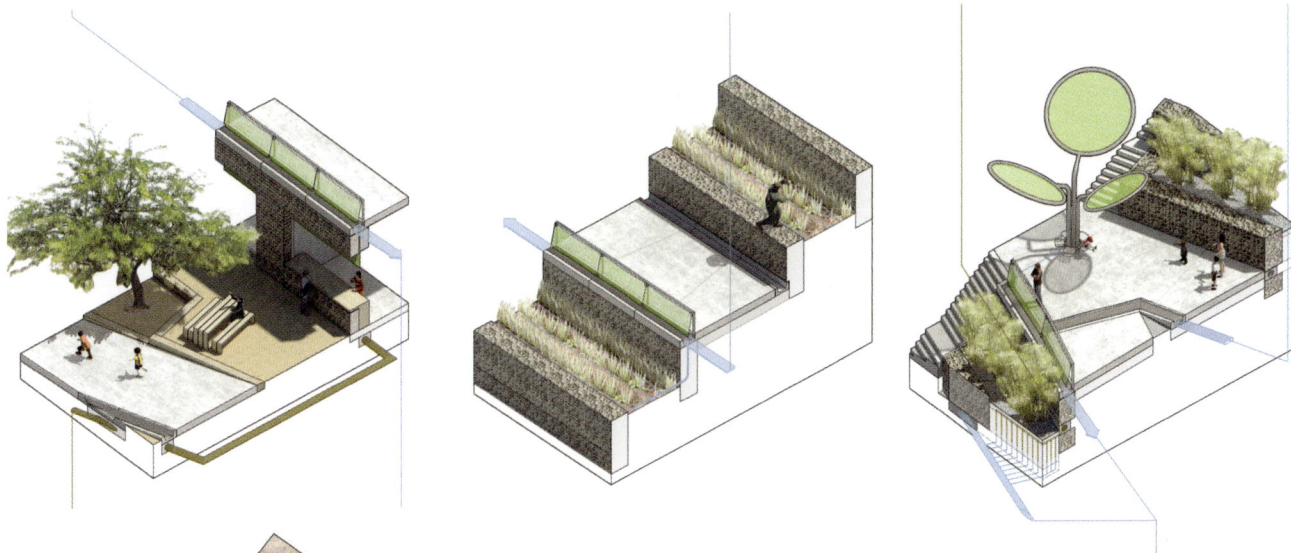

Above:
Design features/ water cycle within
Parque Lomas. Credit: Marius Ege.

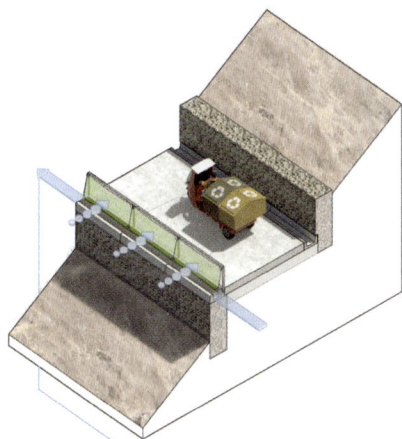

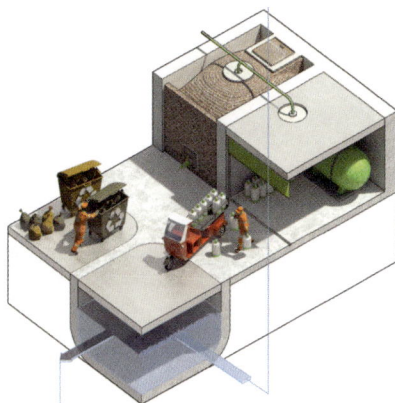

Below:
Extract from LiWa Design Manual:
current reuse of treated wastewater
for urban irrigation.
Credit: Eva Nemcova

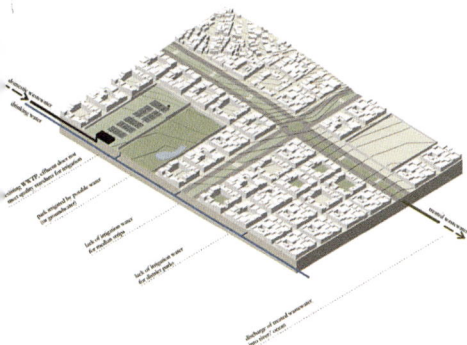

option 1
The effluent of the WWTP is
treated in an open space near by.

option 2
The effluent of the WWTP is
transported to a distend open
space, where it is treated and
reused.

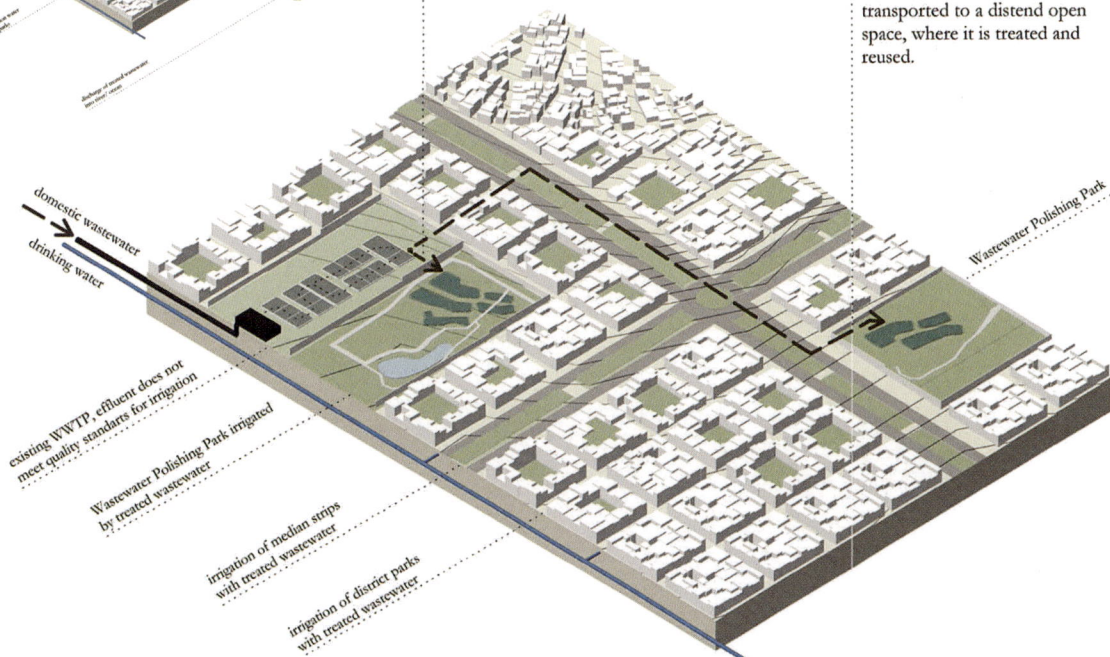

domestic wastewater

drinking water

Wastewater Polishing Park

existing WWTP effluent does not
meet quality standarts for irrigation

Wastewater Polishing Park irrigated
by treated wastewater

irrigation of median strips
with treated wastewater

irrigation of district parks
with treated wastewater

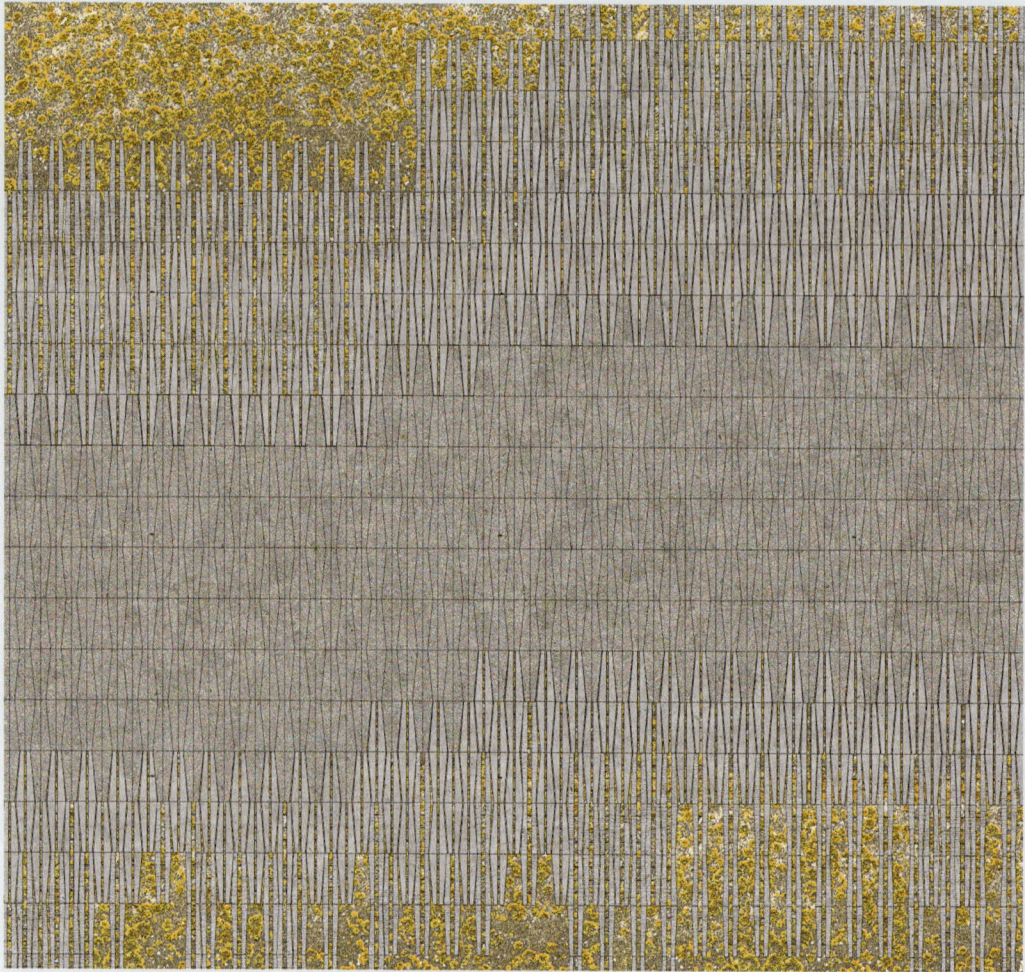

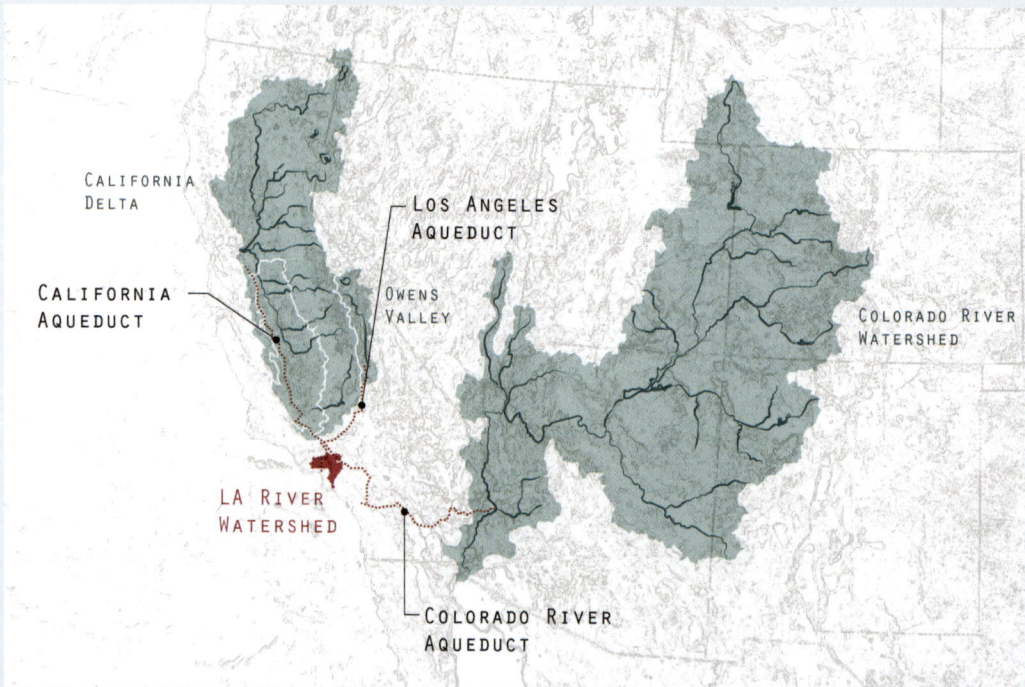

Above:
Gradient Paver System: The Construction of a Living Street.
The proposed system of gradient interlocking pavers and new street topographies supports the integration of ecological performance and social engagement to redefine everyday experience.

Below:
The Actual Watershed of Los Angeles: The city's dependence on distant watersheds adversely affects vast ecosystems.

--- ---

Index of tools | prototypical interventions, layered social and ecological structures, multifunctional watersheds, lawns, greywater reuse, rainwater capture.

WASH: URBAN HYDROLOGICAL NETWORKS FOR RESILIENT CULTURAL ECOLOGIES

Los Angeles, California. Speculative, 2012.

Aja Bulla-Richards

This project proposes prototypical interventions that reconfigure stormwater and greywater infrastructure to initiate layered social and ecological structures in a typical Los Angeles neighborhood. Interventions are designed to impact regional ecosystems. Greywater is utilized to reduce unsustainable water importation and the consequential environmental impacts, and to transform monofunctional infrastructure into multifunctional community watersheds. Integrating ecological performance into the fabric of a neighborhood is critical to redefining urban water infrastructure in relationship to everyday experience.

Arid cities in the western United States are facing an imminent cultural, political, and ecological challenge: dwindling sources of fresh water and a warming climate, coupled with rapid population growth. How can we reimagine and redesign water infrastructure so that monofunctional systems are transformed into resilient socio-ecological cycles that engage and expand everyday experience, promote alternative cultural practices, and reveal latent ecological processes?

Half of the world's cities are located in depleted watersheds. There is a common trend for the growing number of cities facing water scarcity. First, the local water supply is depleted. In the second phase, cities turn to water importation. The third step generally involves centralized wastewater recycling, which, unfortunately, is usually a costly system that can quadruple the price of water. The fourth and final step is to turn to desalination, an expensive and energy-intensive process that carries with it a host of ecological concerns. This trend looks to technological fixes that maintain existing social and spatial structures rather than addressing the design of landscape infrastructure and the potential of alternative cultural practices.

Los Angeles is an iconic example. The small city and surrounding ranches originally depended on local water resources. The promise of a landscape for health and pleasure drew large numbers of people until LA grew to become the second-largest city in the United States. Additional water supplies were secured from distant watersheds to support the growing population.

The consequences of this engineered landscape extend beyond city, state, and national borders. The actual watershed of Los Angeles has expanded to include Owens Valley, the California delta, and the Colorado River. LA has an impact on ecosystems across the western United States and into Mexico. The city has contributed to a dry Colorado River Delta and the endangerment of species in the California delta, and is responsible for the devastation of ecosystems and farming communities in Owens Valley.

Seventeen desalination plants have been proposed along the California coast. Los Angeles is considering

LOS ANGELES WATER CYCLE

RESIDENTIAL WATER REQUIRES
455,600 AFY

EVERY HOUSEHOLD:
200 GALLONS OF WASTEWATER A
DAY

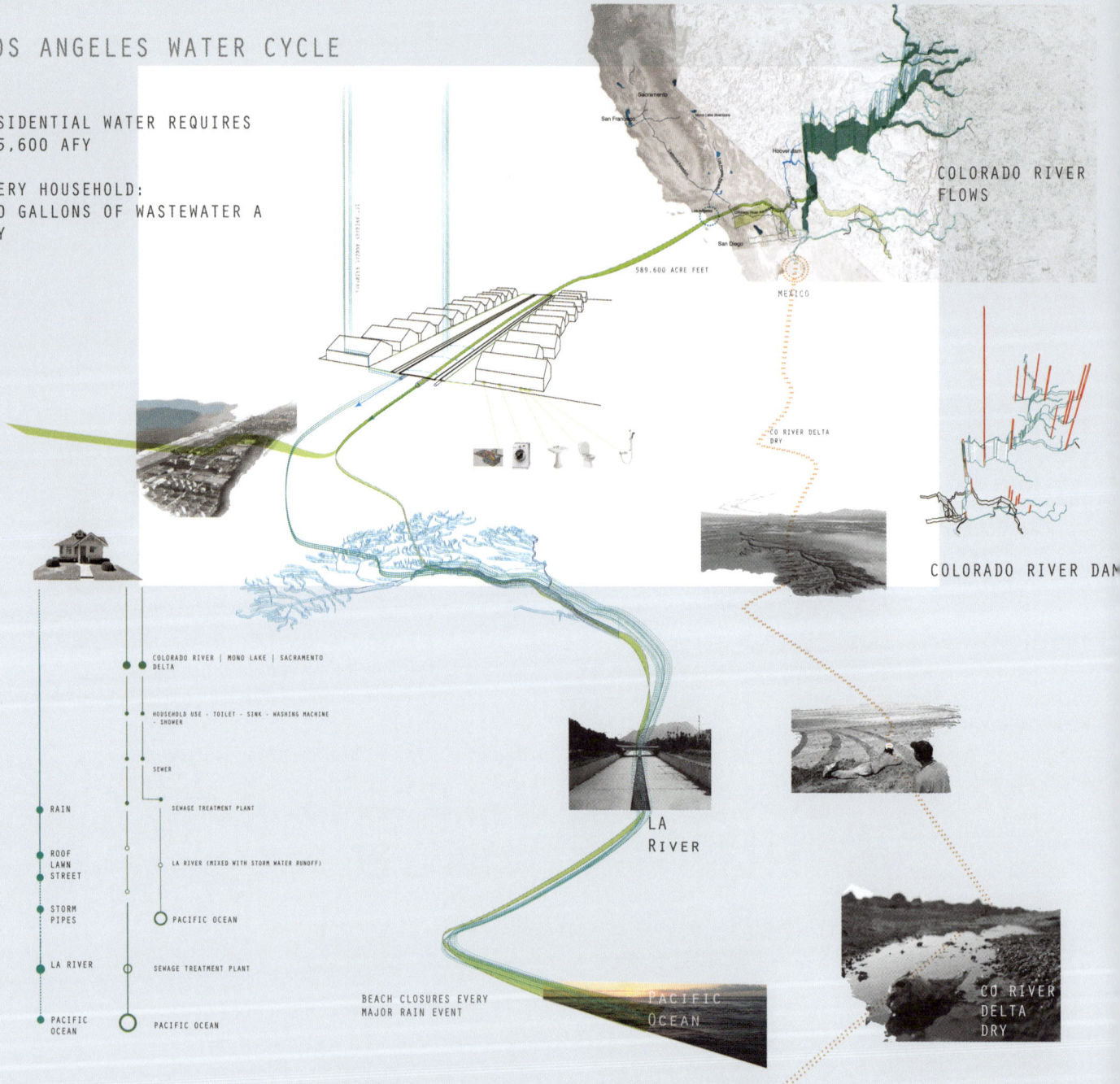

COLORADO RIVER
FLOWS

589,600 ACRE FEET

MEXICO

CO RIVER DELTA
DRY

COLORADO RIVER DAM

COLORADO RIVER | MONO LAKE | SACRAMENTO
DELTA

HOUSEHOLD USE - TOILET - SINK - WASHING MACHINE
- SHOWER

SEWER

RAIN

SEWAGE TREATMENT PLANT

ROOF
LAWN
STREET

LA RIVER (MIXED WITH STORM WATER RUNOFF)

STORM
PIPES

PACIFIC OCEAN

LA RIVER

SEWAGE TREATMENT PLANT

PACIFIC
OCEAN

PACIFIC OCEAN

LA
River

BEACH CLOSURES EVERY
MAJOR RAIN EVENT

PACIFIC
OCEAN

CO RIVER
DELTA
DRY

Above:
The Los Angeles Water Cycle: This
one-directional process is illogical,
treating a precious resource as
waste.

Right:
*Full-scale Prototypes of an
Interlocking Concrete Paver
System.* Various textures provide
subtle coding of the street,
indicating zones for pedestrians,
cyclists, and vehicles without
limiting or isolating functions.

investing in this technology or buying water from companies that include desalination in their water portfolio. Reports produced by nonprofits such as the Pacific Institute warn against turning to this technology because of its impact on water prices, the environment, and land value along the coast.

Water scarcity in the Mediterranean climate of Los Angeles is not primarily a quantitative problem, but instead a mismanagement of resources in support of unsustainable cultural practices and landscape typologies. Water is imported over vast distances, used in commerce, industries, and residences, and then sent to sewers where it is piped to reclamation plants and cleaned through an elaborate three-stage process. The treated water is then either dumped in the Los Angeles River and mixed with dirty street runoff or piped directly into the Pacific Ocean and mixed with salt water, only to potentially be pumped back out and desalinated. Stormwater is handled in a similar manner—streets are engineered to shed rain as quickly as possible, sending it along with thousands of gallons of trash a year into the concrete channel of the Los Angeles River and out to sea. The Los Angeles Basin aquifers are no longer replenished by rainfall, resulting in a diminished local water supply and saltwater intrusion. This one-directional water cycle is illogical because it treats a precious resource as waste. It is time for the city to radically rethink these outdated systems rather than turn to yet another unsustainable centralized infrastructure model, desalination.

Half of LA's residential water is used outside the home, usually to water a front lawn of cool-climate grasses. Approximately 65% of water used indoors for bathing, hand washing, and laundry could be recycled. If biodegradable soaps and detergents are utilized, then this nutrient-rich liquid, called greywater, can be reused to irrigate plants. Greywater is the product of our daily rituals and can connect activities in the home with the public landscape of the street. This project proposes prototypical interventions that reconfigure stormwater and greywater infrastructure to initiate layered social and ecological structures in a typical Los Angeles neighborhood. The monofunctional space of the street is transformed, and compartmentalized daily interactions with water are expanded beyond private space. Between greywater reuse and rainwater capture, LA residences could easily cut in half the amount of imported water they now rely on. This localized water cycle would send more water into the aquifers and about 75% less water

out to sea. If stormwater capture and infiltration were maximized and coupled with greywater reuse, LA could potentially eliminate its reliance on water importation.

Los Angeles is an ideal city to lead a new global trend, one that integrates ecological performance into the fabric of a neighborhood, redefining urban water infrastructure in relationship to everyday experience, and questioning the divide between nature and culture.

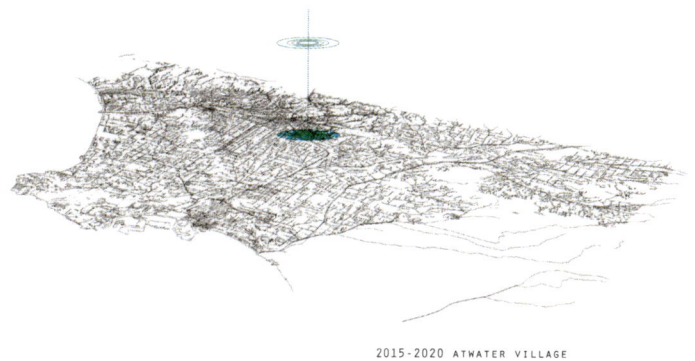

2015-2020 ATWATER VILLAGE

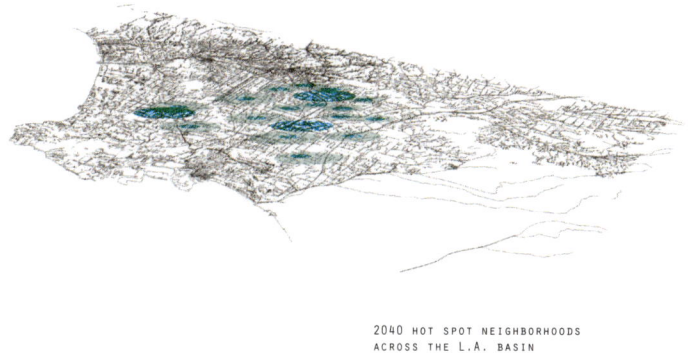

2040 HOT SPOT NEIGHBORHOODS ACROSS THE L.A. BASIN

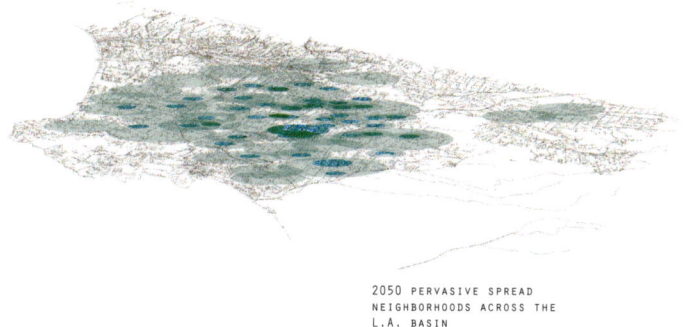

2050 PERVASIVE SPREAD NEIGHBORHOODS ACROSS THE L.A. BASIN

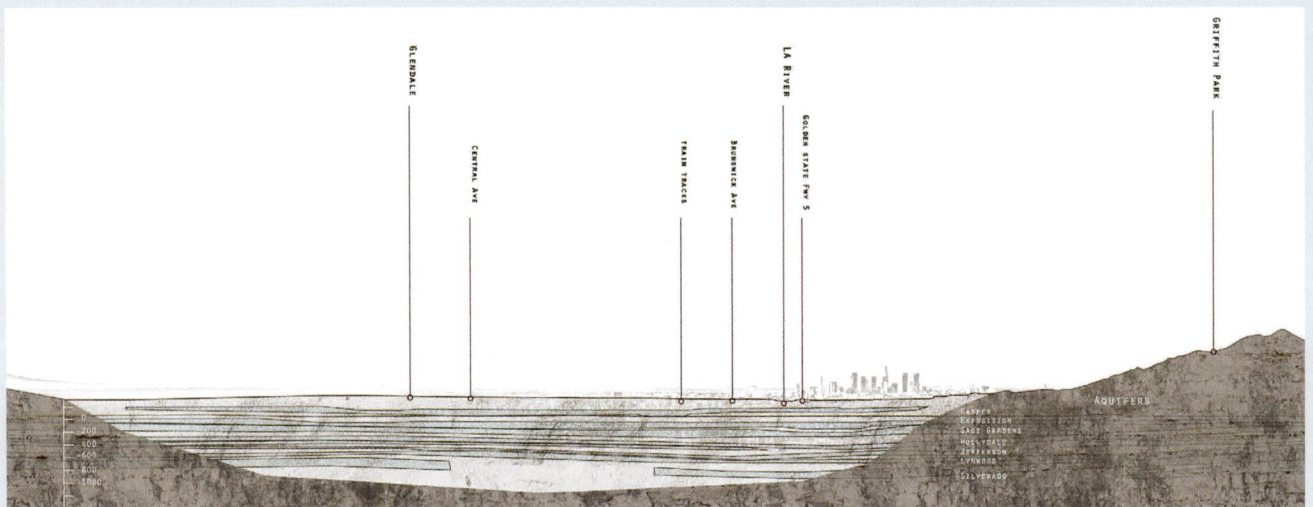

GLENDALE
CENTRAL AVE
TRAIN TRACES
BRUNSWICK AVE
LA RIVER
GOLDEN STATE FWY 5
JEFFERSON
GRIFFITH PARK
AQUIFERS

CAPPER
EXPOSITION
SAGE GARDENS
HOLLYDALE
LYNWOOD
SILVERADO

ATWATER
WATERSHED

SAN GABRIEL

LOS ANGELES

SA

LA Watersheds Transformed: Reading the landscape of sub-watersheds presents new potential identities for LA neighborhoods. Atwater village is proposed as an initial site for this intervention. The emergent nature of this proposal allows for the aggregation of individual street transformations over time, supporting a diverse range of implementation strategies.

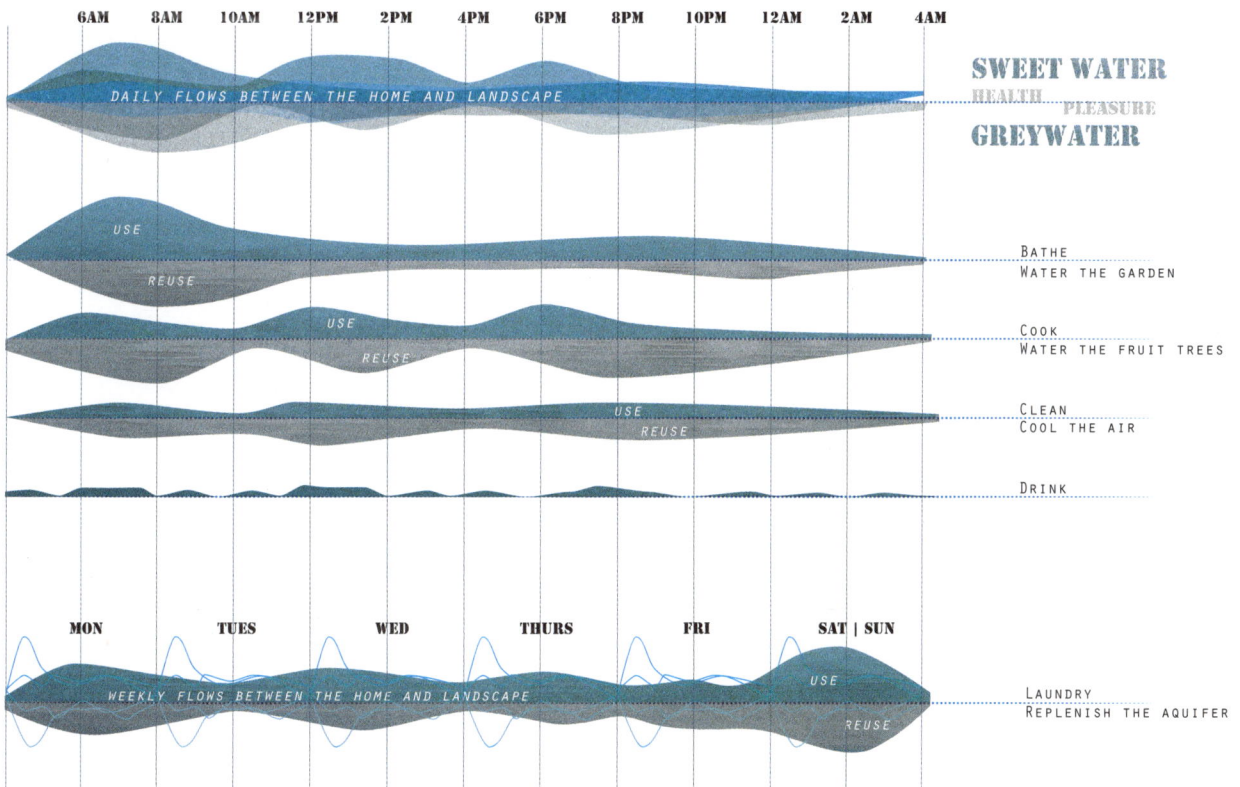

6AM	8AM	10AM	12PM	2PM	4PM	6PM	8PM	10PM	12AM	2AM	4AM

DAILY FLOWS BETWEEN THE HOME AND LANDSCAPE

SWEET WATER
HEALTH   PLEASURE
GREYWATER

USE

REUSE

BATHE
WATER THE GARDEN

USE

REUSE

COOK
WATER THE FRUIT TREES

USE

REUSE

CLEAN
COOL THE AIR

DRINK

MON	TUES	WED	THURS	FRI	SAT | SUN

WEEKLY FLOWS BETWEEN THE HOME AND LANDSCAPE

USE

REUSE

LAUNDRY
REPLENISH THE AQUIFER

Above, Left:
Alternative Futures: As the population increases, residential neighborhoods will need to densify. LA streets can serve as a socio-ecological public space that supports sustainable growth.

Rainwater capture and greywater reuse can eliminate the need to invest in larger sewer systems and expensive desalination plants.

Above, Right:
Greywater Daily Rituals and Rhythms: Nine months out of the year greywater is reused for irrigation. Warm, biologically active soils remove pathogens, filtering water before it enters the aquifer. Water absorbed by plants is transpired through their leaves, creating cool microclimates that promote outdoor social activities.

Below:
Daily Rituals and Rhythms: Flows between the home and landscape promote a new understanding of micro water cycles. Daily rituals may be reconsidered in relationship to the landscape. Water used within the home is transformed and enjoyed again outside the home.

Cultural Practices:
Reimagining underutilized residential streets and front lawns creates public space that can support everyday rituals, expanding the experience of water in the urban landscape.

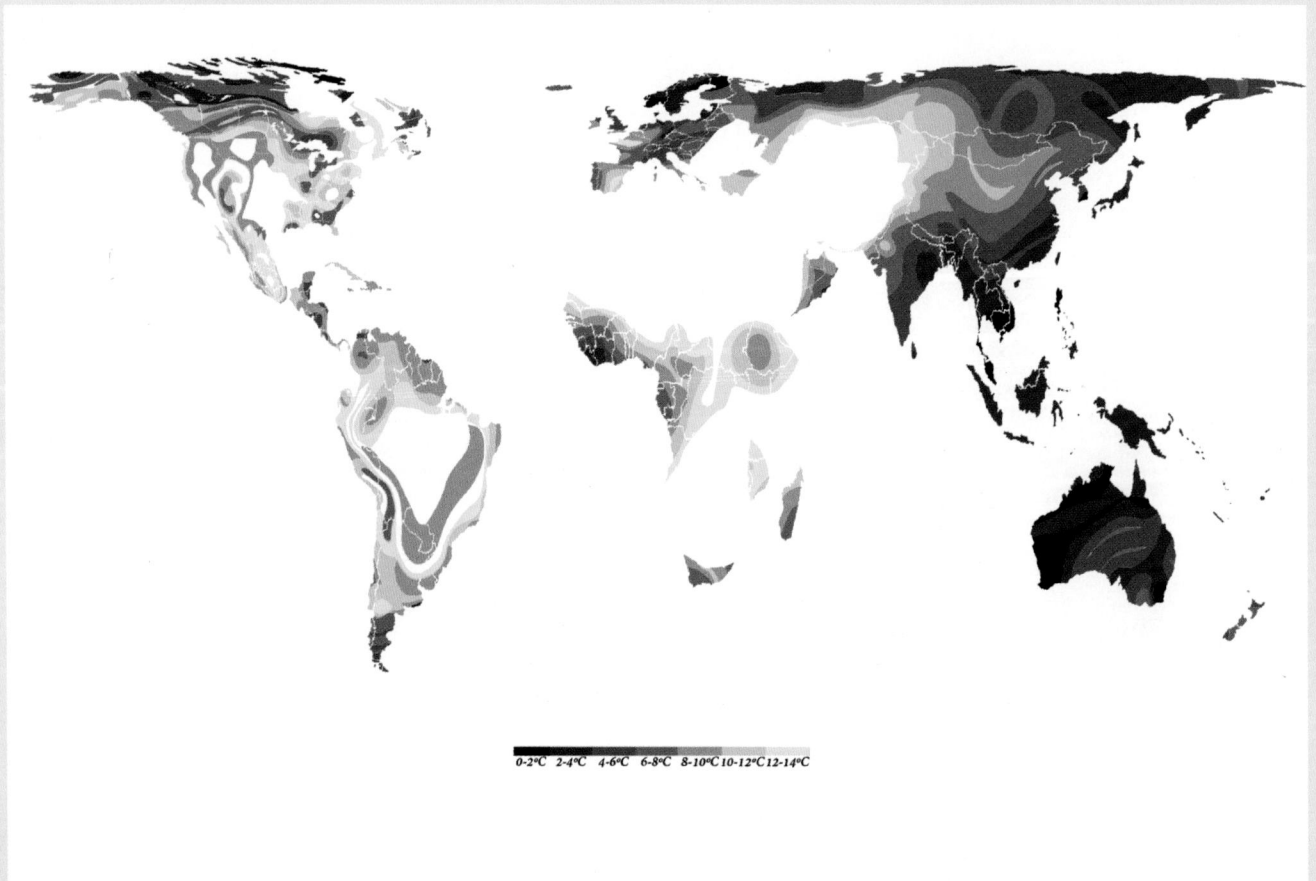

Above:
The vapor house: a moving border allowing a constant definition of intimacy

Below:
Suitable sites for obtaining water from air.
Triple intersection of curves of air temperature, dew temperature and relative humidity above 40%.

0-2ºC 2-4ºC 4-6ºC 6-8ºC 8-10ºC 10-12ºC 12-14ºC

--- ---

Index of tools | watmosphere, Namib desert beetle, refrigerated surface, gadgets.

IMMATERIAL WATER

Anywhere. Speculative, 2012.

Izaskun Gonzalez Barredo and Oriol Valls Guinovart

Water is still a dream for almost 1 billion people. In a planet with around 145 billion liters of suitable-for-consumption water per person, is it possible that water demand overcomes supply? Unfortunately yes. But let's not marvel at this: merely to evacuate 400 to 500 liters of urine and 50 liters of excrement per person annually, we use from

15,000 to 30,000 liters of water. The forecast of world demographic growth for 2050 will just aggravate the difficulty of access to sanitary water resources. Will we let the current urban concentration process saturate the cities with kilometric sanitation nets through which only the water of inequality will be able to flow? We imagine a paradigm capable of questioning the present unequal model of hydric distribution. It is simple: stop basing our lives on it. Water can be obtained from air, which is still an unprivatized resource. There are around 11,340 billion liters of water in air, which means 1.5 million liters of water per person. How many water drops from surrounding air can we store per minute? Will we achieve hydric self-sufficiency? "Immaterial water" generators, our own domestic activities, are the basis for a transition towards the house of zero hydric consumption, in which the primitive dissociation between supply and waste doesn't exist: water is neither created nor destroyed, but transformed.

The principle of Immaterial Water is that water is used only where it is essential. The aim: getting water from atmospheres at 40–60% of relative humidity. The biological example of the Namib desert beetle (Stenocara gracilipes) is applied to the house.

The water vapor rises through the membrane and condensates in the refrigerated ceiling. The membrane behaves like the human skin, allowing transpiration but preventing condensed water from crossing the skin. As the nano droplets begin to accumulate on the refrigerated surface due to condensation, the weight of the accumulated water begins deforming the membrane. A set of water points emerge from the membrane and are ready to receive the gadgets. The stored water is available for moderated consumption under pulverization and vaporization. The gadgets' consumption of water is the minimum amount of water needed. People can place the gadgets everywhere. Since the gadgets share the same diameter, their place is interchangeable. Any place in the house can become a bathroom. By focusing on the bath's most irreducible particle, Immaterial Water optimizes and expands water management.

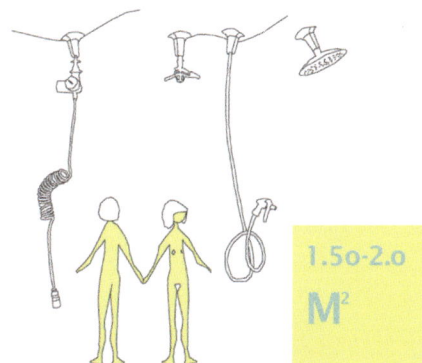

1.5o-2.0
M²

One unit of the Body Surface Area.
The amount of water needed to
adapt consumption to real need.

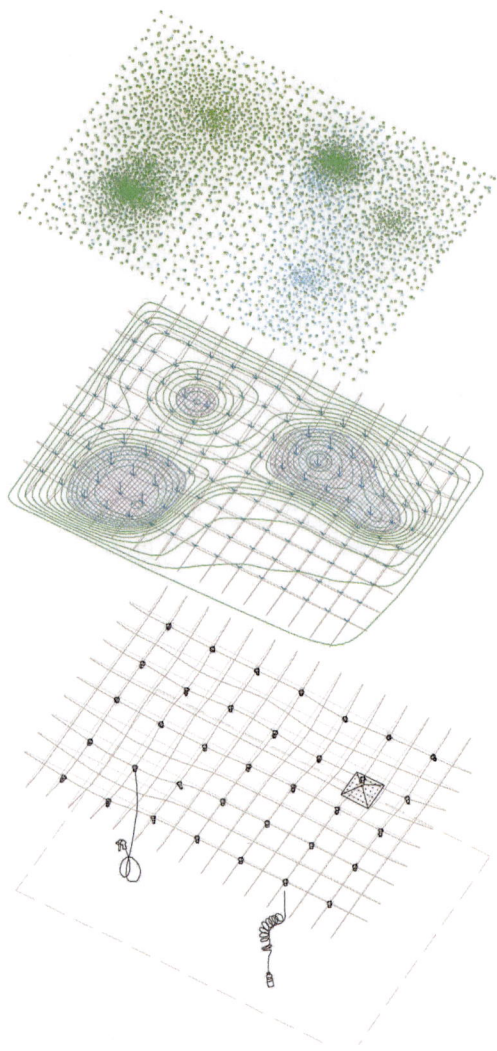

Left:
Immaterial water layers: Water vapor, refrigerated condensation surface and grid of attachment points within attachment area for water gadgets.

Below:
Water generation and water collection inside the house. The steam before and after reaching the membrane is the basis for the house of zero hydric consumption.

immaterial water generators inside the house

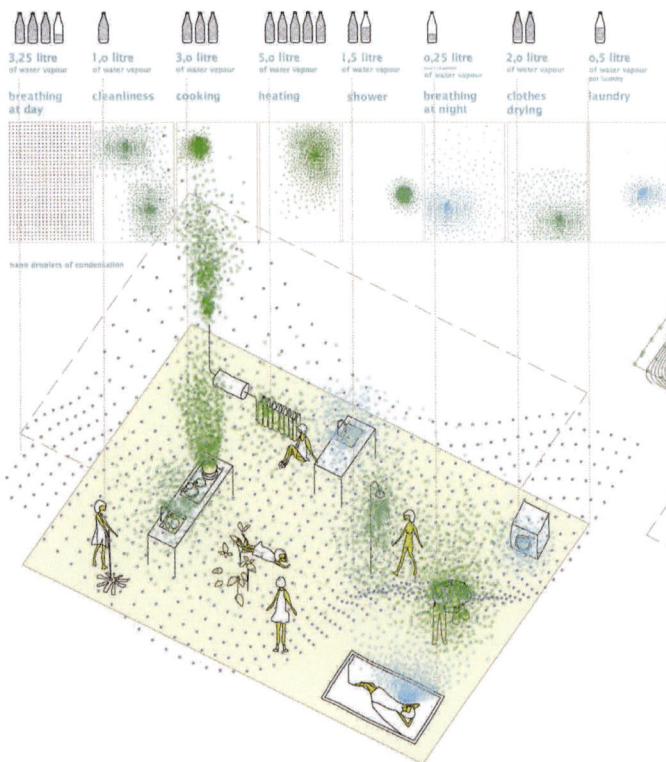

3,25 litre of water vapour	1,0 litre of water vapour	3,0 litre of water vapour	5,0 litre of water vapour	1,5 litre of water vapour	0,25 litre of water vapour	2,0 litre of water vapour	0,5 litre of water vapour
breathing at day	cleanliness	cooking	heating	shower	breathing at night	clothes drying	laundry

nano droplets of condensation

immaterial water
immaterial water grabber in the vapour house

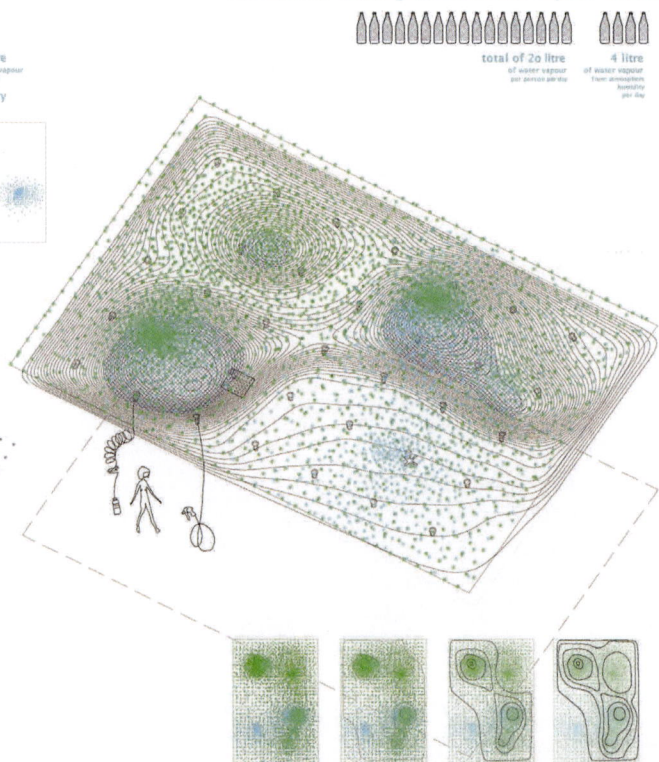

total of 2o litre of water vapour per person per day	4 litre of water vapour from atmospheric humidity per day

271

TERMS OF THE RISING [23] ACCELERATED EROSION OF SHORE-LINES [33] AGING INFRASTRUCTURE [23] AN EXPANDING OCEAN [31] ANTIQUATED STANDARDS [23] APPROACHING MAJOR CENTERS OF POPULATION [23] ATOLL NATIONS [23, 34] AWARENESS AMONG PEOPLE IS VERY VAGUE [43] BATTLE BETWEEN CITY AND COUNTRY [65] BEACHFRONT CONDOMINIUMS [31] BELOW SEA LEVEL [40, 41, 53] BLOCKAGE OF CITY STORM DRAINAGE [36] BREACH [23, 41] BREACHED BY A STORM [36] CATASTROPHE [23, 39, 53, 65, 87] CATASTROPHIC [23, 31, 34, 36, 87] CATASTROPHIC SUBMERGENCE OF THE CITY [31] CHANGING SEA LEVELS [31] CITIES ARE FORTIFIED [24] CLIMATE CHANGE [40, 45, 73] COASTAL EROSION [81] COASTLINE EROSION [69] COLLAPSE [33] CONCENTRATION OF PEOPLE, PROPERTY, HOUSING AND INDUSTRY BEHIND THE DIKES [40] CONTINENTAL ICE SHEETS [23] CORRIDOR OF DESTRUCTION [34] DANGEROUS BEAST [24, 61] DELTAIC COUNTRIES [23, 34] DESTRUCTIVE FRENZY [61] DISASTER [39, 46] DISPLACE [34] DIVERSION OF RIVER SILT [35] DROWNING IN LAWS AND REGULATIONS [47] EARTHQUAKE [23, 31, 32] ECOLOGICAL DANCE [23] EROSION RATE TO ACCELERATE [32] EXPANDING OCEAN [31, 32, 36] EXPECTED INCREASE IN CYCLONE INTENSITY [34] EXTREME WEATHER [24] FALSE SENSE OF SECURITY [24] FEAR [23, 49, 61] FLOODING [23, 33, 34, 35, 36, 39, 40, 41, 42, 43, 48, 50, 58, 59, 65, 69, 75, 85, 87, 93] FLOOD OF UNFOUNDED RUMORS [61] FLOOD PLAIN [53] FLUCTUATING WATER LEVELS [23] FOOD OUTPUT [34] FREQUENT AND VIOLENT STORMS [73] FUTURE DAM CONSTRUCTION [34] GIVE THE LAND BACK TO THE WATER [46] GLACIERS [23] GLOBAL OCEAN EXPANSION [31] HEAD-IN-THE SAND ATTITUDE [35] HEAVILY DEVELOPED COASTAL PLAINS [23, 34] HUGE QUANTITIES OF RAINWATER [49] HUNDREDS OF CITIES AT RISK [23] HURRICANE KATRINA [24, 33, 39, 47, 63] HURRICANES [23, 32, 48] HURRICANE SANDY [73, 77] HYSTERIA [61] IMPENETRABLE BARRIERS [49] IMPROBABLE CITY [48] INACCESSIBLE EVACUATION PLANS [23] INCREASED SALINIZATION OF SURFACE AND GROUNDWATERS [33] INFRASTRUCTURE LOSS [36] INSTANTANEOUS SEA LEVEL RISE [23] INSTANTANEOUS SEA LEVEL RISE [31] ISLANDS MAY COLLAPSE OR DISAPPEAR [33] JUMPED ITS CHANNEL [93] LAND LOSS [33] LANDSCAPE IN FLUX [25, 81] LAND SUBSIDENCE [31, 53] LIQUID INVASION [23] LIVING WITH THE WATER [49] LOSS OF AQUACULTURE [33] LOSS OF BIODIVERSITY [33] LOSS OF PROTECTIVE BARRIER ISLANDS [36] LOSS OF SEDIMENT TRAVELING TO THE DELTAS [34] LOWEST-LYING AREAS OF THE CITY [48] LOW-LYING [23, 24, 34, 35, 40, 41, 48] MAINTENANCE BACKLOG OF THE EXISTING DIKES AND BARRIERS [45] MEGASTORMS [34] MELTING OF BEACH PERMAFROST [32] MONSOON RAINS [33] MORE FREQUENT AND MORE INTENSE RAINFALL [40] NATURAL DISASTER [43, 48] NO NATIONAL EVACUATION PLAN [41] NOWHERE FOR THE EVERGLADES TO GO [33] OCEAN LEVELS [23] OFFICIALS REMAIN UNCONVINCED THAT THE THREAT TO THE NILE DELTA IS REAL [35] ONCE-IN-TEN-THOUSAND-YEARS STORM [40] OUTDATED STANDARDS [41] PESSIMISTIC SCENARIOS [36] POPULATION GROWTH [42] PRECIPITANT OF DISASTER [32] PREDICTIONS HAVE BEEN EXAGGERATED [35] PRESSURE ON THE MECHANICAL

Water Index

Editor
Seth McDowell, Assistant Professor in Architecture

Editorial Team
Benjamin Gregory
Brad Brogdon

Copy Editor
Paula Woolley

UVa Advisory Council
Inaki Alday
Leena Cho
Sheila Crane
Elizabeth Meyer
Suzanne Moomaw
Louis Nelson
Kim Tanzer

Editorial coordination at Actar Publishers
Ricardo Devesa

Published by
University of Virginia
Actar D Inc.
New York, 2016

Grafic Design
Papersdoc SL

Distributed by
Actar D Inc.

New York
440 Parc Avenue South, 17th Floor
New York, NY 10016
T +1 212 9662207
salesnewyork@actar-d.com

Barcelona
Roca i Batlle 2
08023 Barcelona
T +34 933 282 183
salesbarcelona@actar-d.com
eurossales@actar-d.com

Aknowledgements:
Special thanks to the Dean's Office at the University of Virginia's School of Architecture, both Beth Meyer and Kim Tanzer, for the support, advice and resources to complete this book. Both Beth and Kim have been instrumental in initiating and executing this project. We would also like to thank all the architects, designers, landscape architects, planners, historians, writers, engineers, and students who have contributed drawings, images and written material; the many excellent architectural photographers who have consented to the reproduction of their work; and all administrative and technical staffs who have worked with us to collect information.

Printed and bound in China

ISBN: 978-1-940291-40-6
Library of Congress Control Number:
2016931357

UNIVERSITY of VIRGINIA
SCHOOL OF ARCHITECTURE

ACTAR